Leading Edge Techniques in Forensic Trace Evidence Analysis

Cover photos from

On the book's front cover, the left photo shows a Vacuum Metal Deposition (VMD) instrument with a technician removing a test sample showing VMD-developed fingermarks. The right photo shows the handprint that was developed on cloth cut from a pillow case.

Although a simulation, the photo depicts a palm print on a pillow case developed at West Technology Forensics using vacuum metal deposition. The bottom right photo on this page was also developed on a pillow case by VMD. Do you see the impression of a nose?

Photo illustrations created by West Technology Forensics (https://www.west-technology.co.uk/forensic/) and used with permission.

Think of how often a nurse, caregiver, or family member falls under suspicion when a patient dies under uncertain circumstances. Vacuum metal deposition on the pillow might tend to exonerate all as far as death by smothering, or it might actually show the means of death and who did it

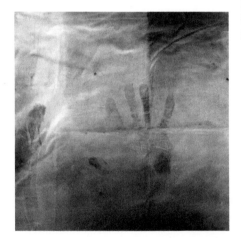

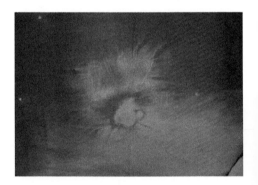

Leading Edge Techniques in Forensic Trace Evidence Analysis

More New Trace Analysis Methods

Edited by
Robert D. Blackledge

Registered Office
John Wiley & Sons, Inc., 111 River Street, Hoboken, NJ 07030, USA

Editorial Office
111 River Street, Hoboken, NJ 07030, USA

For details of our global editorial offices, customer services, and more information about Wiley products visit us at www.wiley.com.

Wiley also publishes its books in a variety of electronic formats and by print-on-demand. Some content that appears in standard print versions of this book may not be available in other formats.

Library of Congress Cataloging-in-Publication Data

Names: Blackledge, Robert D., editor.
Title: Leading edge techniques in forensic trace evidence analysis : more
    new trace analysis methods / edited by Robert D. Blackledge.
Description: First edition. | Hoboken, NJ : Wiley, 2023.
Identifiers: LCCN 2022025481 (print) | LCCN 2022025482 (ebook) | ISBN
    9781119591610 (cloth) | ISBN 9781119591832 (adobe pdf) | ISBN
    9781119591801 (epub)
Subjects: LCSH: Trace evidence–Analysis.
Classification: LCC HV8073 .L33495 2023 (print) | LCC HV8073 (ebook) |
    DDC 363.25/6–dc23/eng/20220714
LC record available at https://lccn.loc.gov/2022025481
LC ebook record available at https://lccn.loc.gov/2022025482

Cover Design: Wiley
Cover Images: © Cover and interior images courtesy of West Technology Systems Limited

Set in 9.5/12.5pt STIXTwoText by Straive, Chennai, India

Contents

List of Contributors

Jocelyn V. Abonamah
Visiting Scientist Program
Research and Support Unit
Federal Bureau of Investigation
Laboratory Division
Quantico, VA
USA

Christoffer K. Abrahamsson
Department of Chemistry and
Chemical Biology
Harvard University
Cambridge, MA
USA

Alina Astefanei
Van't Hoff Institute for Molecular
Sciences
Faculty of Science, Analytical
Chemistry Group
University of Amsterdam
Amsterdam, The Netherlands

Arian van Asten
Van't Hoff Institute for Molecular
Sciences
University of Amsterdam
1090 GD Amsterdam
The Netherlands

and

Co van Ledden Hulsebosch Center
(CLHC)
Amsterdam Center for Forensic
Science and Medicine
University of Amsterdam, Van't Hoff
Institute for Molecular Sciences
1090 GD Amsterdam
The Netherlands

Graceson Aufderheide
Molecular Vista, Inc.
San Jose, CA
USA

Jeffrey G. Bell
Department of Chemistry and
Chemical Biology
Harvard University
Cambridge, MA
USA

Charles A. Bishop
CA Bishop Consulting Ltd.
Consultant on Vacuum Deposition
Technology
Leicestershire
UK

Robert D. Blackledge
Retired, formerly Senior Chemist
Naval Criminal Investigative Service
Regional Forensic Laboratory
San Diego, CA
USA

and

8365 Sunview Drive
El Cajon, CA 92021
USA

Candice Bridge
National Center for Forensic Science
and the Department of Chemistry
University of Central Florida
College of Sciences
Orlando, FL
USA

Christopher Deeks
Channel Manager EMEA – Surface
Analysis at Thermo Fisher Scientific

Joris Dik
Materials Science and Engineering
Delft University of Technology
2600 AA Delft
The Netherlands

Joseph Donfack
Research and Support Unit
Federal Bureau of Investigation
Laboratory Division
Quantico, VA 22135
USA

Brian A. Eckenrode
Research and Support Unit
Federal Bureau of Investigation
Laboratory Division
Quantico, VA 22135
USA

Shencheng Ge
Department of Chemistry and
Chemical Biology
Harvard University
Cambridge, MA
USA

Alwin Knijnenberg
Netherlands Forensic Institute
2490 AA The Hague
The Netherlands

Maria Lawas
Visiting Scientist Program
Research and Support Unit
Federal Bureau of Investigation
Laboratory Division
Quantico, VA
USA

Annelies van Loon
Rijksmuseum
1070 DN Amsterdam
The Netherlands

Roselina Medico
olam food ingredients (ofi)
Department of Plant Science
Koog aan de Zaan
The Netherlands

Cyril Muehlethaler
Department of Chemistry,
Biochemistry and Physics
University of Quebec
Trois-Rivières, QC
Canada

Kandyss Najjar
National Center for Forensic Science
and the Department of Chemistry
University of Central Florida
College of Sciences
Orlando, FL
USA

Padraic O'Reilly
Molecular Vista, Inc.
San Jose, CA
USA

Sung Park
Molecular Vista, Inc.
San Jose, CA
USA

Claude Roux
Centre for Forensic Science
University of Technology Sydney
Australia

Ryan Schonert
VUV Analytics, Inc.
Cedar Park, TX
USA

George M. Whitesides
Department of Chemistry and
Chemical Biology
Harvard University
Cambridge, MA
USA

and

Wyss Institute for Biologically Inspired
Engineering
Harvard University
Cambridge, MA
USA

and

Kavli Institute for Bionano Inspired
Science and Engineering
Harvard University
Cambridge, MA
USA

Foreword

I have known Robert Blackledge for more than 20 years. We started communicating when one of my students began some research on the forensic analysis (and interpretation) of condom lubricants. I was immediately impressed by his willingness to openly share his knowledge and expertise to the overall benefit of forensic science in general and the field of trace evidence (or microtraces, as I would prefer to call it) in particular. This field has always been one of my strong interests since studying at the University of Lausanne, Switzerland, working under Professor Pierre Margot.

Microtraces are the product of a one-off event that occurred in the past. As a result, they are often incomplete, imperfect, or degraded. They are anything but uniform. Further, they can take almost infinite shapes or forms. Finally, microtraces are rarely discovered as stand-alone entities but usually mixed with other materials that were already present on the relevant substrate before the event generating them; sometimes, other materials may contaminate this complex matrix after the transfer event but before the microtraces discovery and collection.

As a result, microtraces detection, recognition, examination, and interpretation are often challenging and require a great deal of critical thinking resting on a sound scientific approach and reasoning. However, these microtraces can often hold the key to solve complex problems, including reconstructing what happened or who was involved. There are ample examples of cases where microtraces brought a breakthrough. Some of them are included in this book.

Unfortunately, over the last 20 years, the value and effective use of microtraces in forensic science has been diluted for a variety of reasons, mostly associated with the focus on high-throughput laboratories increasingly operating in a mechanistic way, following what some call a "pill factory paradigm." At the same time, it has been refreshing to see some great work and long-lasting passions in this field. Anyone who has had the privilege to know Robert Blackledge will agree with my contention that he is a prime example in this category. I will always remember his passionate contributions to the multiple discussions at the National Institute of Justice Trace Evidence Symposia between 2007 and 2011.

With this passion and career-long experience, Robert Blackledge wrote this book, along with the contributions of many leading experts in their field. The title, *Leading Edge Techniques in Forensic Trace Evidence Analysis*, perfectly reflects the purpose of this book. On the one hand, it systematically presents microtrace types that are not covered by traditional forensic science textbooks. Examples include shimmer particles in cosmetic samples, glitter, and other flakes. To some extent, due to their extreme variability and outstanding transfer and persistence abilities, these particles could be considered as cutting-edge microtraces or ideal trace evidence, as has been previously reported.

On the other hand, this book presents cutting-edge technology and methods that, similarly, are not commonly discussed in other forensic science textbooks. These techniques may rarely be found in the average crime laboratory. However, the forensic science community needs to know they exist, and can provide crucial findings in some cases. Therefore, it is essential to improve our knowledge about them and to recognize the case circumstances where they can be exploited to our benefit.

Leading Edge Techniques in Forensic Trace Evidence Analysis is therefore not a handbook on microtraces; there are many existing examples of these. This book deals with selected microtrace types, techniques, and problems. It will be most useful to practicing forensic scientists working in microtrace (trace evidence) laboratories; it will further expand their expertise in this field. It will also be a valuable reference for students, educators, and researchers engaged in forensic science. It is hoped that it will stir some new or renewed passion for microtraces research. Finally, other groups may benefit from reading this book, such as attorneys, judges, and novel writers. They may find critical answers to some questions raised in a case.

Leading Edge Techniques in Forensic Trace Evidence Analysis fills a gap by presenting less commonly discussed forensic science topics. It further illustrates and reinforces the value of microtraces in investigations and court. It is hoped it will inspire and stimulate the reader in showing interest and perhaps garnering more support for microtraces. This field is seminal to forensic science and remains one of its most fascinating areas. This book should well serve this cause.

Distinguished Professor Claude Roux
Centre for Forensic Science,
University of Technology Sydney, Australia
President, International Association of Forensic Sciences

Preface

This book is a sequel to the book published in 2007, *Forensic Analysis on the Cutting Edge: New Methods for Trace Evidence Analysis*, edited by Robert D. Blackledge. As with the previous book, featured are either types of trace evidence having received little previous attention or new and better methods for trace evidence characterization.

As far as new and better methods for trace evidence characterization are concerned, many of the methods although relatively unknown to the forensic science community, have for years been used, studied, and improved by industry, scientific instrument manufacturers, and large university research groups.

Why this antipathy by the forensic science community as far as looking into and implementing new methods and instrumentation? One example that particularly puzzles me is the lack of interest in X-ray photoelectron spectroscopy (XPS). Increasingly today we encounter objects that have one or more very thin surface layers. XPS is a surface analysis method that applies to all elements except hydrogen and helium. With a penetration depth of not much deeper than 10 nm, XPS results are not complicated by the composition of the particle core. It not only is quantitative as far as element concentration is concerned, but also provides an element's oxidation state and identifies those elements to which it may be bonded. Because carbon has so many oxidation states and elements to which it may be bonded, XPS is particularly useful for characterizing very thin surface layers that may be polymers.

In 2010, the paper "The potential for the application of XPS in forensic science," by John F. Watts, Professor of Materials Science, University of Surrey, UK, was published in *Surface and Interface Analysis* 42(5): 358–362 (DOI: 10.1002/sia.3367). A truly outstanding and visionary paper, it included examples and case histories. And yet it was largely ignored by the forensic science community. I should think that among the forensic scientific community it would have received a reaction akin to that of biochemists to advances in DNA profiling.

It would seem that forensic scientists today actually working in crime laboratories prefer to beaver away like clerks in a Charles Dickens novel using validated protocols rather than have the temerity to try something new.

It is my fervent hope this book helps push many of these analytical methods toward acceptance and use by the forensic science community.

Robert D. Blackledge
San Diego, CA
December 2020

1

Forensic Analysis of Shimmer Particles in Cosmetic Samples

Kandyss Najjar[1], Robert D. Blackledge[2], and Candice Bridge[1]

[1]*National Center for Forensic Science and the Department of Chemistry, University of Central Florida, College of Sciences, Orlando, FL, USA*
[2]*Naval Criminal Investigative Service, Regional Forensic Laboratory, San Diego, CA, USA*

1.1 Introduction

Shimmer particles are commonly observed in daily life. From cosmetic products to paint samples, shimmer particles are readily present. However, despite its common presence, it has not been readily considered in the forensic trace evidence community as a form of contact evidence.

In this book's first edition, the chapter on glitter analysis mentioned that glitter and shimmer were often confused with each other [1]. Although both are ingredients in cosmetic products, they are easily distinguished by a cursory microscopic examination. In addition to microscopic analysis, there are fundamental chemical differences that differentiate these items. Regardless of the chemical differences, one property that might be common between these items is the potential to be an ideal contact trace evidence with strong indications of association between two people or a person and a physical item.

The previous glitter chapter in the first edition of this book asked the question: What are the properties of the ideal contact trace? To answer this question, the authors listed seven properties that would make any item an ideal contact trace sample. Based on these criteria, glitter can be considered an ideal contact trace material. The next question is – can shimmer particles be considered an ideal contact trace?

In this chapter, we discuss the chemical properties of shimmer particles, how these particles can be recovered as trace evidence samples, and the most appropriate instrumental methods for analyzing these samples. It will also be discussed how a Questioned and Known shimmer sample can be compared in a forensic casework setting. Through these discussions, it will be demonstrated that shimmer

Leading Edge Techniques in Forensic Trace Evidence Analysis: More New Trace Analysis Methods,
First Edition. Edited by Robert D. Blackledge.

particles can also be considered an ideal contact trace evidence in addition to glitter samples. Although glitter has been demonstrated to be a critical associative evidence in numerous criminal investigations, to date, shimmer has been largely ignored by criminalists.

1.2 What is Shimmer?

Although this chapter focuses on cosmetic shimmer, many other products include shimmer particles. In the automotive paint industry, shimmer is known as an "effect pigment." Kids and adults use them in creating arts and crafts. There are many commercial glue and pen products containing both glitter and shimmer. Moreover, particles have also been incorporated into everyday clothing and shoes, as well as in costumes for special events such as Halloween and Mardi Gras. They are even used for decorating greeting cards and ornaments used during holidays such as Thanksgiving and Christmas.

The focus of this chapter is on the identification of shimmer particles used in cosmetic products and personal hygiene products. Such products include lipstick, foundation, eye liner, hair spray, body lotion, and more. Cosmetic products use mica particles with different layer thicknesses of varying metal oxide coatings to achieve various shades of a certain color [2]. This is done to satisfy the ever-increasing demand for new colors in cosmetics [3]. For instance, some silver pigments are created by coating with titanium dioxide to produce shades ranging from soft silver to dazzling silver. In contrast, other natural or earth-toned colored cosmetics use iron oxide to obtain shades ranging from soft bronze to bright red. Since shimmer can be found in various cosmetic products, and most are intended for everyday wear rather than just for special occasion, as a result, shimmer particles may potentially transfer during close personal assaults and can be used as a form of trace evidence.

1.2.1 Shimmer versus Glitter

Many people assume that glitter and shimmer are the same, when, in fact, they are fundamentally and compositionally different. Cosmetic glitter is a man-made product that is usually composed of either tiny pieces of aluminum foil, plastic without a metallic coating, or plastic that has an aluminum layer. It typically starts off with polyester sheets [1], such as polyethylene terephthalate (PET) or polybutylene terephthalate (PBT) that may have been painted with pigments approved by the U.S. Food and Drug Administration (FDA). These sheets are then cut into tiny pieces with typical shapes of hexagons, squares, or rectangles. Hexagonal shapes are most common, followed by the squared shape. Other manufacturers may create unique glitter shapes such as stars or moons [1].

In her article from The New York Times, *What Is Glitter?*, Caity Weaver described glitter as "aluminum metalized polyethylene terephthalate" [4]. Glitterex Corporation was presented as one of the largest glitter manufacturers in the United States (Cranford, NJ), which sells glitter that is mainly composed of thin Mylar PET film. To achieve a rainbow-like colored glitter, i.e., holographic glitter, a fine layer of vapor-deposited aluminum is placed onto the polyester film and then embossed with a diffraction grating pattern so that light reflects at different directions simultaneously. Finally, some of their glitter is composed of multi-layered plastic of various refractive indices, with each layer more than 230 nm thick, to achieve different colors at different incident angles [4].

On the contrary, shimmer particles are primarily comprised of mica substrates, which are complex aluminosilicate crystals that readily separate into thin flat sheet-like layers. Mica is a naturally occurring mineral but can also be synthesized commercially. Cosmetic shimmer is primarily comprised of mica particles that have been coated with different types and thicknesses of metal oxides to generate different colors or effects. Furthermore, although shimmer particles may fall into a certain size range, their shape is totally irregular and random.

Glitter has been used previously as associative evidence in real-world cases [1, 5, 6], however, shimmer has not been evaluated despite its potential value as associative evidence as well. Little research has been conducted on shimmer analysis as a means of trace/associative evidence. The examination and evaluation of shimmer may indeed expand the scope of forensic particle analysis that is currently available by providing another type of associative evidence to evaluate and a means to compare known and unknown shimmer particles that may have transferred during a close personal contact.

1.2.2 Shimmer Composition and Use

Albeit being mainly overlooked by criminalists, shimmer is very well known by chemists in various industries, including the cosmetics industry, the craft industry, and the automotive industry; and new and improved forms of shimmer are steadily being introduced into the commercial marketplace. The wide variety of shimmer applications increases its potential value as trace evidence, and the fact that many commercial cosmetic products contain several different types of shimmer particles at different relative amounts makes samples with shimmer easier to distinguish from one another.

Shimmer mainly consists of pieces of mica of a certain size range that have been coated with titanium dioxide (TiO_2) of uniform thickness. This chapter will focus on coated mica as it is the most common type of shimmer.

The most common and abundant forms of mica are muscovite and biotite [7]. Biotite mica has the general formula $K(Fe, Mg)_3(AlSi_3O_{10})(OH)_2$, where the potassium (K) can be found bound to either iron (Fe) or magnesium (Mg). Biotite is typically darker in color than muscovite, in usually a black or dark brown color [7, 8]. Biotite is used as a filler or insulator in various electrical and construction applications [7]. The mica used in most cosmetic products is muscovite. Although its general formula is $KAl_3Si_3O_{10}(OH)_2$, depending on geographic location the formula is variable $K(Al, Cr, Mn)_3Si_3O_{10}(OH)_2$, where the potassium could be bound to either aluminum (Al), chromium (Cr), or manganese (Mn). When only K is bound to Al, muscovite is most commonly clear in color, sometimes occurring in light shades of brown, green, yellow, or rose [9]. When Al is substituted with Cr, the mica is referred to as Fuchsite or Chrommuscovite and the mineral is generally green in color [9, 10]. Manganese rich muscovite mica, when Mn is in the place of Al, occurs in colors ranging from pink to red and is known as Alurgite [10]. These chemical differences provide the first manner in differentiating mica substrates.

Once the cosmetic sample has been applied, the shimmer particle is intended to lie flat on the surface. The longest dimension may range from as little as a few microns up to several hundred, but their thickness is typically one micron or less. Although quite small, shimmer particles cannot be considered nanomaterials.

Mica particles use the basics of thin film light interference to achieve their color shifting properties [11]. Light interference occurs in one of two ways. Constructive light interference results in higher amplitudes when two waves are in phase, whereas destructive interference causes the resultant wave to lower in amplitude when two waves are out of phase at a phase difference of half the wavelength. The condition for maximum interference is $\lambda = 2nd\cos\theta$, where λ is the wavelength, n is the refraction index of the spacer, d is the spacer thickness, and θ is the angle of incidence of the light.

Most shimmer particles will be constructed as a sandwich, where the mica, i.e., a semi-transparent spacer, is placed in between two semi-reflective metal oxide layers. The combination of a specific spacer and semi-reflective layers is referred to as a Fabry–Perot interference filter [11] (refer to Figure 1.1). Since a specific color is represented by a narrow wavelength region, various colors would be observed with different incident angles [11]. These optical properties of metal-coated mica shimmer particles can help individualize these cosmetic shimmer particles.

To develop the colors observed via thin film interference, the incident light interrogates the semi-reflective metal-oxide layer first where two outcomes occur. The incident light can be reflected by the metal-oxide layer and due to the semi-reflective nature the incident light can also be transmitted. The thickness of the metal-oxide layer on top of the mica substrate will affect the resulting color observed. At the interface of the metal-oxide and mica layers, the light can be reflected, and/or transmitted into the mica substrate. Based on the thickness of

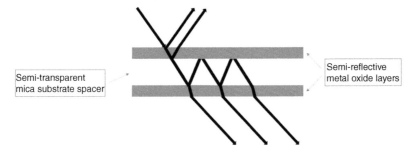

Figure 1.1 Shimmer thin film light interference based on the concept of a Fabry–Perot filter.

the metal-oxide layer and its refractive index, the reflected light at the interface should narrow the wavelength selectivity of the light and as a result influence the color observed by the user. The different colors of shimmer that may be achieved are due to interference between reflected light at the metal-oxide surface and at the interface of the metal-oxide and mica layers. A fuller discussion is presented by Jiang et al. [12] on the basic theory behind optically variable pigments such as metal-oxide-coated mica particles.

Since the reflected color depends on the thickness of the TiO_2, and because mica may not always cleave so that the top and bottom surfaces are perfectly flat (i.e., there are layers that partly overlap), the thickness of the TiO_2 layer may show some variation and the observed reflected color may not be as pure or monochromatic. More expensive, synthetic mica, $KMg_3AlSi_3O_{10}F_2$, often referred to as "synthetic fluorphlogopite," has excellent surface smoothness. Synthetic mica is sometimes preferred over its natural alternative since it may be used to create brighter and more radiant shimmer colors. Other less common substrates include alumina, silica, bismuth oxychloride crystals, and calcium aluminum borosilicate (glass) [13, 14]. Synthetic shimmer substrates are also selected by some companies to avoid human rights implications of using child labor in the mining of natural mica [15].

1.3 Shimmer Detection and Collection

In the 1900s, a French criminologist named Edmond Locard opened the world's first crime lab in France and is well known for his most famous book titled *Treaty of Criminalistics* [16]. Locard believed that "every contact leaves a trace" and developed what is known as Locard's Exchange Principle [16, 17]. In his book *Crime Investigation: Physical Evidence and the Police Laboratory*, Paul Kirk expresses the Locard's Exchange Principle as "Wherever he steps, whatever he touches, whatever he leaves, even unconsciously, will serve as a silent witness against him. Not only fingerprints or footprints, but hair, fibers, glass, paint, blood or others. All of these and more, bear mute witness against him" [18]. This principle suggests that even if perpetrators attempt to mitigate evidence left behind at a crime scene, trace evidence that is often overlooked may be used to connect or link them to the crime scene or the victim. Therefore, since many perpetrators are not aware that cosmetics can be a type of trace evidence, Locard's rule can be applied to cosmetic and/or shimmer transfer as well.

1.3.1 Detection of Cosmetic Stains

Shimmer particles can easily be detected at a crime scene using common methods that are generally present in most collection kits. Cosmetic smears can be found anywhere at a crime scene, depending on the type of crime that was committed, but the most common locations are on clothing, bed sheets, or other physical items (i.e., tables, walls, etc.).

The easiest detection method to locate shimmer particles is to use a flash light. When searching large areas of interest around the crime scene, the shimmer particles will reflect the light and can then be easily detected by the naked eye.

An ultraviolet (UV) light source or light box is another method that can be used to detect cosmetic smears on fabric samples. Most white or light-colored garments bear traces of a fluorescent fabric brightener and when outdoors or under UV-containing light rays, the fabric appears "whiter than white." This is because UV-light rays impinging on the garment cause the fluorescent molecules to emit "white" light. Therefore, if any stains or smears are present on the garment, even if they are colorless and invisible under ordinary light, once placed in a UV light in the dark the smears will appear as a shadowy area on the garment. This is because the smear blocks some of the fluorescent light being emitted from the garment. Although the smear may primarily consist of the vehicle used in the commercial cosmetic containing the shimmer, i.e., lipstick, which may not exhibit fluorescence, this will be a prime location to collect shimmer particles.

1.3.2 Collection of Shimmer Particles

Shimmer particles transferred from cosmetic products may be collected in a manner similar to the suggested collection of glitter particles in this book's first edition [1].

Whether a few individual particles or cosmetic smears were found, the simplest way to collect shimmer samples is to stub the sample with Post-it® notes. The sticky notes could simply be folded and placed into small re-sealable zipper bags to prevent losing the sample prior to analysis. Moreover, all critical information about the collection process can be written onto the sticky note itself prior to being sealed for subsequent analysis. Tape lifts may also be used, but the glue from tape is stronger than the Post-it notes and may contaminate or destroy the evidence upon extraction of the particle from the tape for analysis. One caveat to using Post-it notes is that they can only collect from smaller items or areas, whereas tape lifts may be used on larger items or areas of interest.

If only a smear was located on a hard surface, i.e., a table, the smear may be collected using a cotton swab or scraped off using a spatula. However, with this collection method it will be necessary to have a good extraction method to separate the shimmer particle from the cosmetic vehicle.

1.4   Analysis of Shimmer Particles

It is best practice to first analyze particles using a stereomicroscope or digital compound microscope to obtain size and morphology information. Afterward, other analytical techniques may be utilized to ascertain other chemical information. Techniques commonly used to analyze forensic evidence include, but are not limited to, optical microscopy, Fourier-transform infrared spectroscopy (FTIR), Raman spectroscopy, X-ray diffraction (XRD), and scanning electron microscopy – energy dispersive X-ray spectroscopy (SEM-EDS). Most of the instrumental techniques presented herein are based on the current research conducted by Najjar and Bridge.

1.4.1   Sample Extraction and Preparation

After collection, particles can be removed from the collection media, e.g., Post-it notes or tape lifts, using tweezers and subsequently cleaned. When shimmer particles are in cosmetic products, they are generally contained in a vehicle (e.g., lotion, eyeliner, nail polish, or lipstick) and as a result an appropriate extraction method is needed to separate the shimmer particles from the vehicle. After collecting shimmer particles, it is also possible that there may be traces of the vehicle adhered to the recovered shimmer particles. The particles can then be separated from the cosmetic matrix using a simple hexane wash and filtration process [2]. The particles are then ready for instrumental analysis.

1.4.2   Digital Microscopy

Forensic evidence is typically first analyzed under a stereo-binocular microscope. The stereomicroscope allows for three-dimensional visualization of the item so the analyst can observe its structure, morphology, and size. New advances in the microscopy field allow for a more robust digital analysis of samples where size, color, particle count, and other features are automatically detected via machine-operated microscopy software.

Since shimmer shape, size, and color are very irregular and random, it is quite difficult to gain such information using digital microscopy. Despite the irregularity in size, specifically, it is suggested to conduct analysis based on a size range rather than an average size. Based on recent research completed by the authors Najjar and Bridge, subtle differences in size ranges were observed between different shimmer samples. To demonstrate the subtle differences between different shimmer powders, the size ranges were determined and subsequently compared and presented herein.

Shimmer powder, which had not previously been incorporated into a vehicle, was placed onto a microscope slide and was smeared using a Kimwipe to obtain individual layered particles. Shimmer particles may be very small and are harder to separate based on size simply by visual observation. Therefore, a Keyence VHX-6000 digital microscope was used to obtain size measurements of area, perimeter, minimum diameter, and maximum diameter for the shimmer particles in the field of view of the microscope. Only individual layered shimmer particles were used for the analysis. Some shimmer specks were omitted from size measurements due to overlap with other particles, or because part of the individual shimmer particle was cut off the microscope's image field of view. Table 1.1 shows an example of the size measurements of two red shimmer samples analyzed. In the table, "Low" and "High" signify the low-end and high-end size range limits, respectively. Samples used for this research study were purchased from Just Pigments (Tucson, AZ). Based on their manufacturer details, Red Wine CP-504 is composed of natural muscovite mica coated with iron oxide, whereas Superstar Red CP-7059 is made up of synthetic fluorphlogopite, coated with titanium dioxide, iron oxide, and tin oxide. Table 1.1 displays the difference in size ranges between the natural and synthetic shimmer samples. For instance, Red Wine shimmer had particles with areas ranging from 24.2 to 1895.4 μm^2, while

Table 1.1 Microscopic size measurements obtained for Red Wine CP-504 (natural mica) and Superstar Red CP-7059 (synthetic mica) shimmer samples from Just Pigments.

Shimmer name/ product #	Low area (μm^2)	High area (μm^2)	Low perimeter (μm)	High perimeter (μm)	Low max diameter (μm)	High max diameter (μm)	Low min diameter (μm)	High min diameter (μm)
Red Wine CP-504	24.2	1895.4	17.5	176.1	6.2	58.4	4.0	44.1
Superstar Red CP-7059	76.2	7170.9	30.7	408.3	11.6	175.3	6.9	89.6

Reflectance   Transmittance

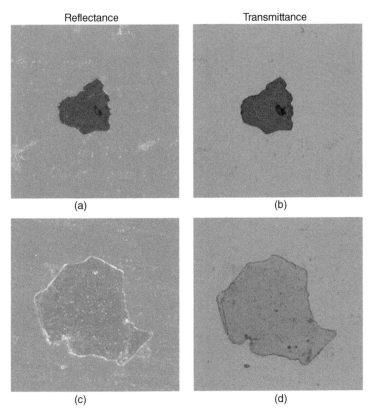

(a)   (b)

(c)   (d)

Figure 1.2   Microscopic images of (a and b) Red Wine CP-504 in reflectance and transmission modes. Microscopic images of (c and d) Superstar Red Wine CP-7059 shimmer samples in reflectance and transmission modes. All images were obtained at 500× magnification.

Superstar Red shimmer had a range of 76.2–7170.9 μm^2. Further comparisons between natural and synthetic mica samples of different colors (e.g., green, blue, and violet) showed similar results, with the synthetic samples always larger in size on average. Figure 1.2 shows the Red Wine and Superstar Red shimmer under the Keyence microscope in both reflectance (Figure 1.2a,c) and transmission (Figure 1.2b,d) light. Both shimmer samples were analyzed at various angles of impingent light to demonstrate the color-shifting properties of mica pigments (Figure 1.3).

Regarding color analysis, although an observer may view both of these samples as red by the naked eye, the synthetic sample appears more translucent under the microscope (Figure 1.2c,d). Thus, natural and synthetic samples may be differentiated using microscopy alone. However, although a difference in color

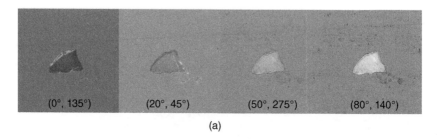

(a)

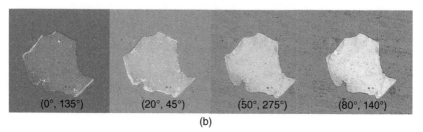

(b)

Figure 1.3 Microscopic images of (a) Red Wine CP-504 and (b) Superstar Red Wine CP-7059 shimmer samples in reflectance mode at different light angles represented by (latitude°, longitude°).

may be visually observed among samples, a reliable objective color measurement, i.e., Red-Blue-Green (RGB) values, may not be obtained per sample. As shown in the figures, there are color differences within single particles of a certain shimmer sample, most probably due to mica's imperfectly flat cleavage.

1.4.3 Infrared Spectroscopy

Infrared spectroscopy is a common technique used in forensic casework. Recent work has evaluated FTIR as an analytical method in the forensic analysis of cosmetic shimmer. Gordon and Coulson aimed to differentiate 53 distinct cosmetic foundations based on FTIR spectral differences and achieved a discriminating power of 98.3% [19]. Moreover, a second group detected and analyzed fingerprints contaminated with various cosmetics. Researchers obtained FTIR spectral imaging of each sample by attaching a focal plane array (FPA) detector to the IR spectrometer [20]. FTIR imaging was achieved because the FPA detector allows for simultaneous collection of one spectrum per pixel [21]. The researchers determined that different cosmetic samples, i.e., body butter and lip gloss, may be characterized based on the mid-IR region (400–4000 cm^{-1}). Differentiation of samples due to the identification of fundamental molecular vibrations was achieved [20]. An example of the vibrations observed in the mid-IR region for muscovite mica shimmer samples is presented in figure 3 from Ref. [22]. This region shows OH vibrations at 3622 cm^{-1} and different SiO$_4$ vibrations (Si–O and Si–O–Si) around 1063, 1028, 993, and 926 cm^{-1}. In comparison, figure 5 from Ref. [22] presents the IR spectrum for biotite mica which has a similar overall spectrum, but there are differences between the two spectra in the fingerprint region [22]. The medium to strong peaks around 500 and 1100 cm^{-1} present in the muscovite mica are less intense and less sharp in the biotite mica sample. Although muscovite mica is most commonly used in the cosmetic industry, other types of mica may be used as well. It is evident that different types of mica may be identified by use of FTIR, as the two mica spectra are different in the *fingerprint* region.

This chapter's authors evaluated the ability to use FTIR to analyze shimmer powders. Comparing shimmer samples from Just Pigments using a JASCO FTIR instrument with an attenuated total reflectance (ATR) attachment, it was found that although the mica samples have the same chemical composition, it is still possible to differentiate them. Karakassides et al. demonstrated that different vibrations, although not the focus of their paper, may be helpful to distinguish mica samples. The region around 950–1200 cm^{-1} represents the stretching of Si–O–Si and Si–O–Al bridges, and the bending vibrations of those bridges occur below 600 cm^{-1}. Additionally, the region around 800–950 cm^{-1} is characteristic of the OH bending vibrations [23]. Figure 1.4 shows the FTIR spectra of four different red samples, three of which are comprised only of mica and iron oxide (Ruby Red, Red Wine, and Flash Red, Figure 1.4c–e), and 1 synthetic red sample, Superstar Red (Figure 1.4f). It is evident that shades of a certain color may be a reason that the samples are differentiated, as shown in Figure 1.4a. Additionally, the natural and synthetic mica samples are easily characterized. The synthetic Superstar Red shimmer sample showed a higher intensity signal around 440 cm^{-1}, which was not present in the three natural mica samples. Interestingly, Flash Red and Superstar Red are marketed as red shimmer samples and appear as different shades of red to the naked eye, but their photomicrographs do not indicate that the sample is red as was observed with Ruby Red and Red Wine (Figure 1.4c,d).

In comparison, two green shimmer samples Blackish Green (Figure 1.4g) and Sparkle Green (Figure 1.4h) were analyzed by FTIR. These two samples, whose spectra are presented in Figure 1.4b, are both comprised of natural mica, titanium dioxide, and chromium green oxide. Despite the similarities in the chemical composition of the shimmer samples, the spectra are individual and distinct from one another.

FTIR does have some limitations when analyzing shimmer particulates. As mentioned previously, metal oxide coatings such as titanium dioxide, iron oxides, and chromic oxides are used in the preparation of shimmer particles. The analysis of simple inorganic materials via FTIR is not particularly characteristic, and the wavelengths at which such peaks are observed are in the far infrared region beyond the detector range of common IR instruments. Additionally, FTIR

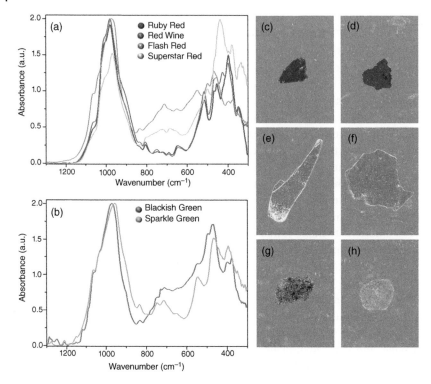

Figure 1.4 Overlay of FTIR spectra for (a) Ruby Red, Red Wine, Flash Red, and Superstar Red and (b) Blackish Green and Sparkle Green shimmer. Spectra shown are an average of five replicates each measured with 64 scans and a resolution of $4.0\,cm^{-1}$. Individual particle photomicrographs collected in reflectance light mode are presented for individual particles: (c) Ruby Red, (d) Red Wine, (e) Flash Red, (f) Superstar Red, (g) Blackish Green, and (h) Sparkle Green shimmer. All images captured at 500× magnification.

microscopy traditionally cannot analyze particles that are only a few microns in diameter. The resolution of the laser is not small enough to interrogate samples that are that small. However, instrument systems capable of beating the diffraction limit have been introduced recently [24, 25].

1.4.4 Raman Spectroscopy

Raman spectroscopy is another useful technique that may aid in the analysis and/or characterization of shimmer. Unlike FTIR analysis, since shimmer particles may vary from 5 μm to over 300 μm in diameter, particle size is not a limitation for Raman microspectroscopy. Raman can also interrogate the sample at wavelengths where peaks from the inorganic material can be observed.

Singha and Singh analyzed muscovite and biotite mica using Raman spectroscopy (refer to figures 4 and 6 in Ref. [22], respectively). In comparison to the muscovite and biotite mica samples that the authors analyzed by FTIR, which appeared very similar, the Raman spectra are clearly different. The muscovite mica spectrum has representative bands which include a strong $1127\,cm^{-1}$ and a medium $914\,cm^{-1}$ peak for the stretching of Si–O–Si and Si–O–Al bridges, respectively. The peak around $579\,cm^{-1}$ belongs to the bending of Al–O–Al bridges, whereas those bands near 407 and $263\,cm^{-1}$ arise from O–Al–O and O–Si–O translations, respectively. Finally, Al–OH translations are observed at $197\,cm^{-1}$. Conversely, the stretching of Si–O–Si bridges in the biotite sample is detected as a medium peak around $1130\,cm^{-1}$. The biotite spectrum also has representative bands for Si–O–Si bending at 685 and $552\,cm^{-1}$ [22]. It is apparent that the two spectra are quite different, and thus if a questioned sample was collected and analyzed, the spectrum should be similar to muscovite mica, primarily because this type of mica is most commonly used in the cosmetic industry.

Raman spectroscopy has recently been included in the analysis of cosmetic products for forensic purposes [26–29]. Gardner et al. used Raman to study and compare both lip glosses and lip balms [30]. They suggested using a $780\,nm$ laser source instead of $532\,nm$ to reduce the likelihood of fluorescence affecting analysis (refer to Figure 1.5). The spectra show a higher baseline using the

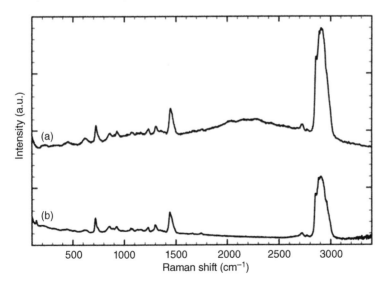

Figure 1.5 Raman spectra of white Claire's cosmetics lip gloss using the (a) 532 and (b) 780 nm laser sources. Source: From Gardner et al. [30].

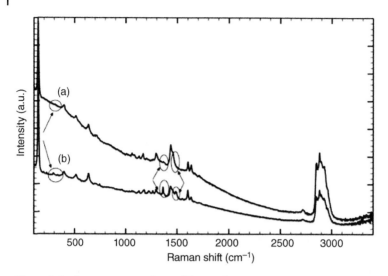

Figure 1.6 Raman spectra of two different lip gloss samples from Clinique Glosswear. (a) Air Kiss and (b) Raspberry Jam. Source: From Reference [30].

532 nm laser, which they believe is due to fluorescence, mainly for the highly colored products. The researchers were able to differentiate among various brands but were not capable of individualizing the products from the same brand. When the sample was translucent and/or lightly colored, i.e., lip gloss or lip balm, it was more difficult to differentiate the samples from one another. Raman analysis was better suited for heavily colored lip cosmetics, where the samples could be differentiated easily due to the more intense colors in the samples. To demonstrate the differences observed via Raman analysis, two colored lip gloss samples from the same product line, Clinique Glosswear, are presented in Figure 1.6 and the key spectral differences between the samples are highlighted [30].

Although the researchers focused on the comparison of the lip gloss and lip balm samples with no emphasis on shimmer at all, they provided a Raman spectrum of a sample containing titanium dioxide and iron oxide (Fe_2O_3). They claimed that the results shown in Figure 1.7 are an example of a typical lip gloss spectrum and that TiO_2 and Fe_2O_3 were regularly detected in lipsticks. The peaks observed for TiO_2 were near 396, 515, and 640 cm^{-1}, and those for Fe_2O_3 were located around 225, 294, and 410 cm^{-1} [30]. The bands detected for these metal oxides are indeed in the far region below 640 cm^{-1}, as mentioned earlier, and thus Raman spectroscopy may be a better option than FTIR for the identification of the metal-oxides in shimmer.

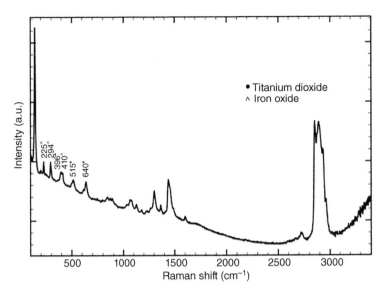

Figure 1.7 Raman spectrum of Clarins Rouge Prodige lip gloss in the color Grenedine, demonstrating a typical lip gloss spectrum. Source: From Reference [30].

1.4.5 X-Ray Diffraction

Although not commonly used in forensic evidence analysis, XRD is beneficial to the study of cosmetic shimmer as a means of trace evidence because the technique allows for the identification of the crystalline lattice structure of the silicate crystals based on unique diffraction patterns. Based on this spectral information, an examiner may establish the presence of mica substrates and/or metal-based pigments in the sample. Information gathered via XRD may be linked to results obtained using more common techniques, e.g., SEM-EDS analysis. While SEM-EDS provides for elemental information, allowing the analyst to evaluate the metal oxides used to coat the mica substrate, XRD provides information about the crystalline structure. The XRD and SEM-EDS are complementary techniques, and thus may both be used by an analyst to chemically characterize shimmer samples. Several research groups have focused on using XRD to examine the crystalline structure of mica samples and naturally found materials.

Singha and Singh analyzed muscovite and biotite mica using XRD, where they demonstrated a clear difference in the diffraction pattern between each mineral (figures 1 and 2 from Ref. [22], respectively). Both samples had a relatively intense peak around 36°. The most significant difference is that muscovite mica had higher peak intensities around 17.9° and 45.6°; whereas, the biotite sample had a base peak at 8.8° [22].

In her dissertation thesis, Elizabeth Kulikov presents the analysis and characterization of 39 cosmetic foundations using XRD. Although not specifically looking into shimmer-containing foundation products, some of her samples contained coated mica. She was able to detect mica, titanium dioxide, as well as other minerals in her samples. Of the 39 samples, 23 of them were classified as mineral-based foundation, and she found that most of them contained muscovite mica, compared to tradition-based foundations which mainly contain talc [31].

Rigaku Corporation, a company known for manufacturing and distributing analytical instruments, examined four cosmetic foundations on a Rigaku MiniFlex benchtop XRD. Of those four foundations, muscovite was detected in three of the samples, and all of the four samples indicated the presence of either iron oxide, titanium dioxide, or both compounds [32]. The results obtained by their study suggested that shimmer is commonly added to cosmetic products to cause a pearlescent effect and thus it is critical that an evaluation and validation of shimmer, in addition to glitter, is conducted to increase the analytical methods that can be used in casework involving cosmetic transfer.

Further analysis of the aforementioned Just Pigments samples analyzed by the authors using an XRD showed that shimmer samples, not previously incorporated into a cosmetic vehicle, are distinguishable. Using the same samples as before, there are spectral differences observed between the natural mica Ruby Red sample and the synthetic Superstar Red (Figure 1.8). The analysis was performed using a PANalytical Empyrean XRD with a copper anode at 40 mA and 45 kV. The main mica peaks were observed around 8.8°, 17.8°, 26.8°, and 45.3°. Such peaks differed between the two different shimmer particles, and in fact, when comparing the spectra to powder diffraction files (PDF) from the inorganic crystal structure database (ICSD), different identifications were obtained for each sample. Results showed that Ruby Red was composed of muscovite mica, while Superstar Red contained potassium magnesium aluminum fluoride silicate, another name for fluorphlogopite, supporting the synthetic designation. The main difference between the two mica substrates is the intensity at 17.8°, where fluorphlogopite typically has a lower intensity band. In addition, iron oxide spectral matches were present for both samples. The peaks consistent with Fe_2O_3 were located at 33.2°, 35.7°, 39.3°, 54.1°, 57.6°, and 72.0°. It is apparent that the natural red sample (Red Wine) had larger amounts of iron oxide. Conversely, identifications were obtained for titanium dioxide in the synthetic red sample (Superstar Red) but not in the natural red sample. The titanium peaks were located at 27.4°, 35.9°, 41.3°, 54.3°, and 69.3°. These results were consistent with the manufacturer details.

Similar results were obtained when comparing a red to orange shimmer sample. The red sample, Ruby Red, showed no presence of TiO_2, as opposed to the orange sample, Coral Reef, where several TiO_2 peaks were detected (Figure 1.9). The peaks for TiO_2 at angles 27.4°, 35.9°, 41.3°, 54.3°, and 69.3° were mainly observed

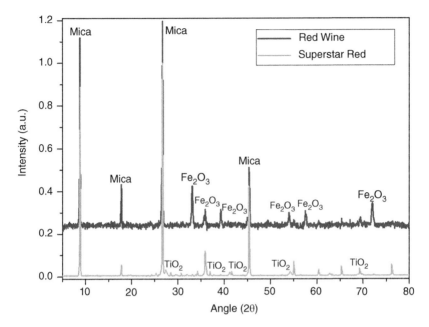

Figure 1.8 Overlay of XRD spectra for Red Wine and Superstar Red. Spectra shown are an average of three replicates.

in the orange sample's spectrum. However, those peaks consistent with iron oxide were present in both shimmer powders. Again, these results were in agreement with the manufacturer composition details for the two samples, and thus demonstrate how XRD may easily identify the presence of shimmer based on the XRD diffraction patterns of mica substrates and metal oxide coatings. Results presented in both figures were created from the average of three replicates per sample.

Cosmetic analysis via XRD has not yet received much attention in the forensic science field, but since most of the ingredients that make up cosmetics are highly crystalline in nature, XRD has the potential to be extremely helpful.

1.4.6 Scanning Electron Microscopy – Energy Dispersive X-Ray Spectroscopy

Scanning electron microscopy with energy dispersive X-ray spectroscopy (SEM-EDS) can be used to obtain black and white images with high magnification and to analyze the elemental composition of the sample. The SEM visualizes individual flaked shimmer particles which allow the analyst to locate the particle of interest for further elemental analysis using energy dispersive X-ray spectroscopy (EDS). The EDS identifies the atomic composition of the

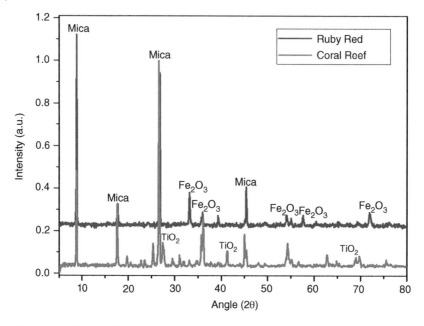

Figure 1.9 Overlay XRD spectra for Ruby Red and Coral Reef. Spectra shown are an average of three replicates.

mica substrates and the different metal-oxide coatings used to generate the certain shimmer color. An SEM image of typical shimmer specks is presented in Figure 1.10.

The elemental information obtained through this technique allows the analyst to differentiate samples based on the composition of the mica substrate and/or the metal oxide coating. Figure 1.11a shows the dissimilarities in the two red samples discussed previously. While both samples showed the presence of K, Al, and Si due to the mica's chemical composition, the synthetic mica Superstar Red sample is characterized by the presence of Mg and Ti, which were elements not present in the natural Red Wine shimmer sample. Another significant difference between the samples was the abundance of Fe from the iron oxide coating. Both samples indicated the presence of iron; however, the Red Wine sample was coated with larger quantities of iron oxide, as the Fe peak at 6.4 keV is significantly smaller for the Superstar Red sample. To demonstrate the differences that are observed between samples of different shimmer colors, Ruby Red and Coral Reef (orange) were analyzed. Figure 1.11b indicates that Ti appeared in the orange shimmer and fluorine (F) was present in the Ruby Red similar to the fluorine observed in the other red samples. Mg was absent in both of these samples, as Mg was mainly characteristic of synthetic mica samples alone. These results are consistent with

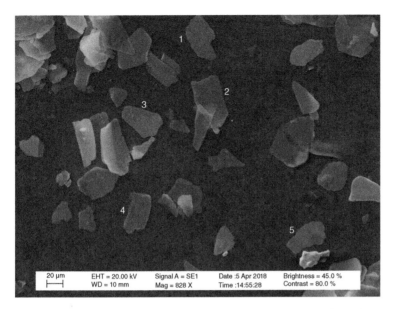

Figure 1.10 SEM secondary electron image of Lemon Yellow CP-6013 shimmer sample from Just Pigments, with numbers representing the particles analyzed by EDS. Image taken using Zeiss (LEO) 1450VP SEM.

manufacturer composition information and are in agreement with the XRD analysis.

1.5 Ideal Contact Trace

In the discussion of glitter in this book's first edition, seven properties were mentioned for a substance to be classified as an ideal contact trace [1]. The properties that make up the ideal trace evidence are that the sample is: (i) nearly invisible, (ii) highly probable to transfer, (iii) highly individualistic, (iv) easily collected, separated, and concentrated, (v) easily characterized, (vi) searchable via computerized database, and (vii) will survive most environmental insults [33]. To support the idea of using shimmer as a form of trace evidence, it is necessary to determine if shimmer meets these criteria to be considered an ideal contact trace. The following sections will demonstrate that shimmer is almost the perfect trace evidence.

1.5.1 Nearly Invisible

Shimmer particles are extremely small and are commonly undetected unless present in a bulk powder form. Albeit difficult, it is possible for an individual to

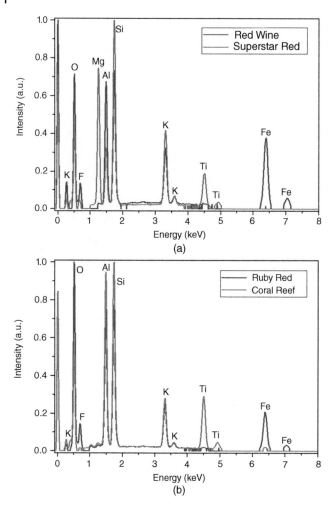

Figure 1.11 Overlay of EDS spectra for (a) Red Wine and Superstar Red and (b) Ruby Red and Coral Reef shimmer samples using a Zeiss (LEO) 1450VP SEM equipped with an Oxford 7353 EDS detector. Each spectrum shown is an average of five replicates.

view a single glitter particle. However, that is not the case for shimmer as it is typically much smaller in size and more transparent than glitter. Single shimmer particles may not be observed by the human eye unless light shined onto it was reflected. Just Pigments offers over 300 shimmer products and claims that their colored mica pigments fall anywhere between 5 and 300 μm in size. If shimmer were to transfer from or onto a victim, suspect, or crime scene, the trace will be nearly invisible due to its small size, i.e., 5 μm or less, and thus the suspect will not attempt to mitigate such evidence left behind, being most likely unaware of its presence.

1.5.2 High Probability of Transfer and Retention

Cosmetic industries include the metal-oxide-coated mica in the chemical make-up of many products in order to provide a shiny or shimmery finish. The shimmer particulates causing this pearlescent effect range in size but most fall below a few hundred microns. Due to the very small size and low weight of these particles, shimmer has a high transfer potential. Moreover, it is quite often recognized that once shimmer specks have transferred onto someone's skin, for instance, the particles linger until noticed and forcefully removed or until the person takes a shower.

1.5.3 Highly Individualistic

As discussed in the opening of this chapter, shimmer may not be characterized based on shape and morphology because such properties vary significantly between individual particles of the same sample batch. Thus, a questioned sample may not be directly compared to a known sample based on shape and morphology alone. However, cosmetic shimmer may be differentiated based on other characteristics. **Size** does differ among samples, but as suggested in Section 1.4.2, it is best to analyze shimmer based on a size range, rather than a size average. This will be difficult if a questioned sample falls within multiple size ranges of different known samples. **Thickness** may be used as another measure of individualization. Shimmer particles have different thicknesses of the mica substrate layer and the metal-oxide layer. Moreover, many manufacturers produce the mica powder in many different **colors** to offer a wide selection to the consumer. Crafter's Choice, a corporation that sells products used in creating soaps, lotions, bath bombs, and cosmetics, offers 19 mica powder assorted sample sets, with each set containing at least 4 different colors of shimmer powder.

Furthermore, shimmer particles may be distinguished based on **density**. For the two red shimmer samples mentioned earlier, Just Pigments claims that the density of Red Wine is between 3.0 and 3.1 g/ml and that of Superstar Red is 0.15 and 0.7 g/ml. Alternatively, a gold sample they offer, called Abstruse Gold CP-307, has a density of 3.2–3.3 g/ml. It is thus obvious that density differs among samples and may help further classify an unknown sample. Lockett et al. describe the use of magnetic levitation to determine the density of trace evidence [34]. They presented the ability to measure the density of glitter particles due to diamagnetic properties and therefore results shall potentially be comparable to shimmer as well. Muscovite mica is a diamagnetic biaxial crystal [35]; however, the presence of iron may potentially make the mica paramagnetic [36] and thus render this method difficult for iron-containing shimmer samples. One other method to measure density is the use of a density gradient column [28]. Particles may be placed into a chemical gradient of different known density standards, and the density level at which the particles fall into will reflect the density of the shimmer [37–39].

Manufacturers also play a major role in differentiating samples. It is evident that there are differences between shimmer samples even within the same company, and thus, it is expected that there are differences among manufacturers as well. Even if mica was purchased from the same source by two or more different brands, manufacturers may use different metal oxides and in varying combinations to achieve similar colors. Finally, additional differentiating information for recovered shimmer may be obtained by analyzing the **vehicle** for the shimmer particles, i.e., lipstick, lotion, eyeshadow, etc. Companies have different product formulations, and therefore, even if the same shimmer particles were used from one source, the adhered vehicle may also prove discriminatory.

In review, shimmer may be individualized based on size, thickness, color, density, manufacturers, cosmetic vehicle, chemical composition, and potentially number of layers.

1.5.4 Easily Collected, Separated, and Concentrated

Particles left behind at a crime scene or transferred onto a perpetrator may be easily collected. Cosmetic smears from foundations, lipsticks, or eye shadows containing shimmer may be located on skin, clothing, bed sheets, furniture, etc. Smears located on an attacker's skin are easily collected using a cotton swab. Otherwise, if the physical evidence is present on clothing or bed sheets, the examiner may cut around the cosmetic smear and particles extracted from the fabric could then be used for further analysis. If scissors are not available, then either a cotton swab or a spatula will work to collect evidence. Make-up could be scraped off using the spatula and placed into an evidence bag. However, if the smear is invisible to the naked eye, a crime scene technician can use natural or UV light to search for shimmer particles. The reflection caused by the light will readily indicate to the technician that a smear may be present for collection. After one or more particles are detected, a simple Post-it note could be used to collect the shimmer. Moreover, as previously mentioned, samples may be separated quite easily from the cosmetic vehicle for subsequent analysis. Griggs et al. suggest a simple hexane wash for the shimmer-containing cosmetic product(s) and subsequent extraction using filter paper [2].

1.5.5 Mere Traces Easily Characterized

Once the shimmer particulates are separated from their matrices, we suggest examining the evidence first using a compound microscope and then proceeding with any of the aforementioned analytical techniques to potentially classify the sample. Shimmer is easily characterized using any of the instruments discussed herein. Some techniques, such as Raman or SEM-EDS, only require a single particle for analysis. Other instruments, i.e., FTIR, only require a small amount of sample. Examples provided in earlier sections of the analyzed particles from Just Pigments demonstrated that shimmer may be differentiated and characterized based on the analytical data collected.

Analyzing 40 shimmer samples of different colors, sizes, and chemical composition from Just Pigments via SEM-EDS has illustrated the classification potential of shimmer. Based on preliminary results, the authors Najjar and Bridge were able to differentiate samples into different classes and achieved a classification rate as high as 100% for the shimmer samples using various statistical models. Moreover, Pearson's correlation coefficients were calculated based on the similarities in the normalized peak ratios of EDS peaks for each pairwise sample comparison of the EDS spectra for each sample. The authors showed that, indeed, mere traces of shimmer may be easily characterized. The 40 shimmer samples generated a total of 400 intra- and 19 500 inter-sample pairwise comparison correlations. As illustrated in Figure 1.12, results from the Pearson's correlation test showed that the mean correlation coefficients between intra- and inter-sample variabilities did not overlap at the 95% confidence level, and thus suggested that samples may be differentiated with little ambiguity. Thus, if an unknown shimmer particle was collected at a crime scene and compared to a known sample, it could be associated appropriately with high accuracy. Results obtained for this study were

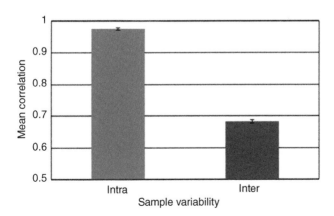

Figure 1.12 Box plot representation of the intra- and inter-sample variability based on Pearson's correlation coefficients at the 95% confidence level for 40 shimmer samples from Just Pigments analyzed via SEM-EDS. The 95% confidence interval is represented by the error bars.

based on the analysis of five particles per sample; however, only one particle is required for SEM-EDS analysis.

1.5.6 Searchable via Computerized Database

No database currently exists for cosmetic shimmer particulates. However, each of the properties discussed above that makes shimmer highly individualistic can be measured via one or more techniques. Results from the analysis of shimmer-containing cosmetic products using multiple instrumentation may be incorporated into a searchable database that could be made available to the forensic science community and to the general public as well.

1.5.7 Will Survive Most Environmental Insults

Similar to glitter, shimmer should fare well in most environmental conditions. Shimmer-containing cosmetic products (e.g., eye shadow, foundation, highlighter, etc.) are designed to survive many environmental insults such as exposure to heat, sunlight, and water/rain, in addition to make-up removers. In fact, titanium dioxide, one of the primary metal oxides used to coat the mica substrates, is commonly used in sunscreens because it is a UV refractor and has high resistance to discoloration due to sunlight exposure. Moreover, when make-up is cleaned using cosmetic wipes, one is able to see the remaining shimmer powder adhered to the towelettes.

1.6   Case Examples

Many criminal investigations have previously gone unsolved due to lack of evidence found at a crime scene, or evidence deemed useless due to lack of technology. However, with the recent advances and improvements in analytical techniques, experts are now examining evidence previously not considered useful. One type of such evidence is cosmetic smears or residue on clothing, furniture, cups, glasses, cigarette butts, and more. The questioned cosmetic smears will most likely contain either glitter or shimmer particles. In this book's earlier edition, several cases were presented for glitter as forensic evidence [1].

Not many forensic cases have involved shimmer-containing evidence, as it is a field just recently being explored. However, in 1991, in his paper published in the *Journal of Forensic Sciences*, Choudhry claimed he was able to aid in solving a case that involved a document containing a lip impression [40]. The questioned document was analyzed against a control lipstick using microspectrophotometry and SEM-EDS with results suggesting that the two samples were from different

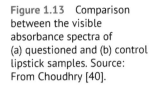

Figure 1.13 Comparison between the visible absorbance spectra of (a) questioned and (b) control lipstick samples. Source: From Choudhry [40].

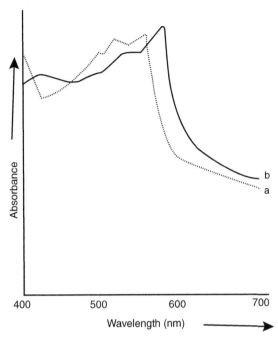

Wavelength (nm)

sources. Figures 1.13 and 1.14 show the results obtained for the questioned and control lipstick samples. Choudhry explained the difference in elemental composition of the pigment particles found in the two lipsticks (refer to Figure 1.14). The questioned sample exhibited main presence of bismuth, while the control sample primarily contained barium [40].

Another case involving pigmented cosmetics was solved in 1912 by Edmond Locard. A woman named Marie Latelle was found murdered in her parent's house. The main suspect was her boyfriend, but he had an alibi. He spent the night playing cards with his friends at a country club. When Locard visited the morgue and examined the corpse, he was convinced that she was strangled to death based on physical markings on her body. Locard scraped underneath her boyfriend's fingernails and the traces collected were skin cells containing pink dust, which Locard believed to be make-up based on the analytical results which indicated the presence of zinc oxide, bismuth, and iron-oxide pigments. Since cosmetics was not mass produced at that time, Locard believed he had a reason to investigate the boyfriend a little further. He found a chemist who had developed a custom face powder for Latelle, which was collected from her room. This face powder was comprised of ingredients similar to those found under the suspect's fingernails. After confronting him about the evidence, the suspect confessed to have changed the time on the wall clock, making his friends believe they were out with him longer

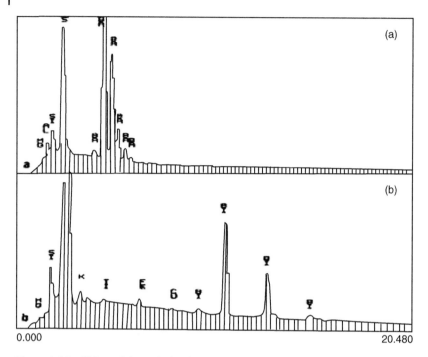

0.000   20.480

Figure 1.14 EDS particle analysis of (a) control and (b) questioned lipstick samples. Source: From Choudhry [40].

than they actually were. Meanwhile, the boyfriend visited the young girl's place and committed the murder [41, 42].

1.7 Conclusion

Close personal attacks are unfortunately common and perpetrators are well aware of typical trace evidence analyzed by examiners. With little to no consideration to cosmetic transfer, shimmer from cosmetic products may prove useful as a new trace evidence. Although glitter has been studied as a source of trace evidence, little to no research exists on the forensic analysis of shimmer. Cosmetic traces containing shimmer particles may be easily collected, separated, and analyzed. Results and discussions presented herein suggest that samples from the same manufacturer are easily distinguished and thus shimmer classification is possible via the various aforementioned techniques (FTIR, Raman, XRD, and SEM-EDS). Shimmer meets all criteria to be considered an ideal contact trace and will potentially aid in cases where cosmetic trace evidence remains after theft, intimate assault, or even car accidents.

Acknowledgments

The authors would like to acknowledge the National Institute of Justice (Grant No. 2017-R2-CX-0005) and the State of Florida for funding the research on cosmetics.

References

1 Blackledge, R.D. and Jones, E.L. (2007). All that glitters is gold. In: *Forensic Analysis on the Cutting Edge: New Methods for Trace Evidence Analysis,* Chapter 1 (ed. R.D. Blackledge), 1–32. Hoboken, NJ: Wiley Interscience.

2 Griggs, S., Hahn, J., and Bonner, H.K.S. (2011). Shimmer as forensic evidence. *Global Forensic Sci Today* 1 (10): 19–23.

3 Argoitia, A. (2007). Technology and applications of microstructured pigments. *JDSU Flex Products GroupAIMCAL Technical Conference Scottsdale, AZ*. October 07–10 2007.

4 Weaver, C. (2018). What is glitter? *The New York Times* (21 December).

5 Grieve, M. (1987). Glitter particles—an unusual source of trace evidence? *J. Forensic Sci. Soc.* 27 (6): 405–412.

6 Jones, Jr., E. (2004). Trace Evidence and Bloodstain Interpretation from the Sanchez/Barroso Case, presented at the Fall Seminar of the California Association of Criminalists, Ventura, CA, USA (October 2004).

7 Sepp, S. Biotite. Sandatlas. https://www.sandatlas.org/?s=biotite&submit .x=16&submit.y=11 (accessed 20 June 2019).

8 Ferry, J.M. and Spear, F.S. (1978). Experimental calibration of the partitioning of Fe and Mg between biotite and garnet. *Contrib. Mineral. Petrol.* 66 (2): 113–117.

9 King, H.M. Muscovite. https://geology.com/minerals/muscovite.shtml (accessed 28 May 2019).

10 Friedman, H. The Mineral Muscovite. https://www.minerals.net/mineral/ muscovite.aspx (accessed 21 June 2019).

11 Argoitia, A. (2002). Pigments exhibiting a combination of thin film and diffractive light interference. *AIMCAL 2002 Fall Technical Conference Meeting Sedona, Arizona October 20–23, 2002*. Flex Products Inc.

12 Jiang, Y., Wilson, R., Hochbaum, A., and Carter, J. (2002). Novel pigment approaches in optically variable security inks including polarizing cholesteric liquid crystal (CLC) polymers. *Optical Security and Counterfeit Deterrence Techniques IV, International Society for Optics and Photonics: 2002*, pp. 247–255.

13 Examination of Cosmetic Pigments, McCrone Inc. 102 slide PowerPoint presentation. https://projects.nfstc.org/trace/docs/final/Cosmetic%20pigment%20presentation.pdf.

14 Pfaff, G. Effect pigments – a successful interplay of chemistry and physics. https://q-more.chemeurope.com/q-more-articles/190/fascinating-displays-of-colour.html (accessed 28 October 2019).

15 Natural VS. Synthetic Mica and Glitters. https://www.renascentbathbody.com.au/blogs/articles/natural-vs-synthetic-mica-and-glitters (acessed 15 March 2022).

16 Edmond Locard. https://www.crimemuseum.org (accessed 23 May 2019).

17 Petherick, W., Turvey, B.E., and Ferguson, C.E. (2009). *Forensic Criminology.* Academic Press.

18 Kirk, P.L. (1953). *Crime Investigation. Physical Evidence and the Police Laboratory.* Interscience, 784 pages.

19 Gordon, A. and Coulson, S. (2004). The evidential value of cosmetic foundation smears in forensic casework. *J. Forensic Sci.* 49 (6): 1244–1252.

20 Ricci, C. and Kazarian, S.G. (2010). Collection and detection of latent fingermarks contaminated with cosmetics on nonporous and porous surfaces. *Surf. Interface Anal.* 42 (5): 386–392.

21 Tahtouh, M., Flynn, K., Walker, S. et al. (2007). FTIR spectral imaging applications in trace evidence. *NIJ Trace Evidence Symposium*, pp. 13–16.

22 Singha, M. and Singh, L. (2016). Vibrational spectroscopic study of muscovite and biotite layered phyllosilicates. *Indian J. Pure Appl. Phys.* 54: 116–122.

23 Karakassides, M.A., Gournis, D., and Petridis, D. (1999). An infrared reflectance study of Si–O vibrations in thermally treated alkali-saturated montmorillonites. *Clay Miner.* 34 (3): 429–438.

24 *Particle Characterization.* Horiba Scientific [At the below website you have access to 46 Application Webinars, 22 Training Webinars, and 30 Technology Webinars, all covering various aspects of Particle Characterization.] https://www.horiba.com/scientific/products/particle-characterization/download-center/webinars (accessed 15 March 2022).

25 Morphologi G3-ID. https://www.malvernpanalytical.com/en/products/product-range/morphologi-range/morphologi-g3-id (accessed 15 March 2022).

26 Salahioglu, F. (2012). Application of Raman spectroscopy to the differentiation of lipsticks for forensic purposes. *Anal. Methods* 5 (20): 5392–5401.

27 Palenik, C.S., Palenik, S., Herb, J., and Groves, E. (2011). Fundamentals of forensic pigment identification by Raman microspectroscopy: a practical identification guide and spectral library for forensic science laboratories. US Department of Justice (237050).

28 Bruce, K. and Went, M.J. (2019). Conference: Kent Researcher Showcase. THE MAKEUP OF MAKEUP, "Raman Spectroscopic Characterisation

of Facial Cosmetics as Associative Trace Evidence". poster. https://www
.researchgate.net/publication/333681922_THE_MAKEUP_OF_MAKEUP_
Raman_Spectroscopic_Characterisation_of_Facial_Cosmetics_as_Associative_
Trace_Evidence (accessed 15 March 2022).

29 HORIBA Scientific (2013). Non-Destructive and In-situ Analysis of Pigments. October 18, 2013. https://www.azom.com/article.aspx?ArticleID=10089 (accessed 28 October 2019).

30 Gardner, P., Bertino, M., and Weimer, R. (2013). Analysis of lipsticks using Raman spectroscopy. *Forensic Sci. Int.* 22 (1–3): 67–72.

31 Kulikov Elizabeth (2013). Spectroscopic analysis and characterisation of cosmetic powders. PhD thesis. RMIT University. http://researchbank.rmit.edu.au/eserv/rmit:160498/Kulikov.pdf (accessed 15 March 2022).

32 Phase identification of common minerals in natural facial cosmetics. Rigaku Products by Application - Application Bytes – XRD. https://www.rigaku.com/applications/bytes/xrd/miniflex/468965291 (accessed May 31, 2019).

33 Aardahl, K. (2003) Evidential value of glitter particle trace evidence. Masters thesis. San Diego, CA, USA: National University.

34 Lockett, M.R., Mirica, K.A., Mace, C.R. et al. (2013). Analyzing forensic evidence based on density with magnetic levitation. *J. Forensic Sci.* 58 (1): 40–45. https://gmwgroup.harvard.edu/files/gmwgroup/files/1171.pdf.

35 Metzger, R.M. (2012). *The Physical Chemist's Toolbox*, 2e. Wiley.

36 Kendall, J. and Yeo, D. (1948). Magnetic susceptibility and anisotropy of mica. *Nature* 64 (2): 135.

37 Webb, P.A. (2001). Volume and density determinations for particle technologists. *Micromer. Instrum. Corp.* 2 (16): 01.

38 ASTM D1505-10 (2010). Standard test method for density of plastics by the density-gradient technique; ASTM International, West Conshohocken, PA.

39 Oster, G. and Yamamoto, M. (1963). Density gradient techniques. *Chem. Rev.* 63 (3): 257–268.

40 Choudhry, M.Y. (1991). Comparison of minute smears of lipstick by microspectrophotometry and scanning electron microscopy/energy-dispersive spectroscopy. *J. Forensic Sci.* 36 (2): 366–375.

41 Nickell, J. and Fischer, J.F. (2013). *Crime Science: Methods of Forensic Detection*. University Press of Kentucky.

42 Ramsland, K. (2012). LOCARD'S VISION: 100 years of crime labs. *Forensic Exam.* 21 (2): 60.

2

Glitter and Other Flake Pigments

Charles A. Bishop[1] and Robert D. Blackledge[2]

[1]*CA Bishop Consulting Ltd., Vacuum Deposition Technology, Leicestershire, UK*
[2]*Naval Criminal Investigative Service, Regional Forensic Laboratory, San Diego, CA, USA*

2.1 Introduction

This chapter updates information on glitter not covered in Chapter 1, ALL THAT GLITTERS IS GOLD!, in the previous book, *FORENSIC ANALYSIS ON THE CUTTING EDGE: New Methods for Trace Evidence Analysis*. Additionally, it covers the wide variety of flake pigments and their manufacture. One publication since that chapter is especially worth mentioning. *Glitter – The Ideal Trace Evidence?* [1], is a 29 slide PowerPoint presentation. A total of 239 glitter samples were examined. Slide 22 illustrates how a database containing the properties of specific brands and glitter types can be put into a searchable database. Properties listed include brand, number designation, shape, size range in μm, area range in μm, color, microspectrophotometry spectra, FTIR ATR spectra (both sides if different), and FTIR Library Search results. Additionally, slide 9 is a breakdown of the sample set by color, slide 12 by shape, and 18 by polymer type based on FTIR ATR spectra. Slide 21 advocates that Raman spectra of inner layers, SEM/EDS spectra, and density by magnetic levitation could also be added to the database. Particle thickness should also be included. The creation and continued addition to the database are quite important. The impression of the general public (jury) is that except for color, glitter is pretty much the same. By reference to such a database, the court testimony of the examining criminalist can dispel that notion.

2.2 Glitter Update

Plastics in general, but especially those plastics that do not readily biodegrade, are increasingly coming under attack. They end up in lakes, streams, and rivers. They

Leading Edge Techniques in Forensic Trace Evidence Analysis: More New Trace Analysis Methods, First Edition. Edited by Robert D. Blackledge.

are carried into oceans and wash up on beaches. They are mistaken for food by water-dwelling species and move up the food chain. Also, tiny fragments serve as substrates that toxic substances in water adhere to. They too are mistaken for food.

As a result of the increasing public awareness of the harmful environmental effects of tiny plastic pieces, the popularity of traditional glitter is likely to decrease. On 28 December 2015 President Obama signed into law H.R. 1321, the "Microbead-Free Waters Act" [2], which prohibits the manufacture, packaging, and distribution of rinse-off cosmetics and nonprescription products such as some face washes and toothpastes.

More recently attention has been given to microplastics that have been contaminating the land [3, 4].

Scientists around the world have been monitoring the amount of microplastics (plastic particles of <5 mm) that are carried by the wind and fall out of the sky to contaminate the land. In the Pyrenees, it was found that 365 microplastic particles per square meter per day were collected. In the Western United States, after a 14-month period of collecting data it was calculated that across 11 protected areas (6% of total United States) 1000 t of microplastics fall per year. One can expect that it is only a matter of time before this too will lead to legislation designed to reduce this problem. Glitter and flake pigments are already close to, or within, the dimensions for microplastics and so could become an endangered product in the near future.

Along with other industries there is an opportunity for switching materials to natural materials or biodegradable alternatives. Therefore, a market exists providing a biodegradable form of glitter can be made inexpensively. It already exists and is being sold. A disadvantage for glitter manufacturers is an advantage for glitter users. That is, some forms of biodegradable glitter may be made quickly and easily at home using primarily inexpensive household materials.

Biodegradable glitter, except for being larger, is more like what this book calls shimmer. It does not consist of hundreds of superimposed very thin polymer layers. But instead of it being particles of mica enclosed in a uniform thickness of titanium dioxide, it is composed primarily of a core of cellulosic material (obtained from hardwoods, mostly eucalyptus), and glycerin (Table 2.1). To obtain its vibrant colors, the core is surrounded by a very thin layer of aluminum, above which is a thin transparent coating that may be clear or colored. Glitter advertised as "edible" is not biodegradable, nor that labeled "vegan"

In the microscopic comparison of Questioned and Known glitter particles, examination under crossed polars may be useful for particles that are translucent rather than opaque [6–8] (Figure 2.1).

Table 2.1   Example of composition of biodegradable glitter.

CAS No.	Chemical name	Conc. (% w/w)
68442-85-3	Cellulose	c. 72.0%
56-81-5	Glycerin	c. 11.0%
7732-18-5	Aqua	c. 6.0%
57-13-6	Urea	c. 3.0%
9010-92-8	Styrene/acrylate copolymer	Max 8%

Source: Data from [5].

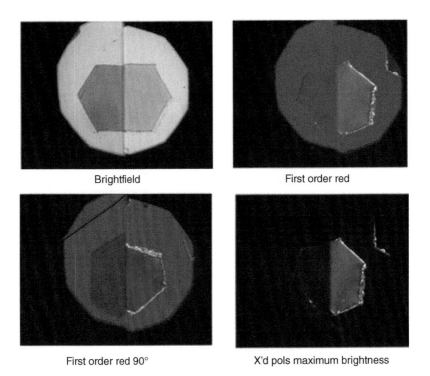

Brightfield

First order red

First order red 90°

X'd pols maximum brightness

Figure 2.1   Two similar but different glitter particles viewed under polarized light microscopy. Source: From Edwin L. Jones, Jr., Retired Forensic Scientist.

2.3 Cutting Film into Individual Glitter Particles

There are three methods of cutting material into glitter. The first is mechanically cutting the sheet; this is like a pastry cutter but on a very small scale. Rather than cut a single glitter flake at a time the cutter is one roll of a nip roll assembly and the other roll is a hard rubber roll. This kiss-cut nip presses the cutter through the web as it passes through the nip so that the whole web width is continuously cut. The rotary die cutter can be manufactured to any shape but it is easier for the shape to be regular with squares or hexagons being the most common. It is critical the cutter is sharp so that it will cut easily as well as release the cut glitter. As the cutter is rigid, the glitter flakes will match the precision with which the cutter was made. The second method for cutting glitter is by laser where the laser is rastered across the web cutting each glitter particle individually. Although this is slower than the die cutter it can be more productive in producing smaller flake sizes, and customized flake shapes. Laser cutting at a speed of 60 m/min on a 1 m wide web would require the laser to move 3 m in distance to cut a row of 50-μm squares. This would require a web speed of only 1 mm/min and would produce 400 000 flakes/min. This is slow compared to the rotary die cutting process [9]. The speed and individual cut for each flake means each flake is likely to have slight variations that are related to the web uniformity and any system vibrations that may affect the laser position. These may be small but when the flake size is small this becomes noticeable (Figures 2.2 and 2.3).

The third method uses a high-speed spinning cylindrical blade and a counterpart fixed blade with the web fed into the cutter at an angle over the fixed blade. The

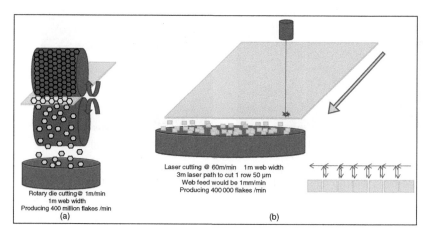

Rotary die cutting @ 1m/min
1m web width
Producing 400 million flakes /min
(a)

Laser cutting @ 60m/min   1m web width
3m laser path to cut 1 row 50 μm
Web feed would be 1mm/min
Producing 400 000 flakes /min
(b)

Figure 2.2 A schematic of rotary die cutting of glitter flakes (a) and laser cutting of glitter flakes (b).

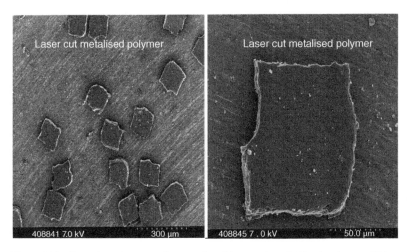

Figure 2.3   Scanning electron micrographs of laser cut metallized polymer glitter.

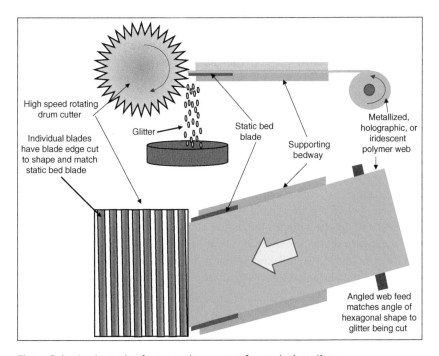

Figure 2.4   A schematic of a rotary drum cutter for producing glitter.

angle of the web feed and the shapes cut into the cylindrical and fixed blades define the glitter shape [10, 11] (Figure 2.4). For this machine square and hexagonal glitter are the easiest to achieve. The fixed and rotating blades have to be brought almost into contact to ensure the glitter is separated during cutting and occasionally teeth are lost from the cutting blades which accounts for some glitter having odd shapes. As with the kiss-cut rotary die cutter the precision of the machining of the blades defines the dimensions of the glitter produced. Punch cutters used to make sequins are inefficient in material use and the mechanics are difficult to miniaturize and so are not used to make glitter. A natural step in the examination of an unknown glitter and comparison to glitter particles from a known source would be to prepare thin cross sections for microscopic examinations as well as Raman spectroscopy. However, the polymers typically present are relatively soft, and biodegradable glitter is reported to be even softer. When making cross sections by hand, as the blade cuts down through, portions of the above layers will tend to slide down with the blade and to some extent obscure layers below. In an effort to avoid this, one of us tried using an ultramicrotome [12]. For this, to provide a support the individual particles were first embedded in an acrylic polymer. The cross sections when cut dropped into water. Once the cross sections had been recovered and dried, microscopic examination showed that the process had resulted in significant swelling, and reliable thickness measurements could not be made. Although not known at the time, we later learned that a typical plastic glitter particle had in excess of over one hundred very thin individual layers [13–18]. These layers were so thin that their thickness was less than the diffraction limit and therefore even at the highest optical magnifications individual layers could not be seen. In recent years, ways have been found to defeat the diffraction limit, but they have yet to be tried on glitter particle cross sections.

Cutting mixed materials where layers of hard and soft materials may be adjacent can cause problems of smearing of the soft material making it difficult to obtain clear thickness measurements. For some materials using cryogenic cooling of the sample may enable a brittle fracture of the material giving a clean view of the layers. For glitter, this would require the flakes to be potted in an acrylic into a sheet and the sheet dipped into liquid nitrogen and then withdrawn and hit with a hammer to shatter the sheet. It is then a matter of luck to quickly find good examples of suitably broken flakes. Similarly freezing the embedded sample can help in producing ultramicrotome samples for some mismatched materials.

2.4 Reflectance

Reflectance from a surface is made up of the combination of specular and diffuse reflectance. A perfectly flat surface with no surface defects or contamination,

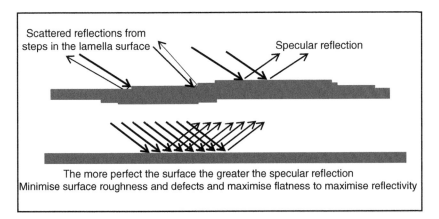

Scattered reflections from
steps in the lamella surface

Specular reflection

The more perfect the surface the greater the specular reflection
Minimise surface roughness and defects and maximise flatness to maximise reflectivity

Figure 2.5 Specular and diffuse reflectance.

if coated with a reflecting metal, such as silver, gold, or aluminum, will have a very high specular reflectance. An example of this would be a glass mirror. By comparison, the surface of untreated paper is rough and when metallized produces a matt metallic surface as a result of the very high diffuse reflectance and low specular reflectance (Figure 2.5).

Individual flakes will be similarly dependent on the quality of the flake surface. Polymer films may be homopolymers with or without the addition of fillers. Fillers will affect the surface profile which helps to control the surface friction and so the handling characteristics of the web.

In addition to the fillers, the polymer surface will have a large amount of surface contamination which also adds to the light scattering as well as contributing to "pinholes" in the metal coating where the particle contamination has been metallized and the particle subsequently moved leaving behind an uncoated area in the coating. Where polymer films are metallized, the coating is very thin usually <40 nm and over time the aluminum coating continues to oxidize and the metal thickness reduces as the metal oxide thickness increases. This will also change the reflectivity as the optical density is reduced and the transmittance increases (Figure 2.6).

Where the glitter is to be metallized and colored there are two options. One is to metallize one side of the polymer web and then coat the metal with a colored lacquer. This does mean that when viewed some of the glitter will be colored metallic and the rest silver metallic depending on which way up the glitter flake rests on the surface. The second option is to either dye the polymer film or to coat the second side with the lacquer so that the same colored metallic appearance is available from each surface (Figure 2.7). This application of a colored lacquer may be a cause of some loss of specular reflectivity. If the coating dries with some stress

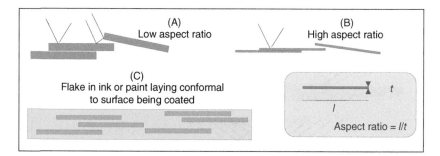

Figure 2.6 Long and thin flakes have a higher aspect ratio than thick flakes.

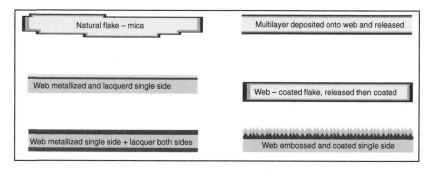

Figure 2.7 Options for metalizing and coating glitter.

in the coating it may cause some curl in the cut glitter flakes as the flake bends to relieve the stress.

Metallized polymer film is produced in very high volumes. The largest metallizers are 4.5 m wide and are capable of running at in excess of 1000 m/min [19, 20] For some small glitter manufacturers, the lowest cost source of metallized film was the offcuts from slit down wide webs. This was a bonus for the metallizers who could sell the offcuts rather than send it for scrap. Large manufacturers who have their own vacuum coater to metallize their own material are likely to have better control over their product which should be more reproducible. Where a glitter manufacturer buys metallized polymer web from a variety of suppliers it is possible that each type will have a different supplier or grade of polymer film and the coating thickness may also vary giving glitter flakes with a wider range of variability making it more difficult to identify a specific source of the flake.

Polymer film can be produced with slight differences but delivering the same mechanical and optical performance. Different polymer manufacturers may use a different catalyst within the process which can be used to identify the source of the film. Different types of filler can be used in the polymer and in different proportions for the different sizes of filler which may also help identify the grade of film being used [21]. If the metallized coating is coated using a colored lacquer either on one side or both sides, the chemistry of the lacquer is also likely to be proprietary and can vary from epoxy, acrylic, acrylic styrene, and nitrocellulose to vinyl. The lacquer in combination with the polymer substrate and metal coating can help to further reduce the number of possible sources of the flakes.

Reflection from metallized flake substrates can also suffer from some loss of specular reflectance due to thickness and flake structure. Few flakes are of uniform thickness. Mica is a common substrate and the way it is mined and refined to make flakes leads to flakes that are substantially flat but with some steps on the flat surface. The edges of the flake, as well as the ends from these steps, disrupt the specular reflection from the whole surface and lead to scattering and reflection losses [22].

Freestanding flakes have the potential to have the highest specular reflectivity. The flake is made from a vacuum coating deposited onto a polymer web which is stripped off the web, ground, and sized. The freestanding flakes are an order of magnitude thinner than flake glitter which reduces the edge effects that cause scattering and lower the peak reflectivity [23]. The flakes as deposited are conformal to the surface they are deposited onto. If the substrate polymer has a high-quality optical surface this will help to minimize the surface defects on the flakes helping to maximize the specular reflectivity.

Comparing the different flake materials, the freestanding flakes are the thinnest and the flakes will be to a very tight thickness tolerance. The flakes made from polymer web will be thicker and will be to the extruded polymer web thickness tolerance. Flakes from coated source flakes, such as mica, will be thicker than the freestanding flakes and the flake thickness will have more variety because the lamella substrate flakes will have steps on each surface. Depending on how well the source material has been split into flakes will depend on how many steps there are on each surface and this will affect the thickness variability.

2.5 Embossed Effects

Polymer web sourced glitter and freestanding flakes can be produced with holographic or grating embossed structures. These structures can produce color or iridescence effects. Both products start with the optical structure being embossed into the polymer surface. There is one anti-counterfeiting flake manufacturer that uses the embossing to define the size and shape of the final flake as well as giving the option of including a logo identifier onto flakes to make batches of flakes company/brand specific [24–26].

2.6 Color

Although in theory, color has great value for discriminating between glitter particles originating from different sources, in practice, it may not be as simple as a side-by-side comparison. Using a scanning goniospectrophotometer, it is possible to plot the color change with changing viewing and illumination angles [27, 28]. When determining color, it is likely that each flake will show a slightly different color and this may differ from the color seen when a larger area is coated with a multiplicity of flakes where the color is averaged out. The range of colors can be plotted and this may be large for low cost products that have a wide manufacturing tolerance. The light reflection is still a key part of the characteristics of the flake helping to give the color a high chroma or saturation or purity of color [29–31]. With security pigments such as the optical variable pigment (OVP) manufactured for use in banknote security, the manufacturing tolerance is very small, the color tightly controlled so that the exact color (hue) and chroma may be used to more confidently confirm the source of supply.

2.7 Specific Gravity

A paper published in 2013 introduced a new method of density determination involving the magnetic levitation of diamagnetic particles [32]. With this method, not only could the density of individual glitter particles be determined, but also particles having different densities could be separated and collected. The method is called MagLev and is described in detail in Chapter 6.

2.8 Is It Glitter or a Flake Pigment?

There are many methods to produce flake, platelet, or two-dimensional materials that can be used as glitter or pigments. The aim is that the material is large in two dimensions compared to the third dimension.

There are various ways to differentiate between glitter and pigment. One is that glitter is where flakes or platelets are used dry. A pigment is where the same flake or platelet is used in a carrier material whether it be a paint, ink, nail varnish, lipstick, etc. An alternative view is that glitter is where individual flakes can be seen by eye whereas a pigment is where the overall effect is seen but not the individual flakes that cause the effect. Whatever definition is used there is an overlap.

Glitter flakes are generally larger than pigment flakes with glitter being >25 μm and may be much larger than this whereas pigment flakes are more typically

<25 µm. The same materials may also be very much larger in size and used to make sequins that are of the order 5 mm diameter.

The flakes may be used for their aesthetic effect but may also be used with overt or covert security features. This does mean that the forensic interest in identifying the structure, size, and materials that make up the flakes may also enable identifying counterfeits.

Glitter flakes are the lower cost products with the covert anti-counterfeiting flakes being the higher cost products. This is also consistent with availability of the different products. Anti-counterfeiting flake materials can be harder to purchase with the increasingly higher security flakes being the hardest to source with scarcity being part of the security strategy. Purchasing the product requires vetting of the company purchasing the material and proof of the need for the increased security. Low cost glitter flakes are widely available from craft stores and online with purchasers and there is no restriction in who buys the product.

The brightness of the material will depend on how smooth the surface of the flake is as well as how it is applied to the substrate. The aspect ratio as well as what proportion of the edge of each flake is showing will also affect the brightness. The more perfect the surface the higher the brightness.

The glitter effect is because the individual flakes are sitting on a surface at very slightly different angles from being parallel to the surface. When viewed only some flakes directly reflect the light source but if you move your viewing point this reflection is lost but other flakes come into line and directly reflect the light source to the eye. This ever-changing bright reflection from different flakes is what we refer to as glitter. If all the flakes were to lie on a surface precisely parallel to the surface then the glitter effect would be lost and the surface would reflect as a whole surface but appear like a cracked mirror (Figure 2.8).

The mirror reflection of flakes is an optical effect and it is possible to make flakes with other optical effects such as colored metallic reflection with a single color. By using multilayer coatings multicolor flakes can be produced or more specific

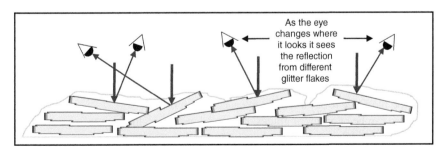

Figure 2.8 Individual flakes sit on a surface at different angles as the eye moves it sees reflection from different glitter flakes.

flakes that change color with viewing angle can be produced [33–38]. Using gratings or holographic structures embossed into a surface can produce flakes that also produce color changing effects [39–41]. A number of these effects are just copying nature [42, 43]. Fish scales are individually transparent but when overlapping become silver in appearance. Some beetles and butterflies also show color changing effects when the viewing angle is changed. All of these are created through multilayer structures that are based on alternating layers of two different materials that each have a different refractive index. For the butterfly this can be alternating protein and air in a Christmas tree-like structure. This precise alternating structure produces color by optical interference. Fish scales have alternating plates of crystalline guanine separated by cytoplasm but the thickness and spacing are variable which means that every color is reflected leading to the silver color that can be as high as 90% reflectivity.

2.9 Materials and Processes that Have Been Used to Produce Flake Materials

Film:

monofilm + liquid crystal coating giving color shift [44]
coextrusion
transparent polymers giving iridescence through refractive index
colored polymers using different thickness layers to give identifiable "barcode" [45]
vacuum coated with metal to give "silver" glitter.
using dyed polymer coating on metal gives colored metal reflective glitter.
specific shaped flakes using die-stamping or laser cutting.

Foil

mirror quality foil
mirror quality foil + dyed coating to produce color

Flake

Lamella natural materials: micas, laponite, wollastonite, muscovite, montmorillonite, etc.
Lamella manufactured materials: micaceous iron oxide
   These can be of different grades: industrial, nontoxic, food grade
   Glass: nanoflake glass of thickness 350, 500, or 750 nm at 30, 120, or 160 μm diameter.
Metal oxides: silica, alumina, and titania coated on any of the above

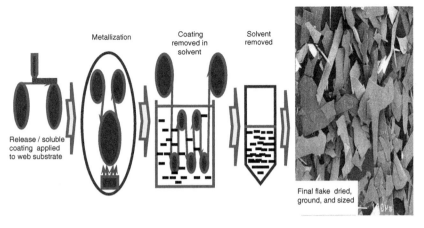

Figure 2.9 The method of production of vacuum metallized flakes (VMF).

Freestanding

self-supporting inorganic coating stripped from carrier film (Figure 2.9)

transparent, semi-transparent, reflective
single colored
color shift, color flop, goniochromic
either of the two above + logo on flake surface
color shifting + magnetic effect to modify the color shift
holographic
customized flake shapes
customized flake identifiers

The effects that can be produced can be produced on every flake or may be modified by how different flakes are used in combination. Embossed logos on individual flakes may be used as a small fraction included in a batch of flakes as a covert method of verifying the authenticity of the source of the pigment (Figure 2.10).

Effects

iridescence, opalescence, pearlescence, nacreous, luster, shimmer
color

Color shift: Changes color with viewing or illumination angle as above but with the addition of a magnetic coating to enable a variation in how the color changes with angle.

Metameric: Different color shift materials that have the same color at one angle but each change to a different color for the same angular movement.

Holographic/grating: Holographic is used as a generic description of many types of effect, this includes kinogram, pixelgram, excelgram, dot matrix, zero order diffraction, gratings as well as holograms or combinations of techniques.

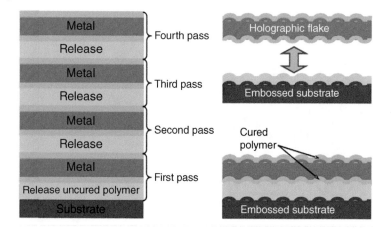

Figure 2.10 Example of Avery patented flake material, four layers to make four flakes and same materials that may be embossed for holographic flakes. Source: Miekka et al. [46]/US Patent/Public Domain.

Thermochromic: Changes color with temperature by adding a liquid crystal coating combination – e.g. color shift + hologram

2.9.1 Post Manufacture Modification

Glitter, reflective, pearlescent, or holographic flakes may be used in combination with other materials to enable a different use.

2.9.2 Polymer Film Flakes – Differences that May Help Discriminate the Flake Source

Polymer films are extruded and then stretched to reduce the thickness [47]. This may be achieved in two ways. The first is to extrude from a narrow slot die and then stretch the film in the forward direction followed by stretching the film in the transverse direction. This biaxial film orientation increases the film length and width whilst reducing the thickness. The biaxial orientation improves the mechanical performance of the film and this process is known as the stenter process (Figure 2.11).

The second method of achieving the same result is to extrude a tube of the polymer and if the end of the tube passes between a nip roller, air can be pumped through the center of the extruding tube which inflates the tube. This combination of nip roll and air pressure also biaxially orients the film and this is known as the bubble process (Figure 2.12).

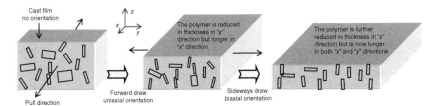

Figure 2.11 Biaxial orientation using sequential stenter process showing how polymer crystallites become oriented.

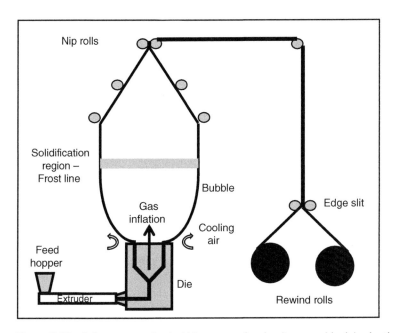

Figure 2.12 Polymer extrusion bubble process for simultaneous biaxial orienting films.

Coextrusion is where two, or more, different polymers are extruded and brought together to form a single web. It is important that there is sufficient adhesion between the polymers that the final film does not separate during use. To make colored or iridescent film the two polymers need to have a different refractive index to each other so that there is a reflection at the interface between each layer. As polymers have similar refractive indices with only a small difference the number of layers needs to be large and the thickness very small in order to make the effect more visible or vivid. To produce such a film, a coextrusion of two polymers is sequentially passed through a series of layer doubling dies where the extrusion is split at the centerline with the flow from one half diverted to be brought under the

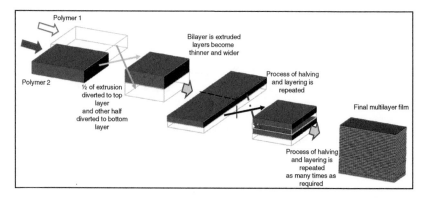

Figure 2.13 Layer doubling extrusion process to produce a multilayer thin film.

other half and so doubling the number of layers. Each of these dies also returns the output to the same width and thickness as the original coextrusion (Figure 2.13). This coextruded film will still undergo biaxial orientation which further reduces the thickness of each layer. Iridescent thin films of 12 μm thick have been produced with >100 layers [48–50]. Since this original layer doubling process for films produced by the stenter process die blocks, spiral coextrusion die blocks have been developed for producing this type of film using the bubble process.

The polymer films used for metallization can vary in a number of ways. The extrusion may be a homopolymer or may be a coextrusion and it may also have a variety of additives included. Different manufacturers also have a different formulation to achieving the same basic polymer. Polyethylene terephthalate (PET), when polymerized, uses a catalyst which can be based on titanium, antimony, germanium, manganese, aluminum, or zeolites [51–53]. Germanium was used but because of cost has been replaced although some Japanese suppliers may still use it for very high clarity PET for bottles. Titanium has been used but it can result in a yellow cast to the films and to reduce this effect additives are used such as cobalt or an organic blue toner. Germanium, and aluminum catalysts can suffer from thermal degradation problems and these may use phosphorus compounds to help thermally stabilize the film. Antimony is probably the most widely used catalyst but there continues to be work to develop a suitable replacement.

If the film was produced as a pure polymer it would not wind well as the surfaces would be too smooth and to wind a good quality roll the surfaces need to be able to slip over each other during the winding process. If the surfaces cannot slip over each other the two surfaces may have no air between the surfaces and they will be in intimate contact and stick together, known as blocking. There are two aspects to preventing blocking [54].The first is to reduce the surface area of the two surfaces that can touch each other. This is done using anti-blocking fillers

The more perfect the surface the greater the specular reflection
minimise surface roughness and defects and maximise flatness to maximise reflectivity

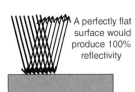 A perfectly flat surface would produce 100% reflectivity

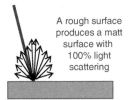 A rough surface produces a matt surface with 100% light scattering

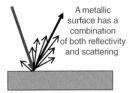 A metallic surface has a combination of both reflectivity and scattering

Figure 2.14 Schematics of light reflection, scattering, and a combination of both.

that distort the surface and provide roughness to the surfaces [55, 56]. Different film will have different proportions of filler included and can have different proportions of different size filler included. Some will have a random spread of size filler others will have a mixture of large and small filler. Anti-blocking fillers do not migrate through the polymer so sufficient needs to be added that there are enough present at the surface to achieve the desired roughness. This does mean that the increase in surface roughness will decrease the specular reflectance and increase the light scattering known as haze. The haze increases by approximately 0.4–1.0% per 1000 ppm of silica filler added to the film (Figure 2.14).

Filler types can vary in shape, size, and hardness. Some common examples are as follows.

Natural silica: a different mixture of shapes and sizes which often contains impurities that the synthetic silica does not. Cheaper than synthetic silica.
Synthetic silica: high surface area, hydroxylated and microporous surface. A good match of refractive index for PE and PP and used to make highly transparent films.
Talc: Magnesium hydrosilicate, a lamellar type of soft rock that has a refractive index that is also a good match to PE and PP.
Limestone: A low cost filler. Calcium carbonate and sometimes a mixture with magnesium carbonate these are used in lower quality film applications.

The second approach to enabling the film surfaces to move over each other is to add slip agents into the polymer [57, 58]. These slip agents reduce the surface energy to the film surface. Slip agents do migrate through the film and over time the amount present on the surface can increase with time or increased temperature. Slip agents can result in a lower metal adhesion when aluminum metallizing. Adhesion can be improved by treating the surface, such as plasma treatment but this is only temporary as following the treatment more slip agent will migrate to the surface.

A group of slip agents are produced by the amidisation of long chain fatty acids. Examples of the most commonly used of these are Steramides, Erucamides, and Oleamides.

Erucamides are longer chain, more heat stable and more oxidation resistant than Oleamides and with a lower vapor pressure create fewer volatiles during high-temperature processing. The Oleamides migrate through to the surface more quickly and are sometimes referred to as "fast blooming."

Stearamide is produced by amidisation of stearic acid (C18:0). Mpt. 98–104 °C
Erucamide is produced by amidisation of erucic acid (C22:1). Mpt. 79–85 °C
Oleamide is produced by amidisation of oleic acid (C18:1). Mpt. 66–76 °C

Waxes can also be used as slip agents. Waxes are similar to oils except they are solid at ambient temperature and generally have a melting point in excess of 40 °C. Microcrystalline waxes have been shown to demonstrate that the harder the wax, the better the slip properties.

Anti-blocking fillers can reduce the coefficient of friction (CoF) down to 0.3–0.4 which may still be higher than required, and so slip additives may be used in conjunction with the fillers. Different manufacturers will use a different balance of fillers and slip agents to provide both ease of handling and the optical performance.

Another option is to use coextruded films with a thin outer layer including the filler to provide the handling performance but because the bulk of the thickness does not contain filler the optical quality is higher than a through filled film.

Things that can be used to help identify the polymer film:

Polymer type and the chemical composition including catalyst, stabilizers, and slip agents.
Polymer structure such as single or coextruded.
Fillers or not. Type of filler both chemical composition as well as size, distribution, and shape.
Film thickness and surface roughness.

Beyond this there is then the metallization which may not provide any characteristic property to identify a specific metallizer. The aluminum when it is evaporated is refined and the residual elemental composition of the aluminum wire used to feed the evaporator is lost as these residuals stay as a crud on the surface of the molten pool and do not get evaporated.

If the glitter is colored it will usually have been achieved by coating the metallized film with a transparent colored lacquer [59, 60]. This will provide chemical information of the coating used including the color additive.

2.9.3 Foil

Aluminum foil is produced by taking thin aluminum sheet and passing it through a series of nip rolls and annealing processes to thin down the sheet [61]. The nip rolls that squeeze the sheet thinner also extend the length of the sheet. This is done repeatedly with the annealing process being done periodically to soften the aluminum as each use of the nip rolls work hardens the aluminum. To reduce the energy of the process lubrication is used which may leave behind residuals in the grain boundaries of the final crystal structure. Some manufacturers wash or heat treat the foil to remove this source of contamination. There will be trace elements present in foil that may differ slightly between batches and suppliers.

2.9.4   Flake Materials

Flake materials fall into two categories, natural and manufactured. There is an overlap in that some mined flake materials can also be manufactured. The natural materials still need to be processed to convert the mineral rock into the flakes. This processing uses both mechanical and chemical exfoliation techniques where absorbed chemicals by swelling or forming gas bubbles can provide enough internal force to split the lamella apart.

2.9.5   Natural Materials

Phyllosilicates covers the groups of layered or sheet minerals that include clays and micas [62, 63]. Mica is a generic name for a group of materials such as muscovite, phlogopite, lepidolite, and biotite. These minerals vary in color depending on the precise chemical composition. Once they have been exfoliated [64] to produce the lamellar flake of required thickness and sized to give the desired aspect ratio, the color is less pronounced. The minerals form a monoclinic crystal structure that has a preferred fracture plane that produces hexagonal-based shaped flakes. Of the range of phyllosilicates, mica is the most well known because when thin it is transparent, electrically insulating, chemically inert, lightweight, and flexible that has been extensively used in the electronics industry. The second highest use for mica has been as a pigment in the paint industry and it is from this route that the metallic flakes used in car paints were developed [65]. The wet-ground mica retained the flat cleaved surfaces that provided a high specular reflection that when coated with a transparent layer such as titania result in a colored metallic reflection [66–68].

Some iron oxide ores have a high lamellar structural content making them suitable for the base flake for some of the coated flake materials. This material is often called micaceous iron oxide [69] and compared to micas the flakes are opaque and dark which, once coated with a transparent oxide, will result in a colored dark metallic flake with the color depending on the oxide coating thickness [70].

Natural pearl essence which is also known as Fish Silver, Essence d'Orient or Mother-of-Pearl essence is a by-product from fish scales [71]. Using organic solvents which dissolve the proteins allowing the purines, guanine, and hypoxanthine to be extracted. These two crystalline materials are platelet shaped and of the order 50 nm thick $\times$ 1–10 µm $\times$ 20–50 µm. The extract is washed to remove the fish oil and residual proteins and then dispersed in a medium appropriate to the final application. This natural material has the advantage that it is more robust than synthetic pearlescent pigments and with a lower density it is less prone to settling in liquids.

Perlite, a hydrated form of obsidian which is an amorphous volcanic glass, is formed of a mixture of oxides with a high (>70%) of silica. This water content can aid the expansion and fracture of the material along with milling in the production of flakes. This material has lower color than mica and when used in pearlescent pigments has less low angle color than mica-based pigment flakes [72].

2.9.6 Synthetic Materials

Synthetic fluorophlogopite or synthetic or artificial mica is produced by smelting natural mica or starting with metal oxide/fluoride mixtures in a furnace at high temperature [73]. The resultant material has the same basic chemical composition of mica except that the hydroxide is substituted by fluorine. The molten material has the stoichiometry precisely controlled and using a seed a single crystal boule is grown under controlled conditions in the same way silicon is made. This crystal has a controlled crystal axis that allows thin mica sheets to be obtained that are more reproducible and transparent than the original mica and with greatly reduced levels of impurities. This material has obtained approval for use in food contact applications.

Micaceous iron oxide can be manufactured from ferric sulphate and sodium hydroxide. The ferric sulphate was a byproduct from pickling of iron and steel products as well as in the production of titania, making the raw material abundant and reducing the cost to manufacture. Heating the solution, increasing the pressure and controlling the alkali concentration encourages the crystals of iron oxide to form. The alkali concentration controls the size of the hexagonal crystals that form. The crystals are finally separated from the sodium sulphate and alkali hydroxide liquid.

Lead-based compounds were an important source of pearl luster flakes with basic lead carbonate being the most widely used. These were all toxic and with the advent of nontoxic alternatives, production was largely phased out [74].

Bismuth oxychloride was the first nontoxic synthetic pearlescent pigment to be manufactured. The primary use for the material is in the cosmetics industry. The manufacture requires very precise control of the pH, temperature, concentration, pressure, and additives to produce flat crystals. This is another flake source that has largely been superseded by materials such as mica.

Glass flake processing has been developed to the point where continuous manufacturing of flakes of thicknesses down to 100 nm is possible [75, 76]. The molten

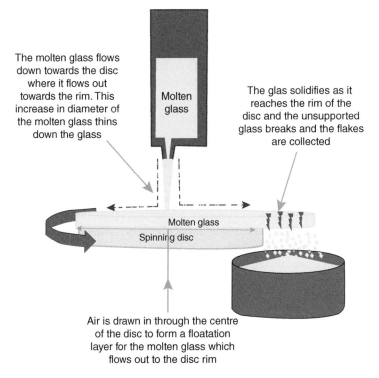

The molten glass flows down towards the disc where it flows out towards the rim. This increase in diameter of the molten glass thins down the glass

Molten glass

The glas solidifies as it reaches the rim of the disc and the unsupported glass breaks and the flakes are collected

Molten glass

Spinning disc

Air is drawn in through the centre of the disc to form a floatation layer for the molten glass which flows out to the disc rim

Figure 2.15 Spinning disc method of pigment flake production.

glass is fed at a controlled rate onto a spinning disc where it flows out to the rim of the disc. The glass does not touch the disc as a flow of air drawn up from the centre of the disc forms a floatation layer between the disc and glass. As the molten glass flows radially out, the thickness reduces and the temperature drops. The aim is that the glass solidifies as it reaches the rim and once over the rim it can be fractured and the flakes collected (Figure 2.15). Flakes produced by this method of 350 nm thickness are used in the production of colored and color shifting pigments [77, 78]. The glass is a borosilicate glass which is also used in cookware and so the flake is taken to be nontoxic but it has not been tested or approved as food safe.

Metal oxides – Silica, Titania, and Alumina. Processes for each of these materials have been developed involving sol–gel as one of the steps. An example of this is titania which is exfoliated from a layered precursor into a colloidal nanosheet that is <1 nm thick and this sheet is reassembled by gelation into thicker sheets that are heated to produce anatase flakes in the range 20–30 nm thick and many microns across [79, 80]. To produce silica, a base catalyzed sol–gel from a precursor such as tetraethyl orthosilicate (TEOS) was used to produce the coating [81]. The sol–gel may be cast onto a surface where it is dried and can be removed and flaked [82].

Alternatively, surfactant can be added enabling the sol–gel to be formed into a bubble that once dried can be broken into flakes without the need to be separated from any substrate [83]. The flake dimension is small compared to the bubble size and the flakes are not noticeably curved and are sufficiently close to being flat that the loss of specular reflectivity is not considered a problem.

2.9.7 Freestanding Flakes

More recently the same process used for making aluminum metal flakes has been used to make the oxide flakes. This comprises vacuum coating the oxides onto a polymer web as it passes around a cooled deposition drum. The polymer web is coated with a release layer that after coating can be dissolved away releasing the coated flakes [84]. This same process is also used to manufacture a variant of aluminum metal flakes. The Aluminum metal flakes can be very flimsy and to make them more robust the process deposits both aluminum oxide and aluminum [85]. The structure is thin oxide, metal, thick oxide, metal, thin oxide where the central thick aluminum oxide provides the stiffness to the flake but does not affect the reflectivity of the aluminum. The quality of all of these flakes depends on the surface quality of the polymer web surface. If optical grade polymer film is used the surface roughness will be minimized which will help minimize any surface scattering from the flakes produced. Flakes produced by this method do not have the steps on the surface that the natural flakes do and so the specular reflectance is higher as there is less diffuse light scattering from each surface.

These oxide flakes are transparent with silica and alumina having a low refractive index and titania, a high refractive index. To make different effects, these flakes are coated with the low index flakes coated with high refractive index material and the high refractive index flakes with a low refractive index material. There are various methods of adding coatings to flakes including sol–gel coating, plasma enhanced chemical vapor deposition and physical vapor deposition [86]. It is preferable that the coating is on both sides of the flake so that it does not matter which way up the flake sits on a surface the effect will be the same. For the vacuum coating processing, the flakes may be rotated in a barrel coater where a magnetron sputtering source deposits downward inside the barrel onto the flakes and as the barrel rotates the flakes tumble down so that both sides may be coated over time [87, 88]. This is an averaging process as at any one time not all flakes will be being coated and some flakes will overlap others and so even on a single flake there may be some variation in the coating thickness. Alternatively, flakes can be turned over by using a fluidized bed reactor where they are fluidized by vibrating the reactor bed and coated either by a magnetron sputtering source or using a chemical vapor deposition process [89–92]. Alternatively flakes can be fluidized by controlling a gas flow through the flakes and an option is to use the

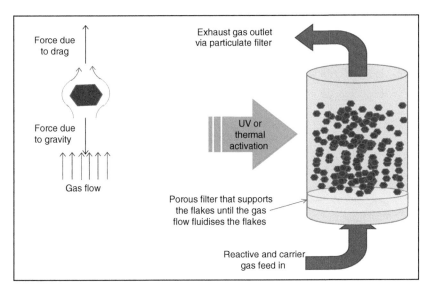

Figure 2.16 Fluidized bed process of coating flake pigments.

chemical vapor deposition gas as the gas floatation keeping the flakes in motion [93] (Figure 2.16).

If the flakes are being vacuum deposited onto a polymer web then then there is the option of producing the whole optical structured flake on the web before it is released from the web and broken into flakes [94]. This has the advantage of producing layers with higher uniformity coating thickness on each layer. This higher precision means that the color can be controlled to a much tighter color specification and the color intensity and reflectivity will be higher than can be produced by other methods.

Aluminum metal flake traditionally was made by atomizing fine droplets from molten aluminum that solidify [95]. These spheres then are loaded into a wet ball mill with steel or ceramic balls and a lubricant and the mill is rotated [96]. The tumbling action, whilst the mill rotates, causes the heavy balls to repeatedly impact the soft aluminum resulting in the aluminum being flattened into flattish shapes known as cornflakes. Converting spheres to flakes the aluminum spreads and flakes and further impacts fracture the large flakes into smaller flakes that are further thinned and the aspect ratio enlarged. The final flakes are not perfectly flat and the edges are irregular. An improvement on these was to take the flakes and further process them to smooth the surfaces to and round the edges leading to an improved surface quality and reduced light scattering. This material is referred to as lenticular or silver dollar flakes and these have a higher reflectivity than the cornflakes and a narrower size distribution. The next improvement in aluminum

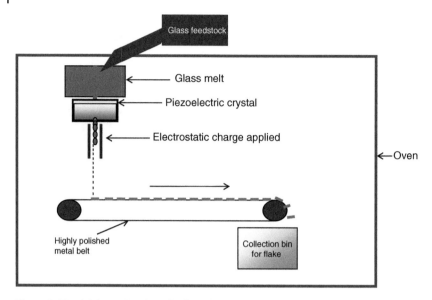

Figure 2.17 Ink jet technology for flake production.

pigment was by vacuum depositing the aluminum onto rolls of polymer whereby the surface roughness and flatness of the flake was directly related to the surface quality of the polymer roll on which it was deposited. This vacuum coated material could be used directly on the polymer film or the metal removed as freestanding flakes. The vacuum deposited flakes are much thinner than the ball milled flakes being as thin as 50 nm thick [97]. The latest technology is to use inkjet technology to atomize molten metal into precise sized droplets that is deposited onto a moving substrate where it solidifies and is released as a flat flake (Figure 2.17). The flakes may be used directly or may be further processed by milling to refine the flake size and thickness [98].

Manufacturing aluminum flakes is hazardous as the aluminum reacts with oxygen in an exothermic reaction [99]. All aluminum has a thin layer of aluminum oxide on the surface that protects the metal from further oxidation. If the aluminum is allowed to oxidize uncontrollably the flake will heat up and can autoignite and so this very thin aluminum flake is always sold in a carrier solvent that is suitable for the final application. The final application usually encapsulates the flake in polymer, ink or paint that protects the flake from further oxidationx.

2.9.8 Effects

The roll to roll production of flakes enables some effects to be created during the flake production process. The deposited coating is thin and fine enough that the coating will replicate the surface it was deposited onto and so if features are included in the surface they will also appear on the flakes. It is possible to emboss structures into the release coating before the vacuum deposition to produce a number of different effects. The embossing can be a simple grating or holographic structure and this will produce the same structure in the deposited coating and, once released from the web, in the flakes [46]. Other features may also be added such as a second structure that creates a weakness in the coating such that when the flake is released it will fracture preferentially along the weakened feature. This can be used to create specifically shaped flakes. Coupled to this method of producing shaped flakes it is also possible to add a feature to produce a logo in the center of each flake enabling flakes to be customized. This identifier flake can be used as an anti-counterfeiting device to add to the portfolio of measures taken to secure the supply chain of materials [24, 25].

2.9.9 Post Manufacture Modification

Homemade glitter has been made from reflective or iridescent polymer sheet by passing the sheet through a micro-cut shredder. Many modern paper shredders are also capable of shredding plastics even as thick as credit cards and compact discs (CDs). The micro-cut shredders produce one shape with the size defined by the security class the shredder falls into as defined by DIN66399 (German Institute for Standardization) [100]. Micro-cut shredders need to be in classes P5–P7 to be micro-cut shredders. The "P" stands for paper and there is a standard for "F" for film or foil but although many paper shredders claim to also cut film, they do not generally refer to the DIN standard for film. The higher the "P" number the finer the cut pieces are and the more expensive the shredder. The most widely available shredders are in the P5 class and these will shred film to an approximate size of $2 \, mm \times 8 \, mm$ [101] producing around 3700 pieces from an A4 sheet. P5 is about half this size and P7 half the size again taking the pieces per A4 sheet to around 12 000 and this is closer to the size of commercially available glitter.

Glitter or other effect flakes are cheap but cosmetics may be expensive. This has driven enterprising people to manufacture their own version of the cosmetics. This is done by buying glitter or pigment flakes and encapsulating them to make new flakes. This can be done by mixing the flakes into a liquid that can be cured or dried. Glitter has been used in glues, gums (spirit gum or gum Arabic), hair lacquer or gel, Vaseline® or clear nail varnish [102–105]. Some are used directly and others are dried or cured and then ground again to return the sheet of encapsulate material into flakes. This material is often brittle and using a coffee grinder [106] a smaller-sized glitter can be produced than by a micro-cut shredder. The encapsulating medium can be colored and so silver glitter can be changed to suit the user's requirement.

There are two broad groups of core materials. The mined materials such as mica and the polymer core materials are usually thicker than the synthetic or vacuum-deposited core materials. The vacuum coated core materials can be as much as two orders of magnitude thinner than the polymer or natural core materials. The internet has enabled many more people to get direct access to glitter as well as the source material manufacturers. This means that the finer flake materials and thinner core materials can be purchased which enables anyone to make more exotic glitter and thinner than previously. In the search for the new effect glitters, different types of glitter are being combined with other materials or coatings. The different flake materials may be blended and using alcohol to help wet the flake powder surface the flakes can be colored before then encapsulating them into a thin sheet and grinding the sheet to make the new effect glitter. An example of this would be by modifying silver glitter by the addition of a liquid crystal powder or leuco dyes which when mixed with a little alcohol will coat the silver flakes with the liquid crystal or dye and when the alcohol evaporates off the silver will be both colored and thermochromic [107, 108].

Analysis of this home-made material should still have a core of manufactured material but in the case of the cosmetic glitter there will be the addition of the encapsulant. This encapsulant could convert a generic core material into something unique which would aid forensic matching.

2.9.10 Color

The eye can be deceived and color is often in the eye of the beholder. To offer a wide range of colors does not mean that a wide range of colors needs to be manufactured. Colored flakes can be blended to modify the perceived color. This blending is more effective when the flakes are of small dimensions and the flake are used to cover a surface. If the flakes are sprinkled very thinly over a surface, the blending is not effective. The other way of having a wide range of colors is that the manufacturing control is not very precise and every batch is different. Those batches that are close in color can be blended whereas others that are easily seen to be different get offered as a color variant.

Some glitters based on metallized polymer film are single side coated. When these flakes are used as silver glitter it does not matter which way the flakes lie. Viewing the flake from the metal side or from the polymer side it will still have the same silver metallic appearance. If the flake has a colored lacquer added on top of the metal it will appear colored metallic whereas from the polymer side it will still appear silver metallic. A bottle of this single side flake will appear colored and if the flake is used en masse it will still appear colored but less intensely colored than if the colored lacquer had been applied to both sides.

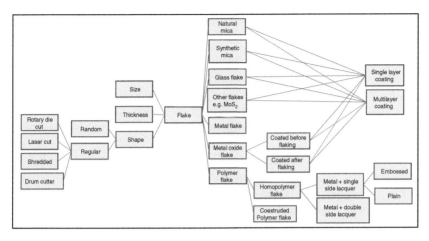

Figure 2.18 Flake production options.

Flakes that have a very high color intensity (chroma) will be manufactured to a close tolerance and are unlikely to be blended and the CIE color characteristics will have a narrow tolerance. As the color tolerance widens the chroma will decrease. Reflection and chroma depend on the flake surface being flat, smooth, and defect free (Figure 2.18).

2.9.11 Discriminating Between Flakes of Unknown Origin

There are a number of questions to answer such as,

1) Are the flakes in a carrier medium or loose?
2) If in a carrier, what is the carrier medium?
3) What size and shape are the flakes? If regular shapes, are they die cut or laser cut?
4) What is the core material? Polymer, metal, natural flake, synthetic flake, glass, metal oxide?
5) If polymer what is the thickness, is the polymer filled, what polymer, what catalyst, what fillers, what additives?
6) If core material is a natural flake, what is the chemical composition including the impurities?
7) Is the core material coated? Single side coated, both sides coated, all sides/edges coated?
8) If coated, how many layers? What material is each layer? What thickness is each layer?
9) Color. CIE for color co-ordinates and chroma (color intensity).

10) What is the type of optical effect of the flake. Reflection, interference, holographic...?

The internet has enabled more individuals to source more materials enabling them to have the opportunity to mix different materials that are not commercially available to the public. The industry continues to work on alternative manufacturing techniques for making the core material, adding the additional coatings, encapsulation as well as working on using sustainable and biodegradable materials [109].

This means that the number of different glitters, commercial, and homemade, can be expected to continue to increase for many years to come. The detailed identification of the composition of multilayer flakes can require the use of different analytical methods. There are many techniques available and choosing which technique to use will depend on many things including the amount and size of the sample available. It is preferable to make a positive identification at the minimum cost. A review of various analytical techniques and their costs has been carried out for the identification of microplastics [110] which is an useful template to follow for identifying unknown particles.

Natural materials contain impurities, some of which may be removed and others which may not. Testing materials for toxicity is expensive and many materials are never tested, others are tested sufficiently to allow them to be labeled nontoxic, which is not the same as food safe but internet buyers may not discriminate between them.

As has been mentioned earlier in this chapter, early glitter flakes were derived from nature from mollusks shells or fish scales. These materials were then copied and industrialized to make cheaper and higher volumes available. This process of copying nature is called biomimetics and is being used in many industries to help develop new materials. Glitters and flakes have also been influenced by bird feathers, butterfly wings, and beetle shells which have shown color and color changing properties. This inspiration leads to inorganic layered or structural materials that exhibited color or color shifting effects [111–113]. The use of inorganic replacements did mean the imitation flakes produced were not biodegradable; however, with a switch to biodegradable film materials, often based on cellulose [114], it is possible to produce biodegradable versions.

2.9.12 Follow the Yellow Brick Road

Concluding this chapter on a humorous note, an actual criminal case where glitter led to the culprits:

To promote the film *The Wizard of Oz*, a theater in Richmond, Virginia had out front a giant glittery foam rubber replica of Dorothy's ruby slippers.

When thieves made off with them all the police had to do was follow a trail of glitter to the apartment of one of the suspects.

References

1 Roux, C. et al. (2011). Glitter – The Ideal Trace Evidence? 29 slide Power-Point presentation, University of Technology Sydney. https://projects.nfstc .org/trace/2011/presentations/Roux-Glitter.pdf#page=1&zoom=auto,726,841 (accessed 15 March 2022).

2 H.R. 1321. The Microbead-Free Waters Act. [Public Law No. 114-114 (12/28/2015)]. https://www.congress.gov/bill/114th-congress/house-bill/ 1321 (accessed 15 March 2022).

3 Brahney, J., Hallerud, M., Heim, E. et al. (2020). Plastic rain in protected areas of the United States. *Science* 368 (6496): 1257–1260. https://doi.org/10 .1126/science.aaz5819.

4 Leahy, S. (2019). *Microplastics Are Raining Down from the Sky*. National Geographic https://www.nationalgeographic.com/environment/2019/04/ microplastics-pollution-falls-from-air-even-mountains (accessed 15 March 2022).

5 Product Safety Data Sheet, Cosmetic Bio-glitter®. https://cdn.shopify .com/s/files/1/0324/3065/files/Cosmetic_Bio-glitter_SDS-RB56.pdf? 14756011220786337448 (accessed 15 March 2022).

6 Vo, C. (2006). Polarized Light Microscopy of Glitter Particles, research project report for Forensic Science Program, Summer 2006. University of Chicago Illinois.

7 Smith, J.M. (2006). Polarize Light Microscope Examinations of Oriented Poly-mer Films. https://www.pstc.org/i4a/pages/index.cfm?pageID=3617 (accessed 15 March 2022).

8 Cosentino, A. (2012). Polarizing microscopy for art examination Pigments, 2:03 minute video. https://www.youtube.com/watch?v=Mmcluf7Kh-0 http:// www.antoninocosentino.it/index.html (accessed 15 March 2022).

9 Custom Printing: Laser vs. Rotary Die Cutting. https://www.printindustry .com/blog/?p=8752 (accessed 15 March 2022).

10 RJA Plastics Glitter Machine Demo. https://www.youtube.com/watch? v=NwjiMneRnWw (accessed 15 March 2022).

11 Image showing drum cutter drum with teeth clearly seen on drum and fixed bed cutter. https://www.alibaba.com/product-detail/Rotary-knife-glitter-cutting-machine-mould_60495038395.html (accessed 15 March 2022).

12 Vernoud, L., Bechtel, H.A., Martin, M.C. et al. (2011). Characterization of multilayered glitter particles using synchrotron FT-IR microscopy. *Forensic*

Sci. Int. 210: 47–51. https://www.academia.edu/10608047/Characterization_of_multilayered_glitter_particles_using_synchrotron_FT-IR_microscopy.

13 Weaver, Caity (2018). What Is Glitter? *The New York Times* (21 December). Found online at https://www.nytimes.com/2018/12/21/stlyle/glitter-factory.htm (accessed 15 March 2022).

14 Schrenk, W.J., Shrum, W.E., and Wheatley, J.A. (1990). Elastomeric optical interference films. Patent US4,937,134, 26 June 1990.

15 Schrenk, W.J., Chisholm, D.S., Cleereman, K.J., and Alfrey, T. (1973). Apparatus for the preparation of multilayer plastic articles. Patent US3,759,647, 18 September 1973.

16 Shetty, R.S. and Cooper, S.A. (1995). Colored iridescent film. Patent US5,451,449, 19 September 1995.

17 Shetty, R. and Allen, S.I. (2001). Flakes from multilayer iridescent films for use in paints and coatings. Patent US6,291,056, September 18 2001.

18 Ponting, M. (2019). Advances in micro and nanolayer multiplying co-extrusion towards the next generation novel commercial products. SPE Flex Pack and AIMCAL R2R Conf Oct 2019 https://www.aimcal.org/conference-proceedings.html (accessed 15 Marc 2022).

19 Applied TopMet™ (2022). Thermal boat evaporation system for Al and AlOx flexible barrier layers. http://www.appliedmaterials.com/products/applied-topmet (accessed 15 March 2022).

20 BOBST (2015). Produces World's Largest Metallizer. www.convertingtoday.co.uk/news/newsbobst-produces-worlds-largest-metallizer-4684389 (accessed 15 March 2022).

21 McManus, M., Lucas, S., and McKinley, B. (2007). Adding up additives. Paper, film & foil converter. January 1st 2007. https://www.pffc-online.com/coat-lam/adhesives/5090-paper-adding-additives (accessed 15 March 2022).

22 Schmidt, C. and Friz, M. (1992). Optical physics of synthetic interference pigments Kontakte (Darmstadt) 1992 (2), pp 15–24 Merck – magazine Pub. Merck.

23 Wissling, P. (2006) Metallic effect pigments – Basics and applications, pp. 23–24 and 52–53 (European coatings literature) Pub. Vincentz ISBN 978-0-81551-532-6.

24 Argoitia, A. (2007). Technology and Applications of Microstructured Pigments AIMCAL Annual Technical Conference USA 2007. https://www.aimcal.org/uploads/4/6/6/9/46695933/argoitia.pdf (accessed 15 March 2022).

25 Argoitia, A., Delst, C.J., Yamanaka, S.A., and Kittler, W.C (2008). Provision of frames or borders around pigment flakes for covert security applications. Patent US 2008/0107856 A1, 08 May 2008.

26 Delst, K.-J. (2012). Advancing Technology for Security Applications Beyond Roll-To-Roll Vacuum Coating AIMCAL Annual Technical Conference USA

2012. https://www.aimcal.org/uploads/4/6/6/9/46695933/delst_pres.pdf (accessed 15 March 2022).

27 Cramer, W.R. and Gabel, P.W. (2001). Measuring special effects. *European Coatings Journal*, 1st July 2001 Issue 7.

28 Li, H., Chen, M., Deng, C. et al. (2019). Versatile four-axis gonioreflectometer for bidirectional reflectance distribution function measurements on anisotropic material surfaces. *Opt. Eng.* 58 (12): 124106. https://doi.org/10.1117/1.OE.58.12.124106.

29 Briggs, D. (2017) The dimensions of colour. http://www.huevaluechroma.com/011.php 8 files to http://www.huevaluechroma.com/018.php (accessed 15 March 2022).

30 Smith, K. (2005–2022) Hue – value – chroma explained. https://www.sensationalcolor.com/hue-value-chroma (accessed 15 March 2022).

31 Abraham, C. (2016) A Beginner's Guide to (CIE) Colorimetry. https://medium.com/hipster-color-science/a-beginners-guide-to-colorimetry-401f1830b65a (accessed 15 March 2022).

32 Lockett, M.R., Mirica, K.A., Mace, C.R. et al. (2013). Analyzing forensic evidence based on density with magnetic levitation. *J. Forensic Sci.* 58 (1): 40–45. https://gmwgroup.harvard.edu/files/gmwgroup/files/1171_0.pdf.

33 Glausch, R., Kieser, M., Maisch, R. et al. (1998, 1998). *Special Effect Pigments*, 33–46. Pub. Vincentz Verlag. ISBN: 3-87870-541-7.

34 Dobrowolski, J.A. (1998). Optical thin-film security devices. In: *Optical Document Security*, Chapter 13, 2e (ed. R.L. van Renesse), 289–328. Edn Pub Artech House. ISBN: 0-89006-982-4.

35 Schmid, R., Lavalle, C., Jones, S.A., and Carroll, J. (2020). Pigment composition with improved sparkle effect. Patent US 2020/0032067 A1, 30 January 2020.

36 Fuller, D.S. and Zimmermann, C.J. (2005). Multi-layer effect pigment. Patent US 2005/0166799 A1, 04 August 2005.

37 Bujard P. and Leybach L. (2011). Interference pigments on the basis of silicon oxides. Patent US 7959727 B2, 14 June 2011.

38 Fuller, D.S. and Zimmermann, C.J. (2006). Transparent goniochromic multilayer effect pigment. Patent WO 2006/088759A1, 24 August 2006.

39 Miekka, R.G., Benoit, D.R., Thomas, R.M. et al. (1997). Process for making embossed metallic leafing pigments. US5624076, 29 April 1997.

40 Argoitia, A. and Bradley, R.A (2004). Diffractive pigment flakes and compositions. US 6692830 B2, 17 February 2004.

41 Rettker, J.P. (2014). Embossed metallic flakes process and product. US 2014/0154520 A1, 05 June 2014.

42 Stavenga, D.G. (2014). Thin film and multilayer optics cause structural colors of many insects and birds. *Mater. Today: Proc.* 1S: 109–121.

43 Hall, S.R. (ed.) (2019). *Bioinspired Inorganic Materials: Structure and Function*. Pub. Royal Society of Chemistry. ISBN: 978-1-78801-146-4.

44 Gailberger, M., Strohriegl, P., Stohr, A., and Mueller-Rees, C. (1998). Effect coating material and effect coating system, especially for vehicle bodies, using liquid-crystalline interference pigments. Patent US5807497, 15 September 1998.

45 Lee, K.P. (1977). Method of tagging with color coded microparticles. US4053433, 11 October 1977.

46 Miekka, R.G., Benoit, D.R., Thomas, R.M. et al. (1997). Embossed metallic leafing pigment. Patent US5672410, 30 September 1997.

47 Breil, J. (2016). Oriented film technology. In: *Multilayer Flexible Packaging*, Chapter 12, 2e (ed. R.J. Wagner). Pub. Elsevier/William Andrew. ISBN: 978-0-323-37100-1.

48 Alfrey, T., Gurnee, E.F., and Schrenk, W.J. (1969). Physical optics of iridescent multilayer plastic films. *Polym. Eng. Sci.* 9 (6): 400–404.

49 Schrenk, W.J. and Alfrey, T. Coextruded polymer films and sheets. In: *Polymer Blends*, Chapter 15, vol. 2 (ed. D.R. Paul and S. Newman), 129–165. Pub. Academic Press. ISBN: 0-12-546802-4.

50 No. 27 - Film/Sheet Coextrusion. Polymer Technology 10/1/2005. https://www.ptonline.com/articles/no-27---filmsheet-coextrusion (accessed 15 March 2022).

51 Ahmadnian, F. (2008). Kinetic and Catalytic Studies of Polyethylene Terephthalate Synthesis. Dr. Ing. Thesis Berlin Technical University.

52 Xiao, B., Wang, L., Mei, R., and Wan, G. (2012). Study of organic aluminum compounds catalysts in poly(ethylene terephthalate) synthesis. *Asian J. Chem.* 24 (1): 42–46.

53 Cholod, M.S. and Shah, N.M. (1982). Catalyst system for a polyethylene terephthalate polycondensation. US4356299, 26 October 1982.

54 Antiblock additives. https://www.ampacet.com/faqs/reasons-for-using-antiblock-additives (accessed 15 March 2022).

55 Keck-Antoine, K., Lievens, E., Bayer, J. et al. Additives to design and improve the performance of multilayer packaging. In: *Antiblock Additives: Multilayer Flexible Packaging*, Chapter 5, Section 5.4.5, 2e (ed. J.R. Wagner), 68–70. Pub. Elsevier. ISBN: 978-0-323-37100-1.

56 Making sure that nothing blocks. Kunststoffe International 8/2014 Additives – Fillers, pp 54–57. Munich: Pub. Carl Hanser Verlag.

57 Keck-Antoine, K., Lievens, E., Bayer, J. et al. Additives to design and improve the performance of multilayer packaging. In: *Slip Additives: Multilayer Flexible Packaging*, Chapter 5, Section 5.4.4, 2e (ed. J.R. Wagner), 68–70. Pub. Elsevier. ISBN: 978-0-323-37100-1.

58 Dapraslip™: slip and anti-block agents in polymer processing. https://www
.italmatch.com/wp-content/uploads/2015/06/brochure-DAPRASLIP.pdf
(accessed 15 March 2022).

59 Nariu, H., Akune, I., Shinohara, T. et al. (1976). Production of reflexible
pigment. Patent US3962397, 08 June 1976.

60 Argoitia, A., Lamar, S., and Kittler, W.C. (2009). Reinforced glitter. Patent
EP2067825, 10 June 2009.

61 Eue, M. (2014). Hydro paper on foil. *Proceedings of AIMCAL Web Coating &
Handling Conference Europe* (9–11 2014).

62 Earle, S. (2015). *Physical Geology*. Victoria, BC: BCcampus. Open Textbook
Collection. ISBN: 978-1-989623-70-1. Retrieved from https://opentextbc.ca/
geology.

63 Phyllosilicates –(Silicate Sheets). http://www.geo.umass.edu/courses/geo311/
phyllosilicates.pdf (accessed 15 March 2022).

64 Hillier, S., Marwa, E.M.M., and Rice, C.M. (2013). On the mechanism of exfoliation of 'Vermiculite'. *Clay Miner.* 48: 563–582.

65 Handa, J., Itou, H., Monohara, T., and Takagi, Y. (1989). Production process
of pigment. Patent EP 0360513, 15 September 1989.

66 Klenke, E.F. and Stratton A.J. (1963). Micaceous flake pigment. Patent
US3087827, 30 April 1963.

67 DeLuca, C.V., Miller, H.A., and Waitkins, G.R. (1977). Rutile coated
mica nacreous pigments and process for the preparation thereof. Patent
US4038099, 26 July 1977.

68 Armanini, L. and Bagala, F. (1979). Iron oxide coated mica nacreous pigments. Patent US4146403, 27 March 1979.

69 Fujiwara, S., Hiroaka, N., and Harada, M. (1991). Titanium dioxide coated
micaceous iron oxide pigments and method for producing the same. Patent
US5002608, 26 March 1991.

70 Pfaff, G. Manufacture and properties of pearl lustre pigments. In: *Iron
(III) oxide mica pigments: Special Effect Pigments*, Chapter 2, Section 2.5.2
(ed. U. Zroll), 40–43. Pub. Vincentz Verlag. ISBN: 3-87870-541-7.

71 Franz, K.D., Emmert, R., and Nitta, K. (1992). Interference pigments.
Kontakte (Darmstadt) 1992 (2) pp 3-14 Merck – magazine Pub. Merck.

72 Bujard, P., Baysang, M., and Bugnon, P. (2010). Interference pigments on the
basis of perlite flakes. Patent WO 2010/066605 A1, 17 June 2010.

73 Nobuoka, S. (1976). Method for manufacture of micaceous α-iron oxide.
Patent US3987156, 19 October 1976.

74 Myers, G.D. (2008). Pearlescent Pigments Making Confections Sparkle.
Pearlescent pigments add a luxury appearance to confections, with fin-
ishes ranging from metallic to silky and smooth. *Professional Manufacturing*

Confectioners Association 62nd Annual Production Conference (April 7–9 2008), Hershey, PA, USA.

75 Watkinson, C. (2015). Glass Flake - Extraordinary Performance Improver. Paint and coating industry, June 2, 2015.

76 Watkinson, C. (2008). Formation of glass flakes. Patent WO 2008/0190141, August 14th 2008.

77 Anselmann, R., Ambrosius, K., and Mathias, M. (2007). Effect pigments based on coated glass flakes. Patent US7226503, June 5th 2007.

78 Horiguchi, H. and Hioki, M. (2018). Luster pigment, method for producing same, pigment-containing composition, and pigment-containing painted product. Patent US 2018/0155551, Jun 7th 2018.

79 Tran, T.Q., Zheng, W., and Tsilomelekis. (2019). Molten salt hydrates in the synthesis of TiO_2 flakes ACS. *Omega* 4: 21302–21310.

80 Sasaki, T., Nakano, S., Yamauchi, S., and Watanabe, M. (1997). Fabrication of titanium dioxide thin flakes and their porous aggregate. *Chem. Mater.* 9 (2): 602–608.

81 Schubert, U. (2012). Chemistry and fundamentals of the sol–gel process. In: *Silica-Based Materials: Synthesis of Inorganic Materials*, Chapter 1, Section 1.2.1, 3e (ed. U. Schubert and N. Hüsing), 4–9. Weinheim: VCH-Wiley Verlag GmbH. ISBN: 3-527-32714-1.

82 Mizuno, T., Yamagishi, T., Yokoi, K., and Doushita, K. (1993). Apparatus for producing flakes of glass. Patent US5201929, April 13 1993.

83 Douden, D.K. and Scanlan, T.J. (1990). Thin silica flakes and method of making. Patent EP0384596, February 2 1990.

84 Schmid, R., Wosylus, A., Kujat, C. et al. (2013). Thin aluminium flakes. Patent WO 2013/127874, 6 September 2013.

85 Coulter, K.E., Mayer, T., Matteucci, J.S., and Phillips, R.W. (2002). Composite reflective flake based pigments, method for their preparation and colorant comprising them. Patent CA2411893, 7 February 2002.

86 Pfaff, G., Warthe, D., Dietz, J., and Foerderer, C. (2006). Effect pigments based on thin SiO_2 flakes. Patent US 2006/0112859, June 1 2006.

87 Baechle, D.M., Demaree, J.D., Hirvonen, J.K., and Wetzel, E.D. (2013). Fluidized Bed Sputtering for Particle and Powder Metallization. Army Research Laboratory Report No. ARL-TR-6435 April 2013.

88 Abe, T., Waatanabe, K., and Honda, Y. (2006). Polygon barrel spattering device, polygonal, barrel spattering method, coated particle formed by the device and method, microcapsule, and method of manufacturing the microcapsule. Patent US2006/0254903, November 16 2006.

89 Takeshima, E. (1989). Process and apparatus for coating fine powders. Patent EP0345795, Jun 8th 1989.

90 Coating of powders. https://www.sidrabe.com/research-development/coating-of-powders.html (accessed 15 March 2022).

91 Ostertag, W. (1985). Preparation of effect pigments coated with metal oxides. Patent US4552593, November 12 1985.

92 Carlotto, J.A. (2006). Vacuum deposition of coating materials onto powders. Patents WO 2006/083725, 10 August 2006.

93 Argoitia, A. (2017). Manufacture of diffractive pigments by fluidized bed chemical vapor deposition. Patent US2017/0306158, October 26 2017.

94 Schmid, R., Wosylus, A., Kujat, C., Merstetter, H.R., Mullertz, C. (2013). Thin aluminium flakes. Patent WO2013/127874.

95 McKay, C.F., McFay, A., and Ringan, E.S. (1997). Metal powder pigment. Patent US5593773, January 14 1997.

96 Ramezani, M. and Neitzert, T. (2012). Mechanical milling of aluminium powder using planetary ball milling process. *J. Achiev. Mater. Manuf. Eng.* 55 (2): 790–798.

97 Seubert, J. (2000). PVD Aluminum Pigments: Superior Brilliance for Coatings & Graphic Arts Paint and coating industry.

98 Wheeler, I.R. (2011). Process for producing metal flakes. Patent US8016909, 03 September 2011.

99 The Aluminium Association (2006). TR2 - Recommendations for storage and handling of aluminum powders and paste.

100 DIN66399. *The New Standard for Top-Security Storage of Information on Paper and modern Data Carriers.* https://international.intimus.com/pub/media/catalog/category/DIN66399_Hompage_en.pdf (accessed 15 March 2022).

101 'Foil shred' and 'Metallic confetti' products. https://www.etsy.com/listing/583710019/gold-foil-shredded-confetti-metallic.

102 Homemade Body Shimmer Lotion. Posted by Kelly. https://simplelifemom.com/2014/06/23/homemade-body-shimmer-lotion (accessed 15 March 2022).

103 How To Make Your Own Body Glitter Gel. https://www.youtube.com/watch?v=G3b27Oc4qS8 (accessed 15 March 2022).

104 Shuntona, B. (2019). 12 Body Glitter Hacks You Need to Copy ASAP Cosmopolitan, October 16, 2019. https://www.cosmopolitan.com/style-beauty/beauty/a9275808/glitter-hacks (accessed 15 March 2022).

105 Helmenstine, A.M. (2020). Easy To Make Glitter Slime. *ThoughtCo.* https://www.thoughtco.com/easy-to-make-glitter-slime-recipe-609154 (accessed 15 March 2022).

106 Archambeault, M.E. (2003). Pigment flakes. Patent US6582506.

107 How to Mix Pigments with Glitter (thermochromic). https://www.youtube.com/watch?v=LhQaMRH2F9o (accessed 20 February 2021).

108 How To Mix Pigments with Glitter (thermochromic). https://www.youtube
.com/watch?v=FvLu56lrNC8 (accessed 20 February 2021).

109 Ecosparkles glitter and shimmer. https://ecosparkles.co/pages/what-is-
ecosparkles-glitter-and-shimmer (accessed 15 March 2022).

110 Primpke, S., Christianson, S.H., Cowger, W. et al. (2020). Critical assessment of analytical methods for the harmonized and cost-efficient analysis of
microplastics. *Appl. Spectrosc.* 74 (9): 1012–1047.

111 Dumanli, A.G. and Savin, T. (2019). Biomimetics of structural colors: materials, methods and applications. In: *Bioinspired Inorganic Materials – Structure
and Function*, Chapter 4 (ed. S.R. Hall), 167–238. Pub. Royal Society of
Chemistry. ISBN: 978-1-78801-146-4.

112 Guan, Q.F., Yang, H.B., Han, Z.M. et al. (2020). An all-natural bioinspired
structural material for plastic replacement. *Nat. Commun.* 11: 5401. https://
doi.org/10.1038/s41467-020-19174-1.

113 The Chemistry and Physics of Special-Effect Pigments and Colorants for Inks
and Coatings. *Paint & Coatings Industry,* June 15, 2003. https://www.pcimag
.com/articles/85016-the-chemistry-and-physics-of-special-effect-pigments-and-
colorants-for-inks-and-coatings (accessed 15 March 2022).

114 Ronald Britton Ltd. Introduction to Bioglitter. www.discoverbioglitter.com
(accessed 20 February 2021).

3

X-ray Photoelectron Spectroscopy

Christopher Deeks[1] and Robert D. Blackledge[2]

[1]*Channel Manager EMEA – Surface Analysis at Thermo Fisher Scientific,*
[2]*Naval Criminal Investigative Service, Regional Forensic Laboratory, San Diego, USA*

3.1 Introduction

X-ray photoelectron spectroscopy (XPS) is a surface analysis technique with a penetration depth of approximately 10 nm. It can provide qualitative and quantitative analysis for all elements except hydrogen and helium. Additionally, it tells an element's oxidation state and the elements to which it may be bonded. It is shown in this chapter that XPS has much to offer the forensic science community. XPS instruments have in recent times evolved such that analysis only takes a matter of minutes, and only requires a sample size of tens to hundreds of microns. In the last few years, argon-ion cluster beam technology combined with XPS has made this analytic method much more useful in crime scene analysis.

3.2 Background and Theory

The beginnings of XPS, as a technique, go all the way back to the discovery of the photoelectric effect by Heinrich Hertz in 1887, prior to its further investigation by Albert Einstein in 1905. The photoelectric effect is the interaction of light and matter; when light is shone onto a material this causes the emission of electrons. The energy of these electrons depends on the element, chemical state, and the electron shell they come from. This is the information that can be extracted from using XPS, making it so useful as a technique.

Using this theory, Kai Siegbahn [1] and his research team in Uppsala, Sweden, managed to record the first ever XPS spectrum in 1954. In the years after, the technique was developed and improved, and became more accessible to users and is now a staple technique in materials laboratories around the world [2–8].

Leading Edge Techniques in Forensic Trace Evidence Analysis: More New Trace Analysis Methods,
First Edition. Edited by Robert D. Blackledge.

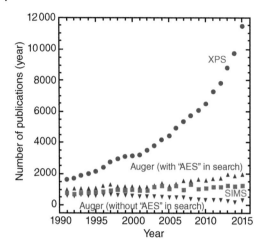

Figure 3.1 Graph displaying number of papers published per year from 1991 to 2015 for a variety of surface analysis techniques. Source: From Reference [9]/Cambridge University Press.

Modern XPS systems have become more and more automated in recent times, allowing high throughput of samples with acquisition taking from just a matter of seconds to a few minutes per sample for basic acquisition. This enables users with even limited knowledge of the technique to get high quality results. The improvements in speed, ease of use, and accessibility of XPS systems have all contributed to a large increase in use of the technique. The graph in Figure 3.1 shows the increase, between the years of 1991 and 2015, in the number of publications featuring XPS [9]. This increase in publications is indictive of the fact that XPS is an easier technique to perform using the modern systems, and this trend is set to continue for the foreseeable future.

In this graph, XPS is compared to other similar surface analysis techniques of Auger and SIMS. Where XPS shows over a 500% increase in papers in this period, the others remain quite stable. From this graph it can be deduced that XPS is increasing in ease of use, and more scientists see XPS as a viable technique for a wider range of applications.

Historically, the use of XPS has not been commonplace in forensics applications [10]. However, the recent increase in ease of use, even compared to that of just a few years ago, has enabled XPS to obtain more definitive data. It is therefore now ideally placed to become a technique of choice for forensics applications.

The mechanism of XPS starts with the illumination of a sample by X-rays. Electrons are then emitted from the sample as they are excited by the incident X-ray energy. These photoelectrons then pass through a variety of lenses to a detector where the kinetic energy of the electrons is measured.

The measured *kinetic* energy from the electrons gives information on the *binding* energy of electrons in atoms in the sample being analyzed. The binding energy of electrons is determined from

$$E_b = hv - E_k - W$$

where E_b is the calculated binding energy, hv is the photon (X-ray) energy, E_k is the kinetic energy of the photoelectron, and W is the work function of the spectrometer.

The X-ray energy is a known constant from the X-ray source in the instrumentation, and the kinetic energy is measured in the instrumentation. Both of these topics are discussed in Section 3.3. The work function of the spectrometer is set as a constant in the XPS system software. The work function is defined as the amount of energy lost between an electron leaving the sample and reaching the detector.

Determination of the binding energy of the photoelectrons now tends to be performed in the system electronics and software, and the user just needs to input parameters required to obtain the data.

XPS offers detection limits down to approximately 0.1 at% of the area being analyzed. This is not as sensitive as some other techniques, but the key objective of XPS is that of determining real surface elemental and chemical information, from the top 10 nm. Once a spectrum is acquired, the peak positions and shape of the spectrum allow information about the elemental and chemical states of the top layers of the sample to be deduced.

Figure 3.2 shows a typical survey spectrum from a fluoropolymer. This covers a binding energy region between 0 and 1350 eV. It shows several peaks, which can

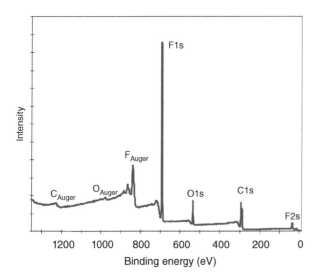

Figure 3.2 An example survey spectrum. Source: [11]/Thermo Fisher Scientific.

be attributed to different elements, chemistries, and electrons from the atoms. For fluorine, three peaks can be observed: F1s, F2s, and F_{Auger}. F1s and F2s show two peaks from two different electron shells of the fluorine atom. Either of these peaks can be used for quantitative analysis, however, the F1s peak is much larger, and so this is normally used. The F_{Auger} peak is due to a secondary ejection in the fluorine atom. As the initial electron is emitted, an electron from a higher energy state in the same atom relaxes and drops into the hole left by the initial electron. This emits energy that causes a secondary emission of an electron. This secondary electron is the Auger electron which causes the F_{Auger} peak. This whole process is referred to as the Auger process and is in fact a whole analytical technique in itself. The Auger peaks can be used in XPS to help with elemental and chemical identification, but often the XPS peaks themselves are sufficient.

It is possible to add a background to the survey spectrum using software, which provides quantitative elemental information. Different peaks have different sensitivities, so even though the F1s peak is larger than the F2s peak, the software and electronics are programmed to know the difference between them. Determining the elemental percentage from each peak requires a different calculation, which most software should be capable of performing. This then would give the same elemental percentage for each peak. Therefore, only one of these peaks needs to be used for determining the atomic percentage of fluorine present in the sample.

There are often different chemical states in a sample, and this can shift the peak binding energy by up to a few electron volts from a typical peak position. This is not often visible in the survey spectrum, due to default parameters being set to lower resolution and higher speed of acquisition, to obtain elemental information to begin with, before obtaining spectra from smaller regions with parameters to allow for better resolution. Figure 3.3 shows an example taken from a high-resolution scan of the carbon 1s region from the same sample as in Figure 3.2. A single carbon chemical state of C—C bonding would be a symmetric Gaussian-shaped curve. However, here it is possible to see that several peaks are present, which means several chemical states are present. By observing the elements present in the survey spectrum, it is possible to get an understanding of the chemical states present. The only elements the carbon can be bonded to are fluorine and oxygen, as no other elements are present in the survey spectrum. Many common chemical states have had their peak positions referenced and so it can be a simple process of determining where the peaks are, and therefore which states are present. In this example, carbon is likely to be bonded as C—C, C—O, C=O, CF, CF_2, and CF_3.

If the carbon, fluorine, and oxygen peaks are all fitted with backgrounds, as they have been here, then the chemical make-up and atomic concentration of the sample can be understood by calculating the area underneath each peak. The background allows software to calculate the area of the peak, and hence calculate the atomic percentage concentration.

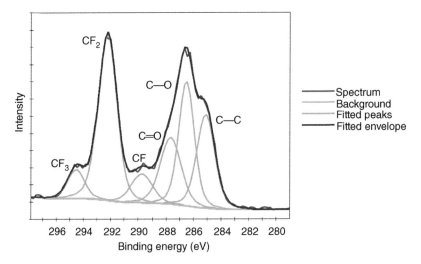

Figure 3.3 An example of a narrow region spectrum of a fluoropolymer. Source: [11]/Thermo Fisher Scientific.

This is a particularly complex example of a single element's spectrum as so many chemical states are present. In other forensic examples, only one or a couple of chemical states may be present for certain elements, but a sample may have many more elements present, which would give the survey spectrum more peaks, and hence more data to be analyzed.

3.3 Instrumentation

An XPS system needs several principal features and components to correctly work and output the best data possible. The main features of a modern XPS system are outlined below.

3.3.1 Ultra-High Vacuum (UHV)

This vacuum is in the region of 1×10^{-9} mbar, which is required by the system for best performance. There are two main reasons for this. Firstly, as a surface analysis technique, any contamination on the top of the sample provides a huge contribution to the collected data. Therefore, during analysis, the contamination could be being analyzed, rather than the intended sample. Even with vacuum it can still provide unwanted gases in the analysis area, causing interaction with the surface, and hence a change in surface chemistry which can give false results, so the higher the vacuum, the more beneficial it is for analysis. A common effect of an unsuitable vacuum level is if oxygen is present in the analysis chamber, which can interact with surface elements and cause an oxide to form.

The second reason for ultra-high vacuum (UHV) is that XPS is used for the precise measurement of the kinetic energy of photoelectrons. If these photoelectrons interact with gas particles within the chamber, this can alter their kinetic energy, and hence affect the final spectrum. Small shifts in the measured kinetic energy will give false results when analyzing the collected data.

To reduce the amount of gas in the chamber where analysis takes places, an XPS system normally has two chambers – the one where all the delicate instrumentation is housed and where the analysis takes place is the "analysis chamber." The second chamber or "entry chamber," is via which the sample is loaded into and out of the analysis chamber. This chamber can be fitted with preparation techniques if desired, to be used before the sample is transferred into the analysis chamber. To transfer the samples between these chambers, either a manual contraption or automated design can be used, depending on the system.

Typically, when the entry chamber is vented to allow a sample to be placed in the system, it is vented with dry nitrogen. As this is an inert gas, it will not react with the sample being analyzed, even if it is not entirely pumped out of the system.

The analysis chamber is kept at UHV at all times, to keep the instrumentation clean and functioning correctly. Having the instrumentation exposed to high pressures can cause contamination build-up on delicate parts and can also cause electrical shorting.

Generally, turbo pumps are used as the main way of creating vacuum inside the XPS system. These are efficient at removing all gases so are favored in the majority of systems. The use of turbo pumps allows vacuum to reach pressures as low as $\sim 1 \times 10^{-8}$ mbar. Ion pumps can be used as an alternative but often suffer when there is the need to remove a large amount of argon from the system, which is the primary gas used for ion etching and charge compensation, which is discussed later in this chapter.

Additionally, a titanium sublimation pump (TSP) is used in the analysis chamber. This is a titanium filament that is set to have a high current run through it every few hours. Titanium is released from the filament, and bonds with particles which may be present in the chamber which are then pumped away by the turbo pumps. This is especially useful for pumping away any water vapor in the chamber.

3.3.2 X-ray Source

To excite the electrons within the sample being analyzed, an X-ray source is needed. This tends to be a monochromated source, which will give a narrower distribution of X-ray energies, so the collected spectra will have narrower peaks with improved resolution. This therefore makes it easier to infer the chemical and elemental information obtained from the peaks.

An aluminum anode is mainly used as the X-ray source in the system. An electron beam is fired onto the anode, which then creates X-rays at an energy of 1486.6 eV. The X-ray spot size on the sample can usually be adjusted to match with the size of the sample being measured. Only the area where the X-ray is being illuminated will cause electrons in the sample to excite and emit from the sample. The size is typically in the regions of a few to several hundred microns, so large samples are not needed. However, the larger the area of interest, the larger the X-ray area can be used, and the more photoelectron emission happens, so analysis time can be improved. The smallest area able to be observed by XPS is in the region of a few microns.

3.3.3 Electron Detector

A detector system is needed to measure the energy of the photoelectrons being collected. An electron multiplier is used for this.

An electron multiplier works as a signal amplifier. Typically, the trademarked Channeltron is used. It functions by each photoelectron impacting the inside of the Channeltron, causing the emission of further electrons, which subsequently impact on another inner surface of the Channeltron, which multiplies further, and so on, until a cascade of electrons reach the end of the "horn," and a signal is output. Figure 3.4 shows an approximation of this.

More recently, Channelplates have been used as well as, or instead of, Channeltrons. Channelplates have a larger, two-dimensional area, causing many smaller electron cascades to occur, behaving as a series of smaller amplifiers. These can also be used to create 2D information from the emitted photoelectrons, rather than just acquiring the energy of the photoelectron, as a Channeltron does.

Figure 3.4 Channeltron. Source: [12]/Paul Gates.

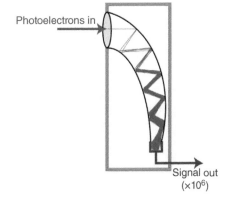

Photoelectrons in

Signal out
($\times 10^6$)

3.3.4 Charge Compensation Source

A charge compensation source is needed when analyzing nonconducting samples. When analyzing a conducting sample, the emitted photoelectrons are replaced in the atom, due to its conducting properties. When an insulating sample is being analyzed, the emitted photoelectrons are not replaced due to the insulating properties. Therefore, the more time the X-ray is illuminating the sample, the more charge builds up.

This charging can happen almost instantaneously, so if analysis is taking place an acquired spectrum shows incorrect energy and the measurements are therefore meaningless. Figure 3.5 shows an example of an insulating sample that has not been analyzed with a charge compensation source turned on.

The charge compensation source is either a continuous flow of electrons, or a combination of both electrons and positive ions to neutralize the sample. These are very low energy, so do not change the chemistry of the sample, and are not measured by the spectrometer. The electrons are provided from a filament within a charge compensation source. If positive ions are used as well, these are created from an inert gas, typically argon. A diagram showing stages of X-ray excitation and charge compensation is demonstrated in Figure 3.6.

The beam of electrons and ions should be larger than the area illuminated by the X-ray to ensure no charging will occur. There is normally a default parameter within the software of the XPS system to ensure this.

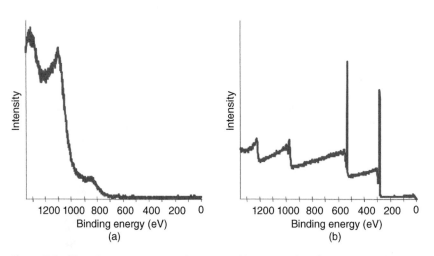

Figure 3.5 Non-charge compensated spectrum (a) compared to the same sample analyzed with a charge compensation source turned on (b). Source: [13]/Thermo Fisher Scientific.

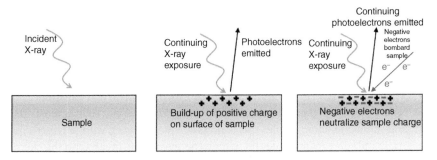

Figure 3.6 Diagram demonstrating charge compensation in XPS.

3.3.5 Sample Cleaning – Monatomic Ions

With XPS being a surface analysis technique, a clean surface on a sample is ideal. However, this is very rarely found on real-world samples, especially ones for forensic applications. Contamination is only avoidable if a sample has been prepared in a regulated and protected environment, often in a vacuum itself.

When analyzing samples, it is often found that a thin layer of carbon–oxygen–hydrogen-based contamination is present on the surface due to interaction with air, water, and commonly due to human interaction which can also cause salt (sodium and chlorine) to be present, even in small quantities.

Therefore, most XPS systems have a third source present which can provide a form of surface cleaning.

The most common method of cleaning is using high energy ions from an ion source. These ions (typically a couple of hundred to a few thousand electron volts) are sputtered onto the sample surface to remove the top layers of atoms, to expose the underlying sample. The ions tend to be created from an inert gas, so during cleaning with the ions, they do not interact with the sample and form different chemical states that were not initially present in the sample. Argon is the most common gas used due to its relatively cheap cost compared to other inert gases.

When ion sources were first introduced, they were much more simple tools with few parameters to choose from. However, now there are generally several sections in an ion source to give the user control over the parameters available. Such parameters can be the energy of the ions, the size of the ion beam and the number of ions emitted per unit of time. This flexibility in a source can help to give optimum settings for different types of samples. For example, if an analyzed sample has a very thick layer of contamination on it, then a high energy beam would be used to remove the contamination in less time, and if the sample is only very small, then the size of the ion beam can be changed to clean just the area of interest.

Ion sputtering [14] is not a straightforward process and can have unwanted effects on the sample. For example, some atoms may be preferentially removed by a high energy ion beam, changing the surface chemistry. Continuous sputtering can also begin to roughen the surface of the sample as well. However, it is necessary to use this method of cleaning if anything within the top ~10 nm requires analysis.

As mentioned, due to the nature of these single ions, they tend to have energies of several hundred electron volts in order to control the path of their emission. This high energy ion then impacts the sample. With harder materials, such as most metals, and semiconductor samples this is no problem. However, with polymer-type materials and some metal oxides these high energy ions can break apart the chemical bonds forming the samples. The chemical states are then altered, meaning when the sample is analyzed the measured chemistry is different from that in the original, non-sputtered sample. This change in chemistry means analysis for softer materials in the forensics field was not feasible, as chemical state measurements are key in this field of study.

However, in the past few years, a second ion sputtering technique has become available, meaning even more in-depth analysis can be performed without causing as much chemical damage, so softer materials can now be cleaned with minimal chemical damage. This is due to the advancement in cluster ion technology. More information on cluster ion sources can be found in Section 3.4.

3.3.6 Depth Profiling

In addition to cleaning with ion sources, depth profiles [14, 15] can be performed on samples. This can then give an understanding of the sample as a function of depth. This involves sputtering the material, analyzing it, and then sputtering again. The cycle is illustrated in Figure 3.7.

Data from a depth down to a few microns can often be collected by performing depth profiles. Beyond this depth, the resolution in the profile becomes poor and the data are more difficult to interpret.

By following this cycle, information from the sample below a depth of the standard 10 nm can be understood. Firstly, a spectrum, or spectra, of the desired

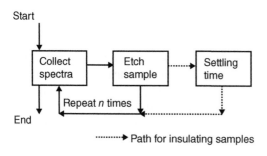

Figure 3.7 The cycle summary for XPS depth profiling. Source: [4]/Thermo Fisher Scientific.

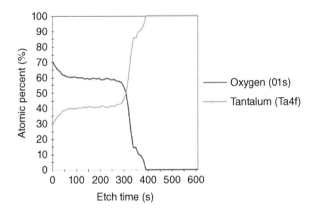

Figure 3.8 Example depth profile of tantalum oxide on a tantalum substrate. Source: [16]/Thermo Fisher Scientific.

regions are collected by using XPS. Once collected, ion sputtering takes places for a user-determined time. This time will depend on how thick the sample is. For example, if a sample has a layer of 500 nm, it will be very time consuming to etch for only a few seconds each cycle, as only a few atomic layers would be removed, meaning many etch cycles will need to be performed before reaching the next layer. However, if a layer is only 30 nm, this etch time would be more appropriate.

On a depth profile, the x-axis is normally set to etch time; however, an estimate of the thickness can be determined by using reference samples of known thickness. The y-axis shows atomic percentage of the element or chemical state. Figure 3.8 shows an example of a simple depth profile of tantalum oxide on a tantalum substrate.

It is important to note here, that once either cleaning or depth profiling has been performed, the sample has been altered, and cannot be set back to its original form. This is particularly important for forensic applications as forensic samples can be very small, so there may just be one chance of obtaining data by using depth profiling. Because of this it is imperative that the user understands the parameters they are using in the profile.

3.3.7 Sample Preparation

Sample preparation is minimal when analyzing with XPS, but an understanding of the sample by the user is useful, if not a necessity, before any analysis takes place.

As XPS is a technique that takes place in vacuum, it is crucial that the sample is vacuum-compatible. This means liquids cannot be analyzed, as they can become gas under vacuum and then be pumped away once the pressure becomes lower. Any solid samples should not be able to sublime in vacuum either – turning from solid to gas as the pressure changes. This is especially important for forensics, as if this occurs, it would be impossible to get the sample back to its original form as the gas would be pumped away through the vacuum pumps.

If the sample is in liquid form, or has a chance of subliming, it can be possible in more advanced XPS preparation systems to cool a sample to freeze it before starting the pumps to create a vacuum. The cooling can often prevent a sample from subliming, as temperatures can be lowered to temperatures of $-190\,^\circ$C by using liquid nitrogen.

Samples should ideally be of a size between a few millimeters and maximum of a few centimeters in width and length, with a thickness of a couple of centimeters. This, again, is because of the vacuum and how the sample has an impact on the pressure. The larger the sample, the more material is going into the system, and hence more gas and liquid is within the sample. This needs to be pumped away by the system before the pressure is low enough for the system to analyze at a safe vacuum level. Most modern systems will have interlocks on the vacuum, meaning analysis can only take place at a safe vacuum level. If the pressure in the system is too high, the analysis equipment such as the X-ray source will not switch on.

For example, a porous material of a size of a few centimeters cubed could take several hours to be ready to be analyzed, as the vacuum would be too high initially. But a solid glass sample of a couple of millimeters thick by a 10 by 10 mm surface could be pumped down and ready to analyze within just a few minutes.

3.4 Argon-Ion Cluster Beam Technology

Ion cluster [6] technology is the latest key development in XPS over the last several years. As previously discussed, single ions have been the main source of sputtering up until this development, but these can cause severe chemical damage to the sample, due to the high energy of the incident ions.

Cluster technology uses the same inert gas as the single ion source, but rather than single ions forming, a cluster is formed as can be seen in Figure 3.9.

This cluster is formed by a high pressure of inert gas (typically argon at a region of 2–4 bar) being fed into the back of the source where a supersonic expansion occurs. This forms clusters of many different sizes. These clusters then move into a segment of the ion source designed to slow them down and ionize them with the use of a hot, ionizing filament. Only a single atom is ionized within the cluster. The next segment in the source can select the size of cluster, from software input by the user. Differing sizes of cluster influence the incident energy of the cluster impacting the sample, and hence have a varying degree of speed when sputtering

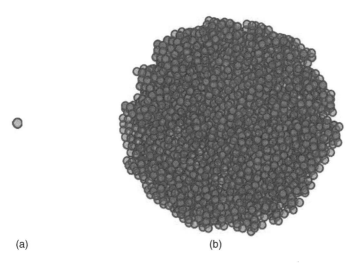

(a)   (b)

Figure 3.9 A single ion used for monatomic sputtering (a), and a cluster ion used for cluster sputtering (b).

the sample, which is why this is of importance as different energies are needed for different samples depending on their properties. Following this, there are typically several stages in the cluster source which can alter other parameters in the cluster such as their energies. The final part of the source is a segment of electrodes to allow direction and rastering of the beam toward the sample to create an etch area.

To be able to create a user-determined etch crater, the energy of the single ions must be in a region of a couple of hundred electron volts. This means when the ion impacts the sample it is hitting it with this amount of energy. However, a cluster needs a couple of thousand electron volts of energy to remain intact, so it does not collapse, but this energy is spread out across the whole cluster. For example, with a cluster size of 2000 atoms and an energy of 2000 eV, the energy is only 1 eV per atom, making it much less energetic than a single ion in the monatomic mode of sputtering. Therefore, the cluster can be utilized for more delicate materials for cleaning and depth profiling.

With a much lower energy across the whole cluster compared to a single ion, much shallower penetration of the sample occurs during impact. While a single ion can penetrate several nanometers, causing breaks in chemical bonds and changing the chemical make-up of the sample which is so important in XPS analysis, a cluster impacts the surface, and breaks up immediately, meaning much less penetration occurs. This penetration can just be the top couple of atomic layers, meaning when analyzed, the top few layers have been removed, but no chemical change is observed in the spectrum. An example of this can be viewed for the polymer polyethylene terephthalate (PET) in Figures 3.10 and 3.11.

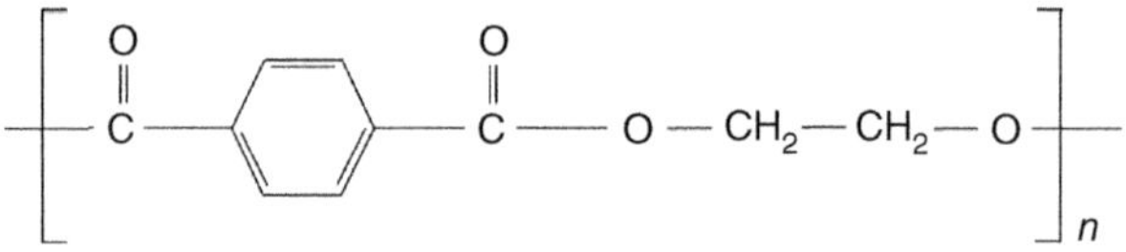

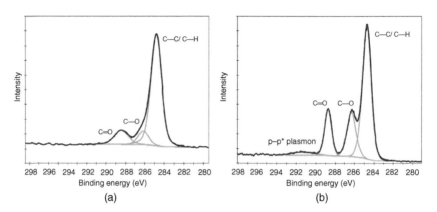

Figure 3.10 Chemical make-up of PET.

Figure 3.11 An example of C1s spectra of a PET sample that has been sputtered with single ions (a) and cluster ions (b). Source: [17]/Thermo Fisher Scientific.

A carbon C1s spectrum of PET should show three distinct peaks from the chemical states (C—C/C—H, C—O and C=O), in a ratio in terms of area of 3 : 1 : 1, with a fourth peak due to a plasmon. The cluster sputtered spectrum shows this. The surface contamination has been removed by cluster cleaning, with just the top ~10 nm of the PET being analyzed, with no chemical damage visible in the spectrum.

The spectrum from cleaning with a single ion shows a very different spectrum. A single peak is still visible from the C—C/C—H, but this has a large tail on the high binding energy side of the peak due to the remaining C—O, which is much lower than expected. The same is found with the C=O. The drop in intensity of these peaks is due to the ions breaking apart the bonds between the C—O and C=O, leaving carbon unbonded, and hence causing a change in the binding energy to 284.8 eV which is characteristic of a carbon peak not bonded to anything else. There is also the loss of the plasmon peak which is characteristic of a phenyl ring. With no evidence of this peak being visible, it is easy to determine that the ions have broken apart the phenyl ring.

This is just a simple example of how monatomic sputtering can alter chemical states of samples and shows the importance of clusters for samples that can be important to forensic applications.

3.5 Evidence Type Examples

3.5.1 Example 1: Surface Modified Fibers

One area that XPS can be of importance in forensics is in analysis of fibers in clothing for criminal investigations [18, 19].

Individual white cotton fibers have very few unique features and can look indistinguishable from one another to the eye. This is where XPS can be of use, as it can distinguish between chemical states that are not optically visible.

Many fabrics nowadays have various coatings and surface-modification treatments to help protect the material beneath. These modifications can include making the material either waterproof, stain resistant, or can give protection from UV rays. These coatings for textiles can comprise a range of chemistries and be applied via a variety of treatments such as wet application or plasma deposition. All these variations can give many different chemical forms, which are distinguishable by the XPS.

At crime scenes threads from clothes, and more commonly fibers, can be left behind by the offender. Often these can be ignored by forensic investigators. However, protective coating on clothing is becoming more common, and so these small fibers can be analyzed and identified using XPS to determine the surface coatings. These small fibers require careful handling and preparation on a sample plate before analysis, but the analysis itself is simple to perform. In this example, several cotton fibers have been analyzed, with different coatings having been applied [11].

Two samples were initially analyzed. One was a standard, plain cotton fiber, and one had been coated with a type of protective layer, which was indistinguishable by eye from the plain fiber. These two samples had survey spectra taken of their "as-received" states; where the samples have been taken, put straight into the XPS system without any modification by the XPS user. The results of this can be seen in Figure 3.12.

This data were acquired on a Thermo Scientific K-Alpha XPS system, by placing the fiber samples across a hole in the sample plate and securing using sticky carbon tape suitable for vacuum. This allowed a large X-ray area to be used for analysis, with no signal being picked up from the sample plate itself. The samples themselves were both approximately 10 mm long. As the sample is insulating, the charge compensation source was required which supplied both low energy electrons and ions to the sample to keep the charge neutral. No preparation was performed on the sample other than extraction from a larger cotton sample and then placement into the system.

From the spectra it is immediately evident that there are differences. The plain sample has two main peaks present; from carbon and oxygen. These are the two main elements that make up cotton, as well as hydrogen, which cannot be detected

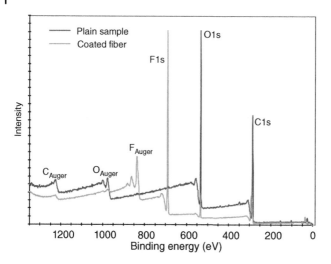

Figure 3.12 Survey spectra comparison of a plain cotton fiber and a coated cotton fiber. Source: [11]/Thermo Fisher Scientific.

by XPS. The coated sample also has a large peak, attributed to fluorine 1s, in addition to the carbon and oxygen. Fluorine Auger peaks can also be identified at ~900 eV confirming this.

If the peaks are measured by adding a background, it is then possible to determine the elemental concentration present on the surface of the samples.

Table 3.1 confirms the differences observed in the spectra. Fluorine is the main element present in the coated fiber, with 50% of the surface (~10 nm) being fluorine. Small amounts of silicon and calcium are present on the plain cotton sample.

Table 3.1 Table showing atomic concentration of plain and coated fiber.

Element	Atomic concentration (%)	
	Plain	Coated
Fluorine	0.0	49.9
Carbon	65.6	41.4
Oxygen	32.8	8.7
Silicon	1.0	0.0
Calcium	0.6	0.0

Source: [11]/Thermo Fisher Scientific.

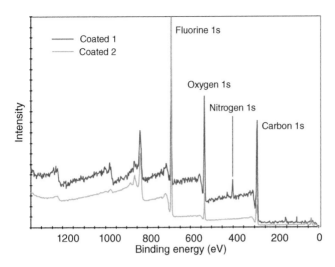

Figure 3.13 Survey spectra comparing two coated cotton samples. Source: [11]/Thermo Fisher Scientific.

Calcium is often used as a whitening agent in cotton, so is not uncommon in this material.

Such differences are easily identified by survey spectra, which provide a broad idea of the elemental information on the surface of the sample. However, with clothing that has been coated by different methods, or by different companies can a difference still be observed?

Figure 3.13 shows two survey spectra acquired from two different cotton fiber samples. Both samples had been coated; one by plasma and one by wet application. Sample "Coated 1" shows more noise in the spectrum as the sample was much smaller than "Coated 2." However, again, it is easy to see differences in the two samples. Percentages of all elements vary between samples, and "Coated 1" has nitrogen present. This is just one example of modified samples being different, and it's clear their elemental information is different, but what can be discovered from utilizing the real advantage of XPS and looking in more depth into the chemical state information?

In Figure 3.14, it is possible to see the chemical state information from carbon 1s spectra. The single line on each spectrum is accumulation of the fitted peaks, with the six individual peaks showing the six different chemical states present. Each of these peaks can be attributed to a different chemical state. In this case (from lower binding energy to higher binding energy): C—C/ C—H, C—O, C=O, C—F, CF_2, CF_3. It is also likely that there is some C—N bonding present, due to the presence of nitrogen in "Coated 1," which would overlap with the C—O peak.

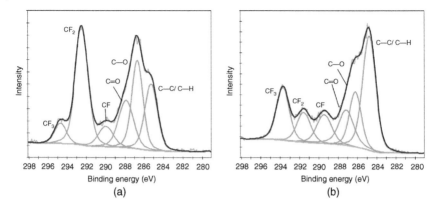

Figure 3.14 Carbon 1s spectra from two modified cotton samples. Coated 1 (a) and Coated 2 (b). Source: [11]/Thermo Fisher Scientific.

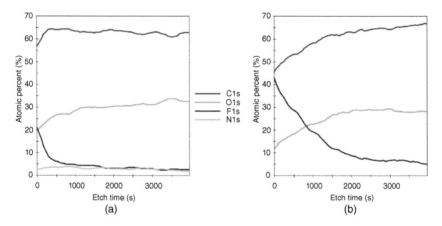

Figure 3.15 Cluster depth profiles of Coated 1 (a) and Coated 2 (b) samples. Source: [11]/Thermo Fisher Scientific.

This data were acquired without making any physical changes to the sample, meaning the sample could be removed from the system and analyzed using other techniques if required.

It is easy to see differences in the top layers by using XPS, however, XPS also allows us to view the change in elemental and chemical make-up of the sample by using cluster depth profiling.

Figure 3.15 shows the collected cluster depth profiles from both samples, showing only their elemental information. This data were collected using a cluster energy of 4000 eV, and a cluster size of 2000, giving each cluster an average energy of 2 eV per atom.

Again, it is easily possible to observe major differences in the two depth profiles, in addition to the fact that nitrogen is present. Coated 1, has less fluorine on the surface, starting at ~20%, and this is soon sputtered away, meaning less fluorine is present in the underlying material. This stabilizes at 8% atomic concentration. There is a steady rise in both carbon and oxygen at the near-surface, which then stabilizes as well.

Coated 2 has a higher concentration of fluorine at the surface, which then is removed at a slower pace than Coated 1, suggesting that the coating is thicker than the one applied to Coated 1. In correlation to this, the oxygen and carbon increase steadily. No nitrogen was found to be present throughout this profile.

All the data shows that no matter if fibers look alike, by using XPS with cluster profiling, significant steps can be made in forensic applications as with XPS it is possible to:

- Differentiate between protective and non-protective coatings on materials, that can be identical to the eye by using both survey spectra and higher resolution spectra
- Compare chemistries of any protective coatings by using high resolution scans of certain elements
- Understand depth and change in thickness of protective coatings by using cluster depth profiling

From this type of data, databases can be built up to hold information of different types of protective coatings, meaning fibers found at crime scenes can be analyzed, (and if needs be, nondestructively) and either matched to coatings from this database, or compared to any suspects' clothing.

3.5.2 Example 2: Glass

3.5.2.1 A

Glass is a vital sample in many forensic cases and is found across a wide variety of objects, including bottles, windows, phone screens, and lighting. In forensic cases, glass is often shattered due to impact and can scatter up to a few meters. During this scattering, pieces of glass can become lodged on suspects' clothes, shoes, or in their hair. These fragments can be analyzed to determine whether the glass has likely come from the crime scene. Luckily, with modern technologies, glass is not always made up of the same material from one type of sample to the next [20, 21].

As with the previously discussed modified fibers, glass also comes in different forms, with many different elements and chemical states combining to make a single sample. For example, windows in buildings may have many different layers on top of the typical silicon oxide substrate that make up the majority of glass samples.

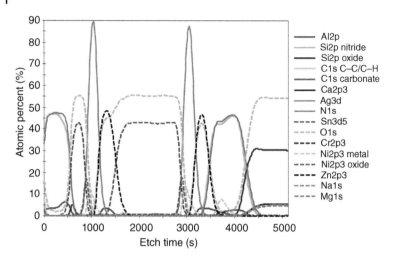

Figure 3.16 Depth profile of a low-emissivity glass coating. Source: Modified from [22].

An example of an XPS depth profile of one of these can be seen in Figure 3.16. Many layers can be observed in the top few hundred nanometers of the sample. This includes metals such as silver, tin, zinc, nickel, and chromium. These layers are *"typically applied to glazing units installed in buildings to assist in controlling room temperatures, reducing the heat loss from the interior in winter months or to control heat entering through the glass in the summer"* [22].

The complex nature of this glass sample can be easily distinguished by XPS using depth profiling to determine the layers on the surface of the glass [23–29]. This is just one example of how glass can be manipulated for human benefit. Other examples of modified glass can be found on phone screens, which contain layers providing protection to the glass substrate underneath, whilst still being thin enough to allow the touch-screen technology to function.

One other example of modified glass is that of self-cleaning glass. Two samples from different manufacturers of self-cleaning glass were analyzed using XPS and monatomic profiling to see if any differences could be observed. These can be found in Figure 3.17. The thickness of the samples was determined from a silicon oxide standard of known thickness that was profiled using the same settings.

From the profiles it is possible to see both samples have a layer of titanium dioxide on the surface, with this being the property behind the self-cleaning result. The TiO_2 on the glass causes a reaction as sunlight is shining, causing photocatalysis which breaks down organic dirt on the surface. As rainwater falls, hitting the glass, the water then runs off and takes the broken-down dirt with it.

Despite both samples having the same elements in the same structure present (TiO_2 on top of a glass substrate of SiO_2 with sodium and calcium), there are

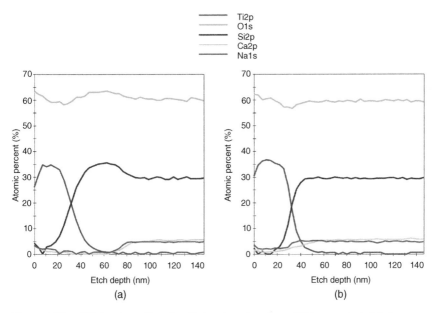

Figure 3.17 XPS depth profiles of self-cleaning glass from Pilkington Activ (a). Source: Based on [23]. PPG Sunclean (b). Source: Based on [24].

several differences which can be observed. The Pilkington Activ™ sample has a layer of TiO_2 of approximately 40 nm, but this tails into the SiO_2 of the glass substrate, whereas the PPG Sunclean® sample has more of an abrupt change between the TiO_2 and SiO_2 substrate. This is likely due to the manufacturing process of the different samples.

During the manufacturing process, titanium dioxide vapor is sprayed from jets onto sheets of molten glass to form the Pilkington Activ glass. Rather than forming a separate layer on the glass surface the titanium dioxide dissolved into the glass at the surface, causing the gradual reduction of TiO_2 in the depth profile.

The PPG Sunclean glass is manufactured differently. Glass panes are made using a traditional technique. A molten separate TiO_2 enriched glass laminate is then poured over pane. As a result of this, the side of the glass coated with TiO_2 has a thin laminate present. This does not mix into the glass itself, which causes a distinction between the TiO_2 and SiO_2 layers which are observed in the depth profile.

Due to these differing manufacturing methods, XPS can easily distinguish between two glass samples that perform the same function. Because of this, glass samples presumed to be from a particular manufacturer can be profiled using XPS and monatomic profiling, therefore determining if they have indeed originated from the presumed source.

3.5.2.2 B

The following data and information are taken from the Application Note [25] "Surface Analysis of Metals in the Glass on Smart Phone Screens" from Kratos Analytical, Manchester, United Kingdom.

An issue with analyzing glass has been the mobility of the sodium and other alkali metals in the glass substrate. With the advancement of phone technology, nearly all phones now have a glass screen. The mobility of these metals is important to the phone manufacturers as they can have a significant impact on the electrical properties of glass. However, these properties in the field of forensics are also crucial.

As previously shown in this example, glass has many layers, and profiling the samples allows the layer structure to be determined. By profiling with monatomic ions, as was regularly done in the past, issues were caused by the mobility of the alkali metals, as is demonstrated in following example.

Figure 3.18 shows two images of the depth profile through the glass of a phone screen. There are several elements found in low concentration, so the profile has been enlarged on the right-hand image. The profile was performed using 5 kV monatomic ions.

Silicon and oxygen are present as SiO_2, which is the main component of glass, and is expected in high concentrations. The lower concentration elements include sodium, calcium, and potassium. After 5000 nm the profile is stable, but in the first 5000 nm a drop in the concentration of potassium is observed, with a rise in sodium. This shows a relationship between the two elements. However, the sodium remains at a fairly constant level of 4%, which is lower than expected for glass (typically ~9%).

This drop in the level of sodium is typical of monatomic depth profiling on glass. However, this restricts its use when analyzing for forensic applications due to the

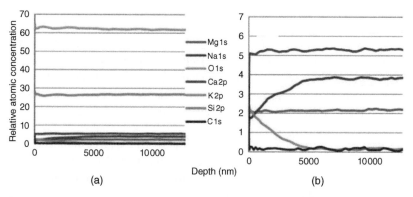

Figure 3.18 Monatomic depth profile of phone screen glass using 5 kV ions. (a) 5 kV monatomic depth profile. (b) Enlarged low concentration elements.

Figure 3.19 Cluster depth profile of phone glass screen using 20 kV clusters of size 500 atoms.

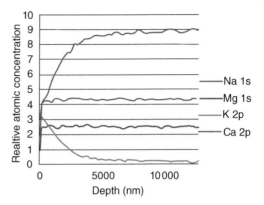

inaccurate data being collected. When monatomic profiling is taking place on glass samples such as this, the sodium is repelled into the bulk of the glass due to the build-up of positive charge in the near surface region, and so the level of sodium in the top layer is lower than expected.

However, now with the availability of cluster ion sources in XPS systems the sodium levels can be analyzed, obtaining correct elemental information. As clusters penetrate less deep into samples compared to monatomic ions, the sodium will not be forced into the bulk, allowing the actual atomic concentration to be analyzed from the glass. Figure 3.19 shows the focused in profile when using 20 kV clusters with a size of 500 atoms.

It is possible to see a large change in the sodium levels here compared to that in the monatomic profile. Figure 3.20 shows a comparison between the sodium and potassium levels between the two profiles.

Here, it is evidently clear that the cluster profile alters the concentrations less than the monatomic sputtering does. The sodium is not forced into the bulk of

Figure 3.20 Comparison of sodium and potassium concentrations between monatomic and cluster profiling. The higher concentrations are from the cluster profile.

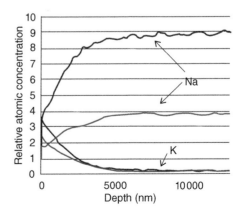

the sample, and so more is present on the surface, hence an increase in the concentration. After the first 5000 nm the sodium level remains stable at 9%, which is expected in a glass sample of this type.

This example again demonstrates the need for cluster technology for forensic examples. Without cluster ions, the acquired data will not be representative of the sample collected and is therefore not appropriate for use in forensics. With the availability of cluster technology, this can give forensic scientists a useful technique that can aid investigations into glass.

3.5.3 Example 3: Hair Fibers with Modifications

Organic material is often left at crime scenes, even in small quantities, but just a single hair left can be analyzed by using XPS to determine from where it originated [30, 31].

The structure of hair is well understood, with a composition of keratin and fatty acids making up the majority of the hair. However, there are many hair products available nowadays, ranging from simple hair gel and wax for styling, to more extreme coloring methods such as using colored hair chalk. These can be applied to hair affecting the surface, meaning a change in elemental and chemical states to the outer layers, which can then be analyzed by using XPS to distinguish between all the artificial products that can be applied.

In this example, both elemental and chemical changes can be observed between different hair products, as well as how this changes with thickness by analyzing depth profiles of different products.

The first example shown in Figure 3.21 has had two depth profiles performed using cluster depth profiling. A cluster energy of 4000 eV was used, with a cluster size of 1000, meaning each atom had an energy of 4 eV. The profiles were taken of a hair, from the same person; one after cleaning with shampoo, and the other after shampooing with a styling gel having been applied afterward. The cluster depth profiles allow the chemical states to be kept intact, meaning it is possible to determine the chemical make-up of the gel.

The two depth profiles show drastic changes between the two hairs and show the affect the hair gel can make on the chemical composition observed.

The hair without gel applied shows a drop in the C—C component from the surface, with a gentle rise in the other carbon components, oxygen, and nitrogen. Sulfur increases slightly to begin with, but then drops away toward the end of the profile.

The hair with gel applied has a much more sustained level of C—C through the beginning half of the profile, with 85% of the composition coming from the C—C component. Both the oxygen and C—O component of the carbon drop initially but increase toward the end of the profile. The C=O, nitrogen and silicon all increase

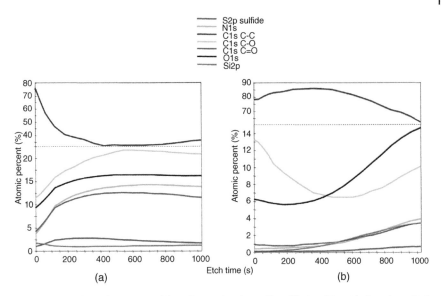

Figure 3.21 Chemical composition cluster depth profile of hair without hair gel applied (a) and with hair gel applied (b). Source: [30]/Thermo Fisher Scientific.

through the profile. At the end of this profile, the atomic levels in chemical states are like that at the start of the profile of just the hair. This profile therefore shows the effect that hair gel has.

Different hair gels are likely to have different compositions and by profiling these, an understanding of the chemical structure of the gels can be reasoned.

Hair chalk has a high probability of transfer and retention if a victim struggles with an assailant during an abduction or sexual assault and could help support an association between an assailant, a victim, and a specific crime scene in a specific case circumstance.

Transfer of hair chalk residues might be ignored by a perpetrator but could easily be noticed on clothing or other articles by an investigator and collected for examination. Even these trace amounts of hair chalk can be readily detected by XPS on material surfaces such as clothing as well as on individual hair samples.

To observe if XPS can determine the difference between hair chalks, the compositional properties of chalks from different manufacturers was measured in a Thermo Scientific K-Alpha. Figure 3.22 shows six example chalks that were analyzed. These hair chalk sample were pressed onto double-sided tape on pieces of aluminum sheet and placed on the sample plate.

Two example survey spectra can be observed in Figure 3.23. These are for white and black hair chalks. It is easy to initially comment on the difference in spectra,

Figure 3.22 XPS sample plate from a Thermo Scientific K-Alpha with six example hair chalks for analysis. The sample plate measures 60 mm × 60 mm Source: [31].

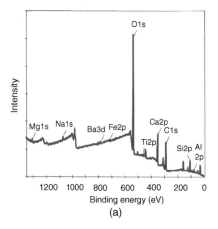

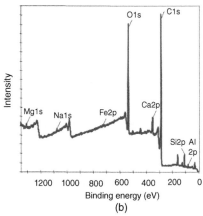

Figure 3.23 XPS survey spectra of white hair chalk (a) and black hair chalk (b). Source: Adapted from [31].

with the white hair chalk having the presence of titanium and showing a large difference in the carbon levels compared to the black sample's spectrum.

All six analyzed chalks showed variance in their elemental atomic concentrations. Once survey spectra were acquired, higher resolution scans of the relevant elements were acquired to determine the chemical states of the samples, and to obtain a better understanding of them.

Table 3.2 shows the elemental composition of the analyzed samples from the acquired higher resolution spectra. All the hair chalk colors contained high amounts of carbon. The black color chalk had the highest amount of carbon, likely due to carbon black or another similar colorant.

Other common major and minor elements in all the hair chalk colors included: oxygen, magnesium, aluminum, silicon, calcium, and iron (most likely as metal oxides and/or carbonate fillers and binders). The orange, red, and violet hair chalk colors contained low amounts (1–3 at%) of nitrogen, which is most likely due to organic dyes and/or pigments.

Table 3.2 Chart showing elemental composition of several hair chalks available from one manufacturer's range of hair chalks.

Color	C	N	O	Na	Mg	Al	Si	Cl	Ca	Ti	Fe	Cu	Ba
White	31.0	–	45.3	0.8	2.3	4.1	9.3	–	6.0	0.7	0.4	–	0.1
Ochre	32.9	–	45.6	0.7	1.4	3.1	8.8	–	6.8	–	0.7	–	–
Orange	39.0	1.3	40.5	0.5	1.2	3.2	9.2	–	4.7	–	0.4	–	–
Red	49.8	1.9	33.3	0.2	1.5	1.5	6.0	0.9	4.7	–	0.2	–	–
Violet	44.6	3.0	37.1	–	0.7	1.6	4.4	–	7.5	–	0.6	0.5	–
Black	68.2	–	21.4	0.3	0.7	1.7	5.4	–	1.9	–	0.4	–	–

Source: Adapted from [31]

Table 3.3 Chart showing surface atomic concentration from three separate manufacturers.

Manufacturer/color/type	C	N	O	Mg	Al	Si	S	Ti
Manufacturer 1. "moon hair" chalk	55.7	–	24.1	0.6	1.6	17.8	–	0.2
Manufacturer 2. "party purple" hair powder	53.8	–	26.4	0.7	2.5	16.6	–	–
Manufacturer 3. "purple swag" liquid hair color	79.5	1.7	18.3	–	–	–	0.6	–

Source: Adapted from [31]

The white hair chalk color contained a low amount of titanium as TiO_2 is a common white pigment and a trace amount of barium (0.1 at%).

These results are from a single manufacturer. However, it is also possible to determine differences in very similar colors between different manufacturers and styles of hair dying. Table 3.3 shows an example of XPS elemental data acquired from three manufacturers styles of hair coloring (one chalk, one powder, and one liquid hair color), all from a very similar purple color.

XPS showed unique surface compositions for the different manufacturer's hair chalks, powders, and liquid hair colors. The chalk and powder had similar levels of carbon present, but the liquid coloring had a ~30% increase. The chalk and powder, as well as having similar levels of carbon, had very similar levels of all other present elements. However, the presence of titanium as TiO_2 allows the chalk to be distinguishable from the powder.

The liquid had nitrogen and sulfur present, which the chalk and powder did not, as well as a lower concentration of oxygen, allowing them to be easily differentiated.

XPS can therefore determine the differences between different hair chalk coloring, as well as different manufacturers and different applying techniques. From these results, it shows that hair chalks are easily measured and determined and can therefore be compiled in a searchable database for any forensic needs when hair coloring has been transferred onto a culprit or indeed on the hair itself.

3.6 Future Directions of XPS and Forensics

Scientific techniques are always evolving, and XPS is no different. Two of the most recent advances could play a large part in appealing to a wider audience for forensic studies. These advances are hard X-ray photoelectron spectroscopy (HAXPES), as mentioned briefly earlier, and near ambient pressure X-ray photoelectron spectroscopy (NAPXPS).

3.6.1 HAXPES

As discussed in Section 3.3, in a standard XPS experiment, an aluminum anode is used that generates X-rays at an energy of 1486.6 eV. This provides a depth of analysis of approximately 10 nm and gives binding energy data between 0 eV and approximately 1300 eV in the spectrum. However, in many systems a second anode is available to be used, either using the same electron source as the aluminum anode, or a separate one. Using this different anode material can alter both the depth of analysis and the XPS peaks that are visible in the spectrum. A typical second anode material is silver, which gives a photon energy of 2984.3 eV, giving approximately double the depth of analysis compared to aluminum, as the photon energy is approximately double. Using a silver anode would additionally give XPS peaks up to a binding energy of 2500 eV, meaning different peaks can be used for analyzing, as well as the original ones from the aluminum anode. This can be particularly useful when two peaks overlap while using the aluminum anode, which can occur when XPS peaks have similar peak energies to Auger peaks. For example, when observing samples of GaN, gallium LMM Auger peaks have an energy similar to that of the nitrogen 1s peak at 398 eV. However, using the silver anode allows analysis of the N1s peak, and the Auger peaks from the gallium are now shifted so that no overlap occurs.

The extra thickness being analyzed by a silver anode also allows an approximation of thicknesses of the material being analyzed without the need for depth profiling using an ion source. If for example an unknown sample has 20% of element A and 80% of element B using the aluminum anode, but when the silver anode is used, element A now has 50% atomic concentration, it is clear that element A is deeper than element B as the silver anode is looking deeper into the sample. Using the silver anode also makes any contamination on the surface less of an issue, as

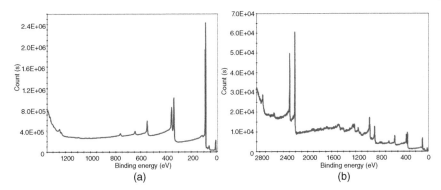

Figure 3.24 Spectrum of gold using an aluminum anode (a) and a silver anode (b). Source: [32]/Thermo Fisher Scientific.

the atomic percentage is much lower if analyzing 20 nm using a silver anode, with contamination just being a couple of nanometers thick.

The main reason that a silver anode is less common than an aluminum anode is due to the sensitivity of the acquired data. The count rate from the silver anode is much less than using an aluminum anode, so a longer acquisition time is required to achieve a decent signal to noise ratio. This is especially true in the lower binding energy region, so the peaks where the aluminum anode may take just a few seconds of acquisition to achieve good quality data, the silver anode may take a couple of minutes. This is the reason HAXPES has been placed into this "future directions of XPS and forensics" segment. If count rate can be increased, it would certainly make HAXPES a more useful technique for forensics (Figure 3.24).

The silver anode is not the only anode available. Table 3.4 shows typical anode materials and the respective photon energies. The higher the photon energy, the deeper the analysis depth.

3.6.2 NAPXPS

While XPS needs to be performed in UHV conditions allowing only vacuum-compatible samples, NAPXPS allows XPS to be performed in much lower vacuum conditions. This is typically in the region of 100 mbar [34].

This higher pressure now allows samples that are not suitable for standard XPS to be analyzed using NAPXPS. For example, samples such as liquids, large porous samples, as well as biological samples [35, 36] can be measured using the technique whereas they were either impossible in standard XPS, or measurements took a long time due to the long pumping times required [37].

NAPXPS typically has the sample analysis area under this near-ambient pressure (in the region of 100 mbar, depending on the sample), whereas the detector

Table 3.4 Frequently used anode materials for XPS and their associated energies.

Anode material	Photon energy (eV)
Magnesium	1253.6
Aluminum	1486.6
Zirconium	2042.4
Silver	2984.3
Titanium	4510.9
Chromium	5417.0

Source: [33]/Thermo Fisher Scientific.

and lens column that collects the photoelectrons from the sample is differentially pumped with a series of pumps [38–40]. This is to ensure minimal gas particulates in the path of the emitted photoelectrons, so the kinetic energies of these emitted electrons are not altered before reaching the detector. However, even with these pumps the sample itself can be in the near-ambient pressures still allowing for a large range of samples to be analyzed [41].

There are still potential difficulties in obtaining decent quality spectra from samples, and this is due to the effectiveness of the charge compensation of the systems. As the area of interest in NAPXPS is typically in applications of life science the need for charge compensation is crucial. As previously discussed, in traditional XPS electrons and positive ions are used from a source to neutralize the sample. The benefit of NAPXPS is having neutral gas particles from the air so close to the sample as these can be used instead. The excitation from the incoming X-ray is enough to interact with these air particles, causing free ions and electrons, creating a "charge cloud" above the sample. This cloud then interacts with the sample surface, and in theory neutralizing it, allowing collection of the photoelectrons by the detector of photoelectrons with the correct energy [42].

While this in theory should mean all data is collected at the correct energy, there has been published data showing incorrectly compensated spectra. With analyzing unknown samples this can pose a problem if it is not known by how much the spectra need to be corrected by. However, as a relatively new advancement this will surely be rectified at a later stage.

One advantage of NAPXPS over traditional XPS is the need for long waiting times in sample pump down. If large porous samples are to be analyzed, a traditional XPS system may need up to several hours of pump-down time to reach the required pressure for analysis. However, with a NAPXPS system, the pressure required to undergo analysis is nowhere near what is needed by a traditional system and so the waiting times are much reduced. In laboratories and facilities that have tight time constraints, and in forensic applications that are time-critical this can be of huge benefit.

3.7 Conclusions

XPS as a technique has grown rapidly from its beginnings in the 1950s, with more and more papers being written including XPS data. This continuing growth in use of XPS has allowed manufacturers to improve the ease-of-use of systems and allowed it to be used in areas where it has not before. As far as additional applications in forensic science, reported use of XPS include investigation of modified and unmodified flax fibers [43], cosmetic trace evidence [31], characterization of the composition of C4 explosives [44], the chemical composition of latent fingerprints [45, 46], silicon dioxide particles [47], near surface composition of alloys [48], fiberglass surfaces [49], photocopier toners [50], gunshot residue [51, 52], and in investigations related to archaeology [53].

From the examples here, it shows XPS can provide vital information on a variety of applications to trace evidence. In the future, XPS is likely to become an even easier technique to use, and with more and more surface elemental and chemical differences being applied in a wide variety of applications it should be considered as a technique to help forensic investigators in their work.

Acknowledgements

The authors would like to thank the entire Thermo Scientific Applications team in East Grinstead, UK, in particular Tim Nunney, Paul Mack, and Robin Simpson, for providing the XPS data and for helpful discussions, as well as Brian Strohmeier of Avery Dennison, USA.

References

1 Siegbahn, K. (1981). Nobel Lecture, Electron Spectroscopy for Atoms, Molecules and Condensed Matter, Kai Siegbahn, Nobel Lecture, December 8, 1981. http://www.nobelprize.org/uploads/2018/06/siegbahn-lecture-1.pdf (accessed 16 March 2022).

2 Watts, J.F. (2008). Applications of XPS, AES and ToF-SIMS for Solving Problems in Materials Research, 46 slide PowerPoint presentation.

3 Watts, J.F. and Wolstenholme, J. (2020). *An Introduction to Surface Analysis by XPS and AES*, 2e. Hoboken, NJ: Wiley. ISBN: 9781119417583.

4 Thermo Fisher Scientific. East Grinstead, UK. http://XPS-simplified.com (accessed 16 March 2022).

5 Wagner, J.M. (2011). *X-ray Photoelectron Spectroscopy*. Nova Science Publishers, forsmat: ebook. ISBN: 1617282405 9781617282409 1617282405.

6 Smart, R., McIntyre, S., Smart, M.S. et al. (1995), X-ray Photoelectron Spectroscopy, 86 slide PowerPoint presentation.

7 Fadley, C.S. (2010). X-ray photoelectron spectroscopy: progress and perspectives. *Journal of Electron Spectroscopy and Related Phenomena* 178–179: 1–83. http://www.sciencedirect.com/science/article/abs/pii/S0368204810000095.

8 Roberts, A., Fairley, C.C., Johnson, B., and Linsford, M.R. (2017). Trends in advanced XPS instrumentation. 6. Spectromicroscopy – a technique for understanding the lateral distribution of surface chemistry. In: *Characterization of Thin Films and Materials* (ed. R. Linsford Matthew et al.). Vacuum Technology and Coating.

9 Powell, C.J. (2016). Growth of surface analysis and the development of databases and modeling software for auger-electron spectroscopy and X-ray photoelectron spectroscopy. *Microsc. Today* 24 (02): 16–23.

10 Watts, J.F. (2010). The potential for the application of XPS in forensic science. *Surf. Interface Anal.* 42: 358–462.

11 Thermo Fisher Scientific (2014). Fibre Sample Plate CCP PowerPoint presentation, May 2014. East Grinstead, UK: Thermo Fisher Scientific.

12 Gates, P. (2014). University of Bristol, School of Chemistry, Cantock's Close, Bristol BS8 1TS, United Kingdom. http://www.chm.bris.ac.uk/ms/detectors .xhtml (accessed 16 March 2022).

13 Data Acquired by Robin Simpson. East Grinstead, UK: Thermo Fisher Scientific.

14 Cumpson, P.J., Portoles, J.F., Sano, N., and Barlow, A.J. (2013). X-ray enhanced Sputter rates in argon cluster ion sputter-depth profiling of polymers.

15 Reconstructed Concentration Depth Profiles from Angle-Resolved XPS using MEM Software, Kratos Analytical Application Note MO454(A). http://www .kratos.com/sites/default/files/application-downloads/MO454%28A%29%20 %20Reconstructed%20depth%20profiles%20from%20ARXPS%20using%20MEM %20software.pdf (accessed 16 March 2022).

16 '*3 kev monoatomic on 30 nm oxide*' Avantage data. Data acquired by Thermo Fisher Scientific, East Grinstead, UK in September 2013.

17 Data acquired by Christopher Deeks, on Thermo Fisher Scientific K-Alpha$^+$ in May 2017.

18 Baily, C. and Nunney, T. (2012). Using X-ray Photoelectron Spectroscopy to Investigate the Surface Treatment of Fabrics. Application Note 52372. East Grinstead, UK: Thermo Fisher Scientific.

19 The effect of washing on the surface chemistry of plasma coated textiles as studied by high resolution XPS. Kratos Analytical Application Note **MO403(1)**.

20 Chapter 2, pages 129–160 in Glass as Forensic Evidence: Purpose, Collection & Preservation. http://study.com/academy/practice/quiz-worksheet-glass-as-forensic-evidence.html (accessed 16 March 2022).

21 Simmons, C.J. and El-Bayoumi, O.H. (1993). *X-ray Photoelectron Spectroscopy of Glass, in Experimental Techniques of Glass Science*. Westerville, OH: American Ceramic Society. ISBN: 0944904580 9780944904589.

22 Nunney, T. (2011). K-Alpha: Characterization of Low-Emissivity Glass Coatings using X-ray Photoelectron Spectroscopy. Application Note 51902. East Grinstead, UK: Thermo Fisher Scientific. http://assets.thermofisher.com/TFS-Assets/CAD/Application-Notes/AN51902-Low-EmissGlass-Coatings.pdf (accessed 16 March 2022).

23 Pilkington Activ™ Self-Cleaning Glass. http://www.pilkington.com/en-gb/uk/householders/types-of-glass/self-cleaning-glass (accessed 16 March 2022).

24 PPG SunClean self-cleaning glass. http://www.glassonline.com/ppg-introduces-sunclean-self-cleaning-glass-for-commercial-applications (accessed 16 March 2022).

25 Smart phone screen – depth analysis of alkali and alkali earth metals Kratos Analytical Application Note MO4427. http://www.kratos.com/sites/default/files/application-downloads/MO427%20applications%20note-smart%20phone%20screen%20profile.pdf (accessed 16 March 2022).

26 Tasker, G., Uhlmann, D., Onorato, P. et al. (1985). Structure of sodium aluminosilicate glasses: X-ray photoelectron spectroscopy. *J. Phys. Colloq.* 46 (8): 273–280.

27 Miura, Y., Kusano, H., Nanba, T., and Matsumoto, S. (2001). X-ray photoelectron spectroscopy of sodium borosilicate glasses. *J. Non-Cryst. Solids* 290 (1): 1–14. http://www.sciencedirect.com/science/article/abs/pii/S0022309301007207.

28 Jiménez, J., Liu, H., and Fachini, E. (2010). X-ray photoelectron spectroscopy of silver nanoparticles in phosphate glass. *Mater. Lett.* 64 (19): 4–5.

29 Teterin, A., Teterin, Y., Maslakov, K. et al. (2019). X-ray photoelectron study of surface-modified silicate glass. *Inorgan. Mater.* 55 (2): 180.

30 Treacy, J. (2016). *Effect of Products on Human Hair CCP*. East Grinstead: Thermo Fisher Scientific.

31 Strohmeier, B.R. (2014). Forensic XPS surface characterization of cosmetic trace evidence. *International AVS Symposium*, Baltimore, MD, USA November 12, 2014.

32 Data acquired by Paul Mack on Thermo Scientific ESCALAB 250Xi. September 2012.

33 Thermo Fisher Scientific, East Grinstead, UK. https://xpssimplified.com/xray_generation.php (accessed 16 March 2022).

34 Patel, D.I., O'Tani, J., Bahr, S. et al. (2019). Ambient air, by near-ambient pressure XPS. *Surf. Sci. Spectra* 26 (2): 1.

35 Jain, V., Bahr, S., Dietrich, P. et al. (2020). Human hair, untreated, colored, bleached, and/or treated with a conditioner, by near-ambient pressure X-ray photoelectron spectroscopy. *Surf. Sci. Spectra* 27: 014001. http://avs.scitation.org/doi/abs/10.1116/1.5109425?journalCode=sss.

36 Avval, T., Hodges, G., Wheeler, J. et al. (2020). Polyethylene terephthalate by near-ambient pressure XPS. *Surf. Sci. Spectra* 27: 014006-1–014006-9.

37 EnviroESCA (2021). SPECS Surface Nano Analysis GmbH, Voltastrasse 5, 13355 Berlin, Germany. http://www.specs-group.com/nc/enviro/products/detail/enviroesca (accessed 16 March 2022).

38 Patel, D., Roychowdhury, T., Jain, V. et al. (2019). Introduction to near-ambient pressure X-ray photoelectron spectroscopy characterization of various materials. *Surf. Sci. Spectra* 26: 016801-1–016801-3.

39 Starr, D.E., Liu, Z., Hävecker, M. et al. (2013). Investigation of solid/vapor interfaces using ambient pressure X-ray photoelectron spectroscopy. *Chem. Soc. Rev.* 13. http://pubs.rsc.org/en/content/articlelanding/2013/cs/c3cs60057b#!divAbstract.

40 Graphene Offers X-Ray Photoelectron Spectroscopy a Window of Opportunity, from NIST *Tech Beat*: December 22, 2014. http://phys.org/news/2014-12-graphene-x-ray-photoelectron-spectroscopy-window.html (accessed 16 March 2022).

41 Near-Ambient Pressure X-ray Photoelectron Spectroscopy (NAP-XPS). University of Manchester, UK: Department of Chemistry. https://www.chemistry.manchester.ac.uk/research/facilities/nap-xps (accessed 16 March 2022).

42 NAP-XPS and NAP-UPS (2020). SPECS Surface Nano Analysis GmbH, Voltastrasse 5, 13355 Berlin, Germany. http://www.specs-group.com/nc/specsgroup/knowledge/methods/detail/nap-xps-and-nap-ups (accessed 16 March 2022).

43 Zafeiropoulos, N., Vickers, P., and Baillie, W.J. (2003). An experimental investigation of modified and unmodified flax fibres with XPS, ToF-SIMS and ATR-FTIR. *J. Mater. Sci.* 38: 3903–3914.

44 Mahoney, C., Fahey, A., Steffens, K. et al. (2010). Characterization of composition C4 explosives using time-of-flight secondary ion mass spectrometry and X-ray photoelectron spectroscopy. *Anal. Chem.* 82 (17): 7237–7248.

45 Bailey, M., Bright, N., Croxton, R. et al. (2012). Chemical characterization of latent fingerprints by matrix-assisted laser desorption ionization, time-of-flight secondary ion mass spectrometry, mega electron volt secondary mass spectrometry, gas chromatography/mass spectrometry, X-ray photoelectron spectroscopy, and attenuated total reflection Fourier transform infrared spectroscopic imaging: an Intercomparison. *Anal. Chem.* 84 (20): 8514–8523.

46 Erdoğan, A., Esen, M., and Simpson, R. (2020). Chemical imaging of human fingermark by X-ray photoelectron spectroscopy (XPS). *J. Forensic Sci.* https://doi.org/10.1111/1556-4029.14483.

47 Quantified imaging of silicon dioxide particles. Kratos Analytical Application Note MO421(1). http://www.kratos.com/sites/default/files/application-downloads/MO421%281%29%20Quantified%20imaging%20of%20SiO2%20particles.pdf (accessed 16 March 2022).

48 Screemay, M. and Ghosh, T. (2001). Near surface composition of some alloys by X-ray photoelectron spectroscopy. *J. Phys.* 57 (4): 809–820.

49 Nichols, G., Hercules, D., Peek, R., and Vaughan, D. (1974). Application of X-ray photoelectron spectroscopy to the study of fiberglass surfaces. *Appl. Spectrosc.* 28 (3): 219–222.

50 Lyter, A. III, (2016). Characterization of photocopier toners by X-ray photoelectron spectroscopy (XPS): how they change with age. *Int. J. Forensic Sci. Pathol.* 4 (4): 231–233.

51 Schwoeble, A., Strohmeier, B., Bunker, K. et al. (2011). Application of X-ray photoelectron spectroscopy (XPS) for the surface characterization of gunshot residue (GSR). *Microsc. Today* 19 (2): 40–45.

52 D'Uffizi, M., Falso, G., Ingo, G., and Padeletti (2002). Microchemical and micromorphological features of gunshot residue observed by combined AFM, SA-XPS and SEM + EDS. *Surf. Interface Anal.* 34: 502–506.

53 Lambert, J., McLaughllin, C., Shawl, C., and Xue, L. (1999). X-ray photoelectron spectroscopy and archaeology. *Anal. Chem.* 88 (17): 614A–620A.

4

Density Determination and Separation via Magnetic Levitation

Christoffer K. Abrahamsson[1], Shencheng Ge[1], Jeffrey G. Bell[1], Robert D. Blackledge[2], and George M. Whitesides[1, 3, 4]

[1]*Department of Chemistry and Chemical Biology, Harvard University, Cambridge, MA, USA*
[2]*United States Department of the Navy, Naval Criminal Investigative Service Regional Forensic Laboratory, San Diego, CA, USA*
[3]*Wyss Institute for Biologically Inspired Engineering, Harvard University, Cambridge, MA, USA*
[4]*Kavli Institute for Bionano Inspired Science and Engineering, Harvard University, Cambridge, MA, USA*

4.1 Introduction

Investigators commonly characterize trace evidence from crime scenes and compare them with objects and materials found in connection to suspects, a process that is often time consuming and labor intensive [1]. Magneto–Archimedes Levitation (which we call magnetic levitation, or, for short, MAGLEV) is a method for measuring and separation using density that has become increasingly widely used in the last decade. MAGLEV is sometimes referred to as fluid magnetic buoyancy [2], magneto–gravimetric levitation, or magneto–hydrostatic levitation [3]. This chapter provides an overview of MAGLEV and its relevance in the analysis of trace evidence, and practical instructions for how to use MAGLEV.

MAGLEV is performed in a transparent cuvette filled with a liquid paramagnetic medium (typically an aqueous solution of manganese(II) chloride [4]) that is placed in the gap between two magnets with like-poles-facing (Figure 4.1). The magnets are held in place with a mechanical fixture, and together they comprise the MAGLEV device. The magnetic field and the force of gravity act to form an apparent linear density-gradient between the magnets in the paramagnetic medium.

To obtain measurements of density or separation in the density-gradient, the material comprising the trace evidence is added to the cuvette and placed in the magnetic field. The particles making up the sample will move and equilibrate according to their specific densities. That is, the levitation height of a particle

Leading Edge Techniques in Forensic Trace Evidence Analysis: More New Trace Analysis Methods,
First Edition. Edited by Robert D. Blackledge.

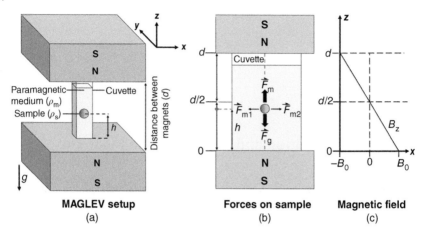

MAGLEV setup
(a)

Forces on sample
(b)

Magnetic field
(c)

Figure 4.1 (a) The MAGLEV device consists of two permanent magnets at the top and bottom of a cuvette filled with a liquid medium that is paramagnetic. The sample is placed inside the cuvette to separate its components. (b) The sample will equilibrate at a specific height (h) above the bottom magnet where the effective gravitational force ($\vec{F}_g$, adjusted for the influence of buoyancy) and magnetic force ($\vec{F}_m$) balance each other. (c) The magnets are placed at an optimal distance d from each other, as determined in past work using simulations in COMSOL Multiphysics Software, to obtain a near-linear magnetic field along the centerline between the faces of the two magnets [5–7].

between the magnets – the distance of the volumetric center of a particle to the surface of the bottom magnet – corresponds to its average density (Figure 4.1) [5, 8].

MAGLEV can levitate and separate diamagnetic trace evidence, such as individual objects or powders. Trace evidence that includes ferro-, ferri-, or paramagnetic matter are attracted to the magnets and can therefore not be levitated in a MAGLEV device without specific techniques [9].

MAGLEV is uniquely suited for the separation of some of the most common types of trace evidence, particularly mixtures of powdered materials. We have published articles on the use of MAGLEV for analysis of particulate materials of particular relevance to forensics, including gunpowder, glitter, and powdered illicit drugs [8, 10].

In the MAGLEV device, the gradient in magnetic field, and thus in density, is used both for sensitive measurements of density and separation of individual objects, or granular matter, such as solid particles, drops of liquids, lumps of gel, or lumps of paste [5, 8]. The method can only separate mixtures of solid particles. A mixture of components that are dissolved in a solvent liquid in the cuvette cannot be separated with MAGLEV. That is, MAGLEV provides useful information about the densities of suspended objects and particles, not dissolved molecules [5, 8].

The search for trace evidence that is rare, out-of-place, and previously unknown is the norm in forensic science. Because all matter has density, MAGLEV can indicate the presence of newly encountered substances. A forensic investigator who wishes to determine the identity of unknown components in a mixture of trace evidence can compare the values estimated by using MAGLEV with densities of known materials.

MAGLEV is uniquely suitable for rapid and convenient comparison of trace evidence, especially of particulate materials [8, 10, 11]. MAGLEV measurements require few components – only a MAGLEV device, a cuvette, and a paramagnetic medium [5]. These simple requirements make MAGLEV a suitable method for analysis of trace evidence, both in forensic laboratories and in low-resource settings such as a crime scenes [8, 10].

Similar to other density-based methods of separation (e.g. float-sink [12]), MAGLEV can significantly reduce the labor needed to separate mixtures of trace evidence compared to manual hand sorting [13, 14]. Other methods of measurement that use density (these methods are listed below) are often time consuming, require expensive non-portable equipment, and well-trained technicians [12, 13, 15]. Additional advantages of MAGLEV include (i) ease of use, (ii) low cost, (iii) portability, (iv) minimal sample preparation, and (v) no use of electricity.

4.2 Objectives of the Work

Comprehensive reviews by Gao and Zhang [16], Ozcivici and Tekin [17], Turker Arslan-Yildiz [18], and ourselves [4] describe the theory and applications of MAGLEV. These applications include diagnostics [19–22], chemistry (especially separation of crystals) [11, 23, 24], biology [25, 26], biochemistry [20, 21, 27], materials science [5, 28], clinical diagnostics [25, 29], and food science [30]. This chapter focuses on the forensic applications of MAGLEV. Here, we provide a brief overview of what MAGLEV is, its working principles, and best practices when using the method. We summarize examples of trace evidence that have been analyzed using MAGLEV, and demonstrate levitation of types of trace evidence that have not been previously reported (i.e. bone, hair, dandruff, and a few types of glass). This chapter provides the most comprehensive, practical instruction to date for how to get started with MAGLEV, and how to design and optimize these experiments. We discuss what types of MAGLEV devices and paramagnetic media are optimal for analysis of different types of trace evidence. The practicalities of using MAGLEV have been previously discussed in numerous separate papers, and are here collected in one place. We have included three types of instructions: (i) a step-by-step protocol that describes how to use MAGLEV, (ii) a troubleshooting protocol to solve the most common challenges, and (iii) a set of schematics of three types of MAGLEV devices that enable laboratories to build (or to commission) their own MAGLEV devices at a low cost. These instructions are suitable as guides for new users of MAGLEV, or as a reference for experienced users.

Specialists in trace evidence who use MAGLEV in their routine work will not need the detailed theory of MAGLEV described in this chapter, but can use it for reference if they will testify to their findings as expert witnesses in court.

4.3   Guidance to the Reader

All sections and subsections of this chapter have the following numerical designations:

4 chapter number
4.1 second number is the section number
4.1.1 third number is the subsection number

Additionally, some section and subsection numbers are followed by a "*" superscript. The sections and subsections with superscripts are not required to be studied unless the reader seeks an in-depth understanding of the physical principles underlying MAGLEV.

4.4   Theoretical Basis*

4.4.1   What Is MAGLEV?

Between the two magnets in the MAGLEV device, an object experiences a weak force from the paramagnetic solution in the gradient of the applied magnetic field (Figure 4.1). The object may levitate depending on factors such as its density and magnetic susceptibility, and the magnetic field strength between the magnets. These factors determine the limiting upper and lower density that an object can have and yet levitate in the MAGLEV device. An object that has a density outside this range of density will not levitate and will instead sink or float. We discuss later how to design experiments to enable levitation of objects with a wide range of densities (see Section 4.6.1).

Common paramagnetic media consist of aqueous solutions of paramagnetic salts, such as manganese(II) chloride or gadolinium(III) chloride. These paramagnetic salts are relatively low-cost and transparent in the visible part of the spectrum. (Ferrofluids tend to be opaque to visible light [4].) There are also paramagnetic media based on solutions of chelated complexes of paramagnetic ions that are soluble in polar or nonpolar organic solvents [8, 30]. Paramagnetic media that use nonpolar solvents (e.g. hexane) do not dissolve polar compounds. Nonpolar paramagnetic media enable MAGLEV of polar compounds or species that would dissolve with (or react with) the components of polar paramagnetic media [8].

"Magnetic levitation" is a term that describes two distinctly different principles of levitation in physics. The first principle is the levitation of vehicles, or contactless electromechanical mechanisms, such as contact- and friction-less bearings in electrical engines and generators [21, 22]. This type of magnetic levitation builds on the repulsion between magnets (or magnets and magnetically active materials), e.g. the repulsion that allows trains to levitate above a MAGLEV train track. The second principle relates to the levitation of diamagnetic materials in a paramagnetic medium. In this type of magnetic levitation, the attraction of the paramagnetic fluid to the magnet causes diamagnetic objects to levitate. For this discussion, we consider the small interaction between magnets and diamagnetic materials negligible. That is, MAGLEV, is due to the *attraction* of the paramagnetic fluid to the regions of the highest magnetic field (often to the faces of the magnets), not the weaker *repulsion* of the diamagnetic objects from them. This chapter will focus on MAGLEV of diamagnetic materials that are – or can be in the context of a possible crime – classified as trace evidence, after suspension in paramagnetic fluids. We will later discuss in greater detail the physical forces that enable MAGLEV.

MAGLEV is distinctly different from immunomagnetic separation, a method that uses superpara- or ferromagnetic beads with surface-bound antibodies to *label* and separate (the magnets *act directly on the magnetic beads*) and enrich proteins, bacteria, and mammalian cells in liquid biological mixtures [31]. In contrast, MAGLEV (as we describe it) is a *label-free* method in which the magnetic field *acts directly on the liquid paramagnetic medium* [4].

MAGLEV effectively reduces the number of potential candidates that can be present (based on density) in a mixture of unknown composition, a commonly encountered situation in forensics [14]. This information can serve as a guide to the selection of additional methods of analysis [8].

Separation by MAGLEV is non-destructive as long as the particles to be separated are insoluble in (and unreactive toward) the liquid medium. This insolubility permits effective enrichment ($\geq 95\%$), and recovery at a high yield of the components of the mixtures [8, 32]. Unlike chromatographic techniques, MAGLEV does not require compounds to dissolve prior to separation. Compounds are separated in their solid state (e.g. crystals or particulates), their original physical characteristics, such as polymorphism of crystals, or three-dimensional shapes of particulates are preserved.

There are different types of MAGLEV devices that have specific applications (Figure 4.2) [4]. In this chapter, we will focus on three types of MAGLEV devices most suited to forensic analysis. (i) Large Working Distance (LWD) MAGLEV that has a relatively large distance between the magnets and that measures density with a high sensitivity (detecting small differences in density, $\Delta \sim 1 \times 10^{-4}\,\text{g/cm}^3$)

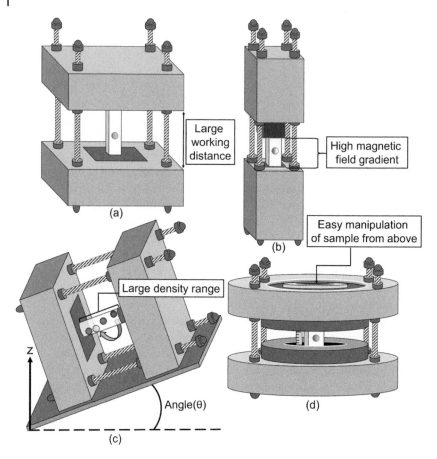

Figure 4.2 Schematics of four types of MAGLEV devices. The colored spheres represent density standards levitating in paramagnetic solutions. (a) A large working distance (LWD) MAGLEV device [5]. (b) A high magnetic field gradient (HMFG) MAGLEV device [23]. (c) A Tilted MAGLEV device [7]. (d) An Axial MAGLEV device [33]. The main benefit of each type of device is described in the figure.

[5]. (ii) High magnetic field gradient (HMFG) MAGLEV, which resolves a larger range of densities ($0.7-3.0\,g/cm^3$); this characteristic makes it suitable for separation of mixtures of unknown composition [23]. (iii) Axial MAGLEV, which has no magnet directly above the cuvette that holds the sample, and therefore enables easy manipulation of the separated components using tools (e.g. pipettes) during levitation [33]. We will discuss the use of tilted MAGLEV in brief; it is described in detail elsewhere [7]. Other types of MAGLEV devices may be suitable for use in forensics under specific circumstances, but these are not discussed in this chapter, e.g. high-sensitivity MAGLEV [6] and high-throughput MAGLEV [34].

4.4.2 Brief Discussion of Trace Evidence Separation Methods and their Limitations

The forensic methods currently used to analyze trace evidence are well established and validated. There are, however, unsolved problems that remain, such as efficient extraction of evidence from complex matrices (e.g. soil or other debris). In fact, trace evidence in complex mixtures is routinely collected at outdoor crime scenes, suspected arsons, and bombings [35]. Common methods used to extract trace evidence from complex matrices, such as hand-sorting or sieving, suffer from poor rates and yields of recovery, and require significant amounts of time and labor [35].

A method used to separate large amounts of particulate trace evidence based on density is the "sink-float" method. This method can separate large amounts of materials but is labor intensive and requires multiple steps of separation in a series of solutions of distinct densities [12, 36]. In contrast to the sink–float method, MAGLEV can separate multiple fractions that differ in density in a one-step procedure, and this property saves time and reduces the volume of liquid medium required [8, 10, 11, 33].

4.4.3 Brief Discussion of Density and Determination Methods

In forensics, density can be used to compare two pieces of evidence, typically to determine if two objects of interest originate from the same source, e.g. two pieces of glass or plastic [1]. Macroscopic objects are straightforward to weigh. It is, however, difficult to determine the weight of objects of small mass (<0.2 mg) because this weight is below the detection limits for most laboratory scales. The volume of a solid macroscopic object is often determined by measuring the weight of a pycnometer (a vessel with a well-known internal volume and weight), containing the object of interest, before and after the vessel is filled with a liquid of well-known density [12, 37]. These measurements enable the volume, weight, and hence the density of the object to be calculated. Standard pycnometry is only practical in determining the volume of relatively large objects (>1000 µl). The density of larger objects can also be determined using different methods, including the sink–float method [12], density-gradient columns [13, 38], and hydrometers [39].

The densities of individual small objects (millimeters to centimeters in size) are measured using techniques, such as micropycnometers of small volume (i.e. 148 µl) [40], liquid density-gradient columns [41, 42], density-gradient centrifugation [43], and image analysis for volume estimation combined with micro-balance measurements [44–46]. These techniques differ in cost, precision, ease-of-use, and portability. The method of density-gradient columns is closest in performance to MAGLEV, particularly in terms of simplicity and sensitivity of measurements, $\Delta\rho \sim 10^{-4}\,\text{g/cm}^3$, and has been successfully used to analyze trace evidence; examples include pieces of glass [1] and polymer fibers [38]. Density-gradient columns have four drawbacks in comparison to MAGLEV: methods using these columns (i) require trained personnel, (ii) are not portable, (iii) require minutes to hours for separation and physical removal (tens of minutes to hours) of the separated components, and (iv) lose its density-gradient over time (hours to days) [41, 42, 47].

4.4.4 Detailed Discussion of Theory*

A diamagnetic object of a density (ρ_s) that is suspended in a paramagnetic medium of density (ρ_m) will either sink $(\rho_s > \rho_m)$ or float $(\rho_s < \rho_m)$ because of the interaction of gravity and buoyancy. The object may, however, be levitated in the paramagnetic medium in the magnetic field of the MAGLEV device, if the applied magnetic field is of appropriate magnitude and shape. When the object reaches its equilibrium position, the magnetic force the object experiences as a result of the direct interaction of the magnetic field and the paramagnetic medium is in balance with the forces of gravity and buoyancy [5]. The object will levitate along the magnetic centerline (see dashed gray line in Figure 4.1b) between the two magnets in the MAGLEV device.

The arrangement of magnets in MAGLEV devices generates a magnetic "bottle" (Figure 4.3) between the two magnets. The magnetic "bottle" is of a shape and magnitude that encourages stable levitation of objects along the central axis between the two magnets, and thus aligns automatically trace evidence in the MAGLEV device. The change in magnetic field strength between the magnets is approximately linear to facilitate a simple interpretation of the results by establishing a linear relationship between the density of the object and its levitation height in the MAGLEV device. The map of the magnetic field strength in Figure 4.3 was simulated using COMSOL Multiphysics software.

Equations (4.1) and (4.2) summarize the physical principles of the method. Eq. (4.1) describes the balance of physical forces acting on the object suspended in a paramagnetic medium in an applied linear magnetic field. Eq. (4.2) describes the relationship between the density of the sample, ρ_s, and the levitation height, h, which we measure experimentally.

$$\vec{F}_g + \vec{F}_m = (\rho_s - \rho_m)V\vec{g} + \frac{(\chi_s - \chi_m)}{\mu_0}V(\vec{B} \cdot \nabla)\vec{B} = 0 \tag{4.1}$$

$$\rho_s = -\alpha h + \beta \tag{4.2}$$

In Eq. (4.1), $\vec{F}_g$ is the gravitational force that acts on the suspended object. $\vec{F}_g$ is corrected for the force of buoyancy, i.e. accounting for situations where an object has a density that is different from that of the solution [7]. $\vec{F}_m$ is the indirect magnetic force experienced by the suspended object as a result of the interaction

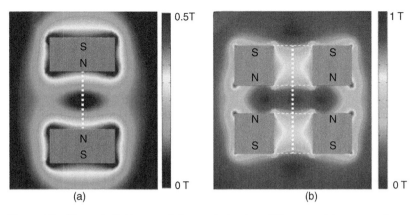

Figure 4.3 Magnetic field strength (see scale bars) at different positions around two magnets for two types of MAGLEV devices. (a) The LWD MAGLEV device (the gap between magnets is 45 mm). (b) The Axial MAGLEV device (the gap between the upper and lower ring magnets is 15 mm) [33]. The figures show the magnetic field strengths in cross-sectional views of the MAGLEV devices. Source: (b) is modified and reprinted with permission from Ge et al. [33]. Copyright (2018) American Chemical Society.

between the magnetic field and the paramagnetic medium. V is volume of the object. $\vec{g}$ is the acceleration due to gravity (where $|\vec{g}|$ is 9.80665 m s^{-2} on earth). χ_s is the magnetic susceptibility of the object. χ_m is the magnetic susceptibility of the paramagnetic medium. μ_0 is the magnetic permeability of free space. $\vec{B}$ is the magnetic field. h is the levitation height of the object above the bottom magnet. Small variations in gravity at different places on earth will not cause significant changes in the measured densities (especially when density standards are used for calibration).

Equation (4.2) is in the slope–intercept form [48], where ρ_s is the density of the object, α is the slope of the curve, h is the levitation height of the object, and β is the intercept of the density-axis. For most density measurements, we first calibrate the system using density standards, e.g. glass beads of known densities, to determine the coefficients (α, β) of the linear regression; we then determine the density of the trace evidence by measuring its levitation height.

4.5 Preparation for Density Determination Via MAGLEV

4.5.1 Choosing the Type of MAGLEV Device: Precision, Accuracy, Working Distance, Sensitivity, and Range of Density

All MAGLEV devices measure the density of trace evidence with high precision (within the limits of their sensitivity), where precision is generally defined as how close independent and repeated measurements of density (levitation height) cluster for a specific object for a set of conditions [49]. A well-calibrated MAGLEV device measures density with good accuracy (1-10%) [4, 5], where accuracy is generally defined as the agreement between the measured value and a true value. (The true value is defined as a standard value of density or the average density calculated from a large set of measurements [49].)

The range of density ($\Delta\rho_{range}$) that a MAGLEV device can measure is determined experimentally for each set of conditions. The range of density can be adjusted by varying four factors: (i) the magnetic field strength of the magnets, (ii) the distance between the magnets (or in practice, the distance between the bottom and the top surface of the liquid pillar of paramagnetic medium inside the cuvette), (iii) the type of paramagnetic species in the paramagnetic medium, and (iv) the concentration of the paramagnetic species in the paramagnetic medium. Levitation of the density standards enables the determination of the slope of the curve (α) using Eq. (4.2). The slope of the curve is used in Eq. (4.3) to determine $\Delta\rho_{range}$

$$\Delta\rho_{range} = (\alpha h_{lower} + \beta) - (\alpha h_{upper} + \beta)$$
$$\Delta\rho_{range} = \alpha(h_{lower} - h_{upper}) \tag{4.3}$$

Where h_{lower} and h_{upper} are the positions at the surface of the lower ($h_{lower} = 0$) and upper magnets (e.g. $h_{upper} = 25$ mm in the HMFG MAGLEV). The difference in measured density between these two positions is the range of density ($\Delta\rho_{range}$). In practice, some of this range is not accessible: (i) the thickness of the bottom of the vessel that holds the paramagnetic liquid (e.g. 1–2 mm for a regular cuvette) and (ii) the gap of air between the top surface of the paramagnetic liquid in the vessel and the surface of the upper magnet.

MAGLEV is uniquely suited for measuring the density of small objects (the approximate diameter of the object = Ø = 10 μm–1 mm) with high sensitivity, where the sensitivity of the MAGLEV device is defined as the smallest difference in density that can be observed. This sensitivity is determined by the smallest difference in levitation height of an object that can be observed. For example, if the range of density ($\Delta\rho_{range}$) is 0.1 g/cm^3, and the distance between the magnets (d) is 45 mm, and the smallest difference in distance that can be resolved (d_{limit}) is 0.1 mm, then the smallest difference in density that can be resolved (ρ_{limit}) is $\sim 2.2 \times 10^{-4}$ g/cm^3 ($\rho_{limit} = \Delta\rho_{range}/(d/d_{limit})$). The large working distance MAGLEV (LWD MAGLEV) device has a large gap (45 mm) between the magnets, a high sensitivity ($\pm 1 \times 10^{-4}$ g/cm^3), and excellent accuracy (using dilute paramagnetic media); it is therefore especially suited for comparing the densities of two objects or types of particulate compounds, for example, to determine whether the materials originate from the same source [5]. High-sensitivity MAGLEV has an even higher sensitivity ($\pm 2 \times 10^{-6}$ g/cm^3) [6], but a detailed description of this device is outside the scope of this chapter.

The MAGLEV devices discussed in this chapter have an optimal distance (LWD MAGLEV: 40–47 mm [5], HMFG MAGLEV: 20–26 mm [23], Axial MAGLEV: 12–17 mm) [33] between the magnets that allows for a linear relationship between an object's density and its levitation height. A larger distance between the magnets, beyond the listed ranges, permits a larger working distance, but results in a nonlinear correlation between height and density. The nonlinearity may complicate quantitative density measurements and analyses, but can still be useful for separations. Among the MAGLEV devices this chapter focuses on, the LWD MAGLEV device has the largest working distance, i.e. the largest distance between the magnets with maintained linear magnetic field lines (and therefore a linear gradient in density). A large working distance is beneficial for several reasons. (i) For a specified range of measured density, a larger working distance allows for a higher sensitivity than shorter distances. The smallest difference in height of an object that is resolved in camera images (without microscopy) is 0.1–0.3 mm for all devices. (ii) Larger working distances allow levitation of larger objects, and of a greater number of objects at the same time. (iii) A larger physical separation of components of granular samples allow for a higher purity of separated materials and an easier extraction of the different fractions using a pipette. A larger working distance, however, results in a MAGLEV device that measures a smaller range of density.

The working distance between the magnets in the MAGLEV device should be appropriate for the volume of the sample being tested. The working distance, generally, should be at least 5–10 times the largest dimension of the levitating object. The amount of sample should be small enough to allow the fractions to levitate freely without any of the fractions touching the walls of the cuvette after the particles have equilibrated. The Axial and HMFG MAGLEV devices have smaller working distances (15–25 mm), but they can measure larger ranges of densities compared to the other MAGLEV devices, which is beneficial for simultaneous levitation of samples that contain multiple components with larger differences in density.

4.5.2 Testing the Compatibility Between Trace Evidence and Paramagnetic Media

Forensic investigators who intend to use MAGLEV for density-based analysis of a new type of trace evidence should consider three factors. (i) The trace evidence should be diamagnetic (para-, ferri-, or ferromagnetic trace evidence cannot be levitated). Ferromagnetic (e.g. iron metal), and ferrimagnetic (e.g. magnetite [Fe_3O_4]) particles can be separated from the mixture prior to levitation in the MAGLEV device using other methods described elsewhere [50, 51]. (ii) The trace evidence should be compatible with the paramagnetic medium. MAGLEV can only levitate objects and particles that do not dissolve in or react with the paramagnetic medium used. (iii) The combination of the MAGLEV device and the paramagnetic medium should generate a density-gradient with a $\Delta\rho_{\text{range}}$ that brackets the density of the trace evidence.

The compatibility between the paramagnetic medium and a sample of unknown composition is easily investigated by immersing part of the sample in a polar solvent (e.g. water) and another part in a nonpolar solvent (e.g. hexane) for 20 minutes. If the sample dissolves, swells, or becomes otherwise changed (e.g. it changes in color, or cracks) by contact with the solvent, the sample is incompatible with this type of solvent.

The separation of trace evidence is generally simpler operationally than the determination of density using MAGLEV. For example, the attachment of small air bubbles to a minority of the particles in a mixture is, in most cases, of little significance to the separation of the mixture as long as it *does separate*. It is often not necessary to calibrate the density-gradient when separation is the only objective.

It is crucial to monitor for drift in calibration of the density-gradient during prolonged experiments because the evaporation of solvent will change the density and the magnetic susceptibility of the paramagnetic solution. Ideally, the experiment should be conducted with a glass-bead density standard levitating in the paramagnetic medium, together with the sample. Any change in the levitation height of the standard during the course of the experiment will indicate a drift in the calibration.

4.5.3 Analysis of Nonpolar Compounds Using MAGLEV

Paramagnetic media based on aqueous solutions of paramagnetic ions are the most cost-effective options for analysis of trace evidence that does not dissolve, react, or swell in aqueous media. The paramagnetic ions are commercially available as inorganic salts and complexes of metal ions and chelate compounds. Table 4.1 lists the prices of the paramagnetic species that we prefer to use. Manganese(II) is the most cost-efficient (moles of ions per US $) aqueous paramagnetic medium, followed by dysprosium(III) and gadolinium(III). Manganese(II) chloride is, however, slightly more toxic (if swallowed), corrosive, and harmful to the environment [55] than to the other two metal ions [56, 57]. We still consider manganese(II) valuable for experiments – and in fact routinely use it – because of its low cost and because its solutions can reach high magnetic susceptibilities.

Manganese(II) chloride can be handled without serious concern for toxicity as long as common sense and standard personal protective equipment (safety glasses, gloves, and lab coat) are used, and the liquid waste is disposed of in accordance with Material Safety Data Sheet guidelines for the material.

Chelation of the paramagnetic ions to form metal ion chelate complexes tends to reduce the toxicity and corrosiveness of the paramagnetic medium compared

Table 4.1 The cost of the most common paramagnetic compounds used for MAGLEV.

Chemical name	Abbreviated name (Type of solvents that dissolve the compound)	Cost (US$) per gram compound (US$ per moles of ion) (Supplier)[a]	Cost (US$) per separation[b]
Manganese(II) chloride tetrahydrate	$MnCl_2 \cdot 4H_2O$ (polar)	0.1 (13.7) (Beantown chemical)	0.02
Gadolinium(III) chloride hexahydrate	$GdCl_3 \cdot 6H_2O$ (polar)	1.6 (602.2) (Beantown chemical)	0.68
Dysprosium(III) chloride hexahydrate	$DyCl_3 \cdot 6H_2O$ (polar)	0.8 (301.6) (Oakwood Chemical)	0.34
Holmium(III) chloride hexahydrate	$HoCl_3 \cdot 6H_2O$ (polar)	2.7 (1009.2) (Beantown chemical)	1.14
Diethylenetriaminepenta acetic acid gadolinium(III)	Gd(DTPA) (polar/nonpolar)	2.2 (1212.9) (Analabs Ltd)	1.37
Dihydroxy-hydroxymethyl propyl-tetraazacyclodode cane-triacetic acid gadolinium(III)	Gadobutrol (polar)	14.4 (8696.7) (Drugs .com)	9.78
Diethylenetriamine triacetic acid didecyldiacetamide gadolinium(III)	Gd(DTAD) (nonpolar)	9.41 (10 584.2) (Precursors from Carbosynth Ltd. and VWR)[c]	11.91
Tris(dipivaloylmethanato) trioctylphosphine oxide gadolinium(III)	Gd(DPM)$_3$TOPO (nonpolar)	2.9 (3218.6) (Precursors from Strem Chemicals [52] and Santa Cruz biotechnology [53])[c]	3.62
Acetylacetonate trioctylphosphine oxide gadolinium(III)	Gd(acac)$_3$TOPO (nonpolar)	3.7 (3105.9) (Precursors from Alfa Aesar [53] and Santa Cruz biotechnology [54])[c]	3.49

a) The costs were based on pricing information of bulk sales (≥ 25 g) accessed on the websites of the suppliers on 01 June 2020.

b) The cost of filling a cuvette (2.5 ml) with paramagnetic medium (0.45 M) for levitation in a HMFG MAGLEV device. The listed costs assume that the paramagnetic medium is only used once.

c) These compounds are not commercially available. The costs include the two main precursors (commercially available) used for their synthesis. The table lists suppliers that retail compounds that are priced in the lower range commercially available.

to those of the same ions in the form of their dissolved inorganic salts (e.g., manganese(II) chloride in water), although the latter is more convenient, and is usually acceptable [55]. For example, metal ion chelate complexes (Table 4.1) dissolved at low concentration in aqueous media are sufficiently biocompatible to permit retrieval of live cells (mammalian cells and microorganisms) after separation with MAGLEV [25, 58].

Figure 4.4 shows the experimentally determined levitation height for density standards in different MAGLEV devices, paramagnetic media, and ion concentrations. When it is possible to approximate roughly the density of trace evidence (e.g. by using literature values), these standard curves are useful guides when choosing a combination of a MAGLEV device and a paramagnetic medium composition that will be able to levitate the samples.

4.5.4 Analysis of Polar Compounds Using MAGLEV

Polar compounds dissolve in aqueous paramagnetic media, precluding the use of MAGLEV. The use of nonpolar paramagnetic media enables levitation of polar compounds. The nonpolar paramagnetic solutions consist of metal ion chelates of relatively large molecular size, such as $Gd(DPM)_3TOPO$ and $Gd(acac)_3TOPO$ that are dissolved in nonpolar solvents [8]. These compounds are not commercially available, but synthesizing them from commercially available precursors, i.e. $Gd(DPM)_3$ and $Gd(acac)_3$, is relatively easy [8]. The $Gd(DPM)_3TOPO$ in its concentrated state (0.9 M) is an oil with a viscosity similar to liquid honey. We therefore dilute this compound to 0.45–0.50 M to obtain a lower viscosity that allows for MAGLEV. $Gd(acac)_3TOPO$ by contrast, is an oil of a low viscosity (in between water and liquid honey) in its concentrated state (1.1 M) that may be used for MAGLEV without dilution.

We prefer to use gadolinium(III)-based compounds to manganese(II)-based compounds, even though the costs of the latter compounds are lower. The use of gadolinium(III) compounds, instead of manganese(II) compounds, slightly increases the cost of the experiments, but the higher magnetic susceptibility of the gadolinium(III) compounds significantly boosts the range of densities that can be measured in a MAGLEV experiment.

Figure 4.4e, f contains standard curves that provide good starting points when developing protocols for the separation of new types of trace evidence.

4.5.5 Calibration

The way we prefer to calibrate MAGLEV devices is by using commercially available glass-bead density standards. We buy our density standards from American Density Materials, Inc. (Staunton, VA). These standards are made

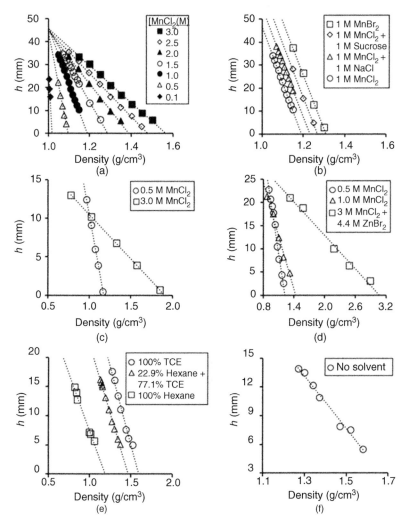

Figure 4.4 The levitation heights of glass-bead density standards in various paramagnetic media and in different MAGLEV devices: (a) LWD MAGLEV (45-mm gap between magnets) with levitation in aqueous solutions of manganese(II) chloride (MnCl$_2$) [5]. (b) LWD MAGLEV (45-mm gap between magnets) with levitation in aqueous solution of MnCl$_2$ (1.0 M) with the addition of different diamagnetic solutes to modify the density of the solutions [5]. (c) Axial MAGLEV (15-mm gap between magnets) with levitation in aqueous solutions of MnCl$_2$ [33]. (d) HMFG MAGLEV (25-mm gap between magnets) with levitation in aqueous solutions of MnCl$_2$, with and without the addition of diamagnetic solutes [23]. (e) HMFG MAGLEV (25-mm gap between magnets) with levitation in a paramagnetic medium consisting of Gd(DPM)$_3$TOPO (0.45 M) and nonpolar mixtures of hexane and tetrachloroethylene [8]. (f) HMFG MAGLEV (25-mm gap between magnets) with levitation in paramagnetic medium consisting of Gd(acac)$_3$TOPO (1.1 M) as obtained after synthesis [8].

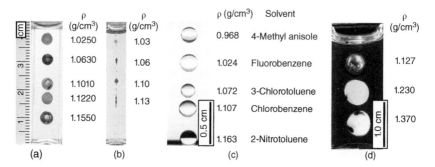

Figure 4.5 Calibration of MAGLEV devices using density standards. (a) Glass-bead density standards levitating in an aqueous solution of manganese(II) chloride (1.00 M) in a LWD MAGLEV device [34]. (b) Microbead density standards made of polyethylene and additives. The particles in (b) were levitated in the same device and medium as in (a) [34]. (c) Liquid droplets (3 µl) of nonpolar organic solvents levitating in an aqueous manganese(II) chloride (0.50 M) solution in an axial MAGLEV device. The specified densities of the solvents are the ones reported by the suppliers [33]. (d) Glass-bead density standards levitating in a solution of Gd(DPM)$_3$TOPO (0.45 M) dissolved in a solvent mixture of 23 vol% hexane and 77 vol% tetrachloroethylene in a HMFG MAGLEV device [8]. (Source: (a)–(c) are modified and reprinted with permission from Ge et al. [33]; Ge et al. [34]. Copyright (2018) American Chemical Society. (d) is modified and used with permission from John Wiley & Sons, Inc. Copyright 2020 Wiley [8].)

of glass, and they enable accurate (sensitivity: $\pm 1 \times 10^{-4}$ g/cm^3) calibration of density (Figure 4.5a, d). The use of glass makes them inert to most solvents and easy to remove from the medium after calibration [5].

Solid materials of known density, such as plastics or crystals of chemicals, can also be used as standards, but the densities of these materials are less well defined than commercially available density standards.

Microbead density standards made of a range of materials (primarily plastics) are available from companies, such as Cospheric (Figure 4.5b), and are provided as powders (Cospheric LLC, Santa Barbara, CA). The particles have a distribution of densities around a specified value. The densities of individual particles can vary considerably, but they provide a fair estimate of the density as a collective [33]. These are especially suitable for density calibration inside small tubes, or for use as disposable standards. The plastic microbead density standards are incompatible with most nonpolar paramagnetic media because the plastic will swell and alter their densities.

Density standards can also be made from droplets of solvents of known density. The solvents should be immiscible with the paramagnetic medium to form a droplet that is stable over time. For example, drops of nonpolar organic solvents are suitable for use in aqueous paramagnetic media (Figure 4.5c), and aqueous salt solutions in nonpolar paramagnetic media. The density of the droplets can be adjusted in at least three ways, such as (i) by using solvents of different, but known, densities, (ii) by dissolving a compound (such as salt) in a solvent, or (iii) by adjusting the proportion of two miscible solvents of different density. The size of the droplets is tuned by pipetting a specified amount of solvent into the paramagnetic medium using a micropipette [30, 33].

The density standard will levitate if it has a density that is within the range of what the device can measure. The levitating glass-bead and its height of levitation are captured with a camera. Ideally the levitation height is determined for 6–12 standards of different density to produce a statistically relevant standard curve. If multiple standards are levitating at the same time, they should not be in contact with each other, as contact can affect the accuracy of the measurement. Handle density standards with care and wash them properly after use, in order to ensure there is no alteration in their density.

If a density standard touches the bottom (does not levitate) of the cuvette, a bead of lower density should be tried to allow for levitation. If the bead is in contact with the liquid meniscus at the top, a density standard with a higher density must be used to separate the effect of density and trapping at the cuvette surface. Note, the standard can become stuck at the interface of the air and the paramagnetic solution, and gently shaking the cuvette helps the beads detach from the meniscus and levitate freely in the medium. Ensure that no bubbles are attached to the standard when it levitates, because this type of attachment can change the apparent density of the bead. The bubbles can be removed by rinsing the standard with pure solvent and by gently rolling it with a gloved-hand on a tissue that is wet with solvent.

We recommend that the MAGLEV devices are calibrated and recalibrated at the beginning and the end of experiments (at a minimum), to ensure there was no significant shift in the calibration.

4.6 Protocols for Measurement of Density, and Separation Using MAGLEV

Caution: Large neodymium magnets exert significant forces when attracted to certain metals (especially iron and steel) and other magnets. These forces can seriously damage objects, or hurt people that hold, or are in close proximity to, the magnets. The magnets can pinch or crush any part of the body (most commonly hands and fingers) trapped between them. The magnets are less likely to cause damage when fixed in a MAGLEV device than those handled as independent objects. The magnets can cause loss of data in electronic devices. As a rule, the MAGLEV devices described in this chapter will not damage electronics that are more than 25–40 cm away from the magnets. We advise users of MAGLEV devices to educate themselves on the dangers associated with large neodymium magnets before starting any experiments.

The devices are simple to construct. (See Section 4.8 *for schematic diagrams that detail their construction*.) The HMFG MAGLEV requires 25-mm-tall cuvettes, which can be made by cutting standard, 45-mm-tall plastic cuvettes with a hand saw with fine teeth (use sandpaper to smooth cut edges if needed).

The protocol below is designed to setup a procedure for the separation or density measurement of a new type of trace evidence using MAGLEV. The setup of an optimized protocol for a new type of trace evidence can take a few hours. After the procedure is established, however, MAGLEV-based analysis of that particular type of trace evidence becomes simple and rapid.

4.6.1 Basic Protocol for Typical Use of the MAGLEV Device

1. To start an experiment, a MAGLEV device should be placed on a level plane. If the paramagnetic solution used contains volatile organic solvents, it is preferable to place the MAGLEV device in a fume hood. (*See* Section 4.5.1 *for how to select the type of MAGLEV device*.)

2. A cuvette should be filled with paramagnetic medium to within 1–2 mm from the top of the cuvette. (*See* Sections 4.5.2–4.5.4 and 4.6.2 *for how to optimize the composition of the paramagnetic medium*.) Cover the top of the cuvette with a lid or a small piece of aluminum foil or Parafilm to reduce evaporation of solvent from the paramagnetic medium. Aluminum foil is preferred for apolar paramagnetic medium.

3. To generate data for a standard curve, one should place a glass-bead standard (*See* Section 4.5.5 *for how to select density standards*) in the cuvette and place the cuvette between the magnets in the MAGLEV device. A self-supported ruler may be constructed by gluing a piece of a ruler to a cuvette. The length of the ruler should be 0.5–1.0 mm shorter than the distance between the magnets in the MAGLEV device. One should place the ruler next to the cuvette with the paramagnetic medium for reference of height. To optimize the visualization of the sample, illuminate the cuvette with external lamps.

4. Density standards that consist of a solid material can be rolled back and forth for a few seconds on a tissue paper wetted with the same solvent used to prepare the paramagnetic medium. This step reduces the risk of bubbles forming on the standards when they levitate.

5. The standard should levitate along the vertical centerline of the MAGLEV device. If the standard touches the walls of the cuvette, the position of the cuvette may be adjusted to ensure contact-free levitation of the standard in the cuvette.

6. After the position of the standard has reached equilibrium, the standard should be imaged with a camera. The cuvette should always be imaged with the camera from the same position in order to avoid parallax error (i.e. error in the measurement of the levitation height caused by viewing the cuvette from different positions). After imaging of the standard, it should be removed from the cuvette using tweezers. To avoid damaging the surface of the standard, it is preferable to use metal tweezers without serrated tips, or plastic tweezers.

7. To avoid build-up of dried components from the paramagnetic medium on the surface of the standard (which would alter its density), wash it thoroughly with solvent after each use.

8. To construct a standard curve, steps 3–5 are repeated for (ideally) 6–12 density standards of different densities. Fewer standards can be used, but this reduction will decrease the accuracy of the results.

9. The levitation height of the density standard should be measured from the surface of the bottom magnet to the middle of the bead (which we approximately define as the middle of a spheroid superimposed on the image within the outer boundary of the bead).

10. To determine the density of the sample one begins by placing the sample in the cuvette. One should allow time (2–60 minutes) for the levitation height of the sample to equilibrate. After the sample has equilibrated, it should be imaged. Objects that are larger than a millimeter in diameter often equilibrate in height in seconds. Particles that are 30–1000 μm in diameter (1000 μm = 1 mm) will equilibrate within minutes to tens of minutes. Particles that are 10–30 μm in diameter, or smaller, take hours to equilibrate (if they reach a stable equilibrium at all).

11. One can accurately determine the levitation height of the samples in the images by placing a ruler next to the cuvette for reference. We approximate the levitation height as the geometric center of the beads or the aggregates of the particles. The simplest way to measure the levitation height of a sample accurately in images is using software, such as ImageJ [59] or Microsoft PowerPoint. Use the standard curve to find the density of the sample (Figure 4.6). The separated (in density) components of the samples may require individual examination with independent techniques to confirm their identity.

12. The levitating fractions of different densities are carefully extracted from the cuvette in the MAGLEV device using a pipette. For accuracy, and to avoid plugging, the inside of the tip of the pipette must be significantly larger than the diameter of the largest particles to be extracted. The fractions of different density can easily be accessed from above using the Axial MAGLEV device (Figure 4.2). For the LWD and HMFG MAGLEV devices, however, we recommend the use of a custom-made cuvette that has a shape that is optimized for the extraction of the components of the levitating samples using

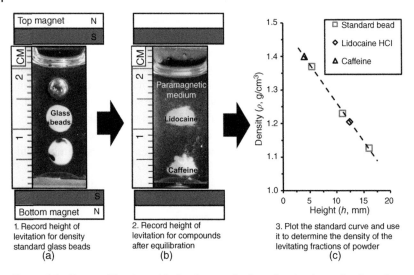

Figure 4.6 From calibration with density standards to density determination of samples. The MAGLEV device is calibrated using (a) glass-bead density standards. (b) The densities of the two components in the mixture are determined using the standard curve (c). The ruler is used as a reference of distance (levitation height). The beads and the powders in the images are levitated in a solution of Gd(DPM)$_3$TOPO (0.45 M) in a solvent mixture of hexane (23 vol%) and tetrachloroethylene (77 vol%). Lidocaine and caffeine are used as an example because they are readily available and they separate cleanly.

a pipette from the upper side of the magnet (Figure 4.7). (The construction of the custom-made cuvette is detailed in the Supporting Information of our previous publication [8].) The extracted samples should be rinsed repeatedly with solvent under suction filtration on a filter to remove any residual paramagnetic medium. The sample can be dried rapidly from residual solvent by pulling air through the filter using suction filtration.

13. After use, it is important to clean the MAGLEV device of any spilled paramagnetic medium. The magnets of the MAGLEV devices will corrode after long-term exposure to the media, and the plastic fixtures that hold the magnets will become discolored and sometimes will crack.

4.6.2 Troubleshooting the Experiments

In the event that a sample does not levitate (i.e. the sample floats or stays at the bottom) in the MAGLEV device, we recommend the use of these four strategies to obtain accurate measurements:

(i) Boost the magnetic susceptibility of the solution by increasing the concentration of the paramagnetic species. The type of paramagnetic ion can also be changed to one of higher magnetic susceptibility, e.g. gadolinium(III) has a

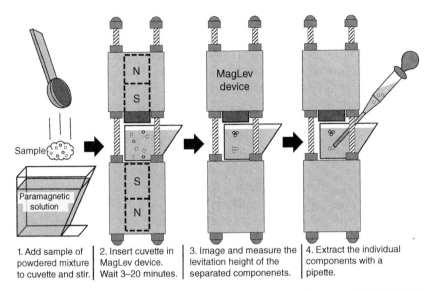

1. Add sample of powdered mixture to cuvette and stir.	2. Insert cuvette in MagLev device. Wait 3–20 minutes.	3. Image and measure the levitation height of the separated componenets.	4. Extract the individual components with a pipette.

Figure 4.7 Separation and extraction of powdered compounds using a HMFG MAGLEV device and a custom-made cuvette with one slanting side to allow easy access with the pipette tip. The schematic shows the separation of two compounds followed by the extraction of one of the levitating fractions using a Pasteur pipette.

higher susceptibility than manganese(II), to increase the magnetic susceptibility of the solution. Both of these changes will expand the range of density over which MAGLEV can levitate samples but, if made, the changes will also lead to a decrease in the sensitivity of the system [5, 30].

(ii) Modify the density of the paramagnetic medium to match closely that of the sample, and thus to require a smaller magnetic force for levitation. Add soluble diamagnetic compounds to increase the density of the paramagnetic medium. Examples of diamagnetic compounds include dense inorganic water-soluble salts (e.g. $ZnCl_2$), water-soluble organic compounds (sucrose), or polar solvents of a higher density than water (e.g. D_2O). Reduce the density of the paramagnetic solution by adding miscible solvents of lower densities (e.g. adding methanol or ethanol to aqueous solutions) [5, 30].

(iii) If the composition of the paramagnetic media is changed using these two methods and the particles (or mixtures of particles) of interest still do not levitate, Tilted MAGLEV (Figure 4.2c) can be used [7]. The density measurements of the Tilted MAGLEV are lower in sensitivity than the other MAGLEV methods, but enable the estimation of the complete range of densities observed in matter at ambient conditions (from ~0 to ~23 g/cm³) [7]. Rotate the cuvette slowly until the different fractions of the sample equilibrate at specific positions – some samples will rest on the walls of the cuvette. The wall of the cuvette provides a force perpendicular to the wall that counteracts the force of gravity along this direction, and effectively reduces gravity along the central axis of the Titled MAGLEV device [7].

(iv) The use of aqueous paramagnetic solutions tends to be accompanied by the formation of bubbles on the levitating objects and the walls of the cuvette. The bubbles can alter the measured density of the objects, and their formation on the walls of the cuvette can also obscure the view of the levitating objects. Reduce the problem of bubble formation by degassing the solution. For example, by reducing the pressure above the solution using vacuum for 20–30 minutes, or by heating up the solution close to the boiling point, and then cooling it down to room temperature. Several cycles of freezing and thawing (under reduced pressure) can also be used to remove dissolved gases from the medium. The addition of small quantities of a surfactant (e.g. one drop of a dilute [0.1 wt%] Tween-20 solution) to the paramagnetic solution in the cuvette is an effective way of reducing the adhesion of bubbles formed at surfaces in the polar medium. An alternative is to use nonpolar paramagnetic media that through solvation of the hydrophobic surfaces pose no, or few, problems with adhesion of hydrophobic bubbles of air to the sample.

4.7 Trace-Evidence-Like Materials That Have Been Analyzed with MAGLEV

MAGLEV can separate and estimate the densities of a range of types of common trace evidence. Table 4.2 provides an overview of the types of materials that have been levitated using MAGLEV with a focus on materials most relevant for forensics.

4.7.1 Bone

The shape and morphology of skeletal remains provide criminal investigators with clues that tell a story about the identity of the victim: for example, sex, age, lifestyle, and nutritional status [63]. Skeletal parts of human or animal origin found in the environment are in more or less deteriorated states, ranging from complete skeletons to small fragments of bone [64]. The breakdown of the bone is accelerated by warm weather, direct sunlight, frost heaves, ice thaws, animals, and soil acidity [63, 65]. The bone can also be ancient or a result from a recent crime. Bones buried in soil tend to deteriorate more slowly [63, 65]. After the larger bones are collected, smaller pieces are found by *screening* (i.e. collection of material that is large enough to not pass through the screen) the soil and litter of the surrounding

Table 4.2 Substances relevant to forensics that have been levitated with MAGLEV.

Substances levitated with MAGLEV	Size	Densities measured (g/cm³)	Type of MAGLEV device	Paramagnetic media	References
Human and bovine bone (dried)	mm	>1.75	LWD	$MnCl_2$ or $MnBr_2$ in water	This book chapter
Human hair and dandruff	μm–mm	1.12–1.33	LWD	$MnCl_2$ in water	This book chapter
Glitter and gunpowder	μm–mm	1.226–1.395	LWD	$MnCl_2$ in water	[10]
Illicit drugs, adulterants, and diluents: e.g. fentanyl, acetyl fentanyl, cocaine, heroin, caffeine, lactose, lidocaine, sucrose, methamphetamine	μm	1.10–1.58	HMFG	Gd(DPM)TOPO and Gd(acac)TOPO complexes in hexane/tetracholoroethylene	[8]
Polymers[a] ABS, HDPE, Nylon, PC, PCL, PDMS PET, PLA, PMMA, PP, PS, PU, Teflon, Torlon	μm–mm	0.9–2.3	LWD HMFG Axial	$MnCl_2/GdCl_3$ in water and alcohol	This book chapter and [6, 7, 23, 34, 60–62]
Polymorphs of crystals: *trans*-Cinnamic acid, carbamazepine, sulfathiazole, S-/RS-ibuprofen	μm–mm	1.093–1.580	LWD	$MnCl_2$ in water	[11]
Cocrystals: carbamazepine/salicylic acid and carbam azepine/camphoric acid systems	μm–mm	1.219–1.443	LWD	$MnCl_2$ in water	[32]
Medication: Microspheres from drug capsules	mm	1.3284	High sensitivity	$GdCl_3$ in water	[6]

(continued)

Table 4.2 (Continued)

Substances levitated with MAGLEV	Size	Densities measured (g/cm³)	Type of MAGLEV device	Paramagnetic media	References
Food: cheese, grains, milk, peanut butter, vegetable oils	mm	0.9–1.4	LWD	$MnCl_2$ and $GdCl_3$ in water; Gd-DTAD in nonpolar organic solvents	[5]
Minerals and semiconductors: aluminum oxide, cerussite, diamond, silicon, silicon nitride, stibnite, zirconium silicate	μm–mm	2.33–6.55	LWD Axial Tilted	$MnCl_2$, $GdCl_3$, and $MnBr_2$ in water	[7, 23]
Metals: aluminum, brass, copper, gold, indium, iridium, lead, mercury, osmium, silver, tin	μm–mm	2.7–23.0	HMFG Axial Tilted	$MnCl_2$, $GdCl_3$, $DyCl_3$, and $MnBr_2$ in water	[7, 34]
Glass: ALON, borosilicate, Gorilla glass, sodium-lime	μm–mm	2.23–3.96	LWD HMFG Tilted	$MnCl_2$, $MnBr_2$, and $ZnBr_2$ in water	This book chapter and [7, 34]

a) Polymers and their acronyms: acrylonitrile butadiene styrene (ABS), high-density polyethylene (HDPE), polycarbonate (PC), polycaprolactone (PCL), polydimethylsiloxane (PDMS), polyethylene terephthalate (PET), polymethylmethacrylate (PMMA), polypropylene (PP), polystyrene (PS), and polyurethane (PU).

area and the material in the burial mound. The screening may results in finds of small pieces of bone, hair, buttons, jewelry, and bullets. The bones are analyzed for, e.g. macroscopic anatomical features by eye [63], microstructure characteristics using histology [66], or DNA or mitochondrial DNA using genetic analysis [67]. MAGLEV is suitable as a secondary method of density-based sorting after the primary mechanical screening by size. MAGLEV can separate the material into several fractions according to differences in their density. There are two main types of bone, both are natural composites consisting of the protein collagen and the calcium–rich mineral hydroxyapatite. Cortical bone is the first type and it is compact in character, and covers the external surfaces of skeletal parts [68]. "Cancellous" or "trabecular" bone is the second type and it is a spongy and porous type of bone found on the inside of bones. Cortical bone that is wet is reported to have a density of approximately 1.85 g/cm [68–70] and 1.58 g/cm^3 when it is dry [68]. The corresponding values of density for cancellous bone are 1.02 g/cm^3 when wet and 0.34 g/cm^3 when dry [68].

In Figure 4.8a, b, human bone dislodged from a bullet (125 g, 0.38 Special +P, jacketed hollow point with a terminal weight of 117.9 g) was levitated using MAGLEV. The bullet was fired with a Ruger 357 MAGNUM revolver (4″ barrel). MAGLEV separates the bone from the inorganic mineral apatite based on density.

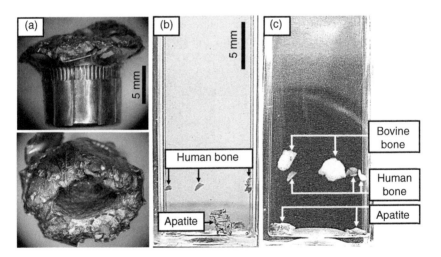

Figure 4.8 Separation of bone from mineral using MAGLEV (LWD). (a) A fired bullet after a gunshot to the head of an adult female. Three pieces of bone (red dashed circles) are lodged in the bullet. (b) The fragments of bone that were extracted from the bullet were levitated in a solution of saturated manganese(II) chloride (aq., ρ = 1.45 g/cm^3). (c) Levitation of pieces of bone extracted from bullet, and pieces of bovine bone, in a solution of manganese(II) bromide (4 M, aq.). Saturated potassium bromide (aq.) was added to attain a density of 1.75 g/cm^3. Source: CREDIT: Mr. Lucien C. Haag/Forensic Science Services, Inc., Carefree, AZ.

It is not possible to conclude if human and bovine bone fragments can be differentiated considering the limited dataset and the similarities in the density (Figure 4.8c). The exact density of the bone in the figure is not known because the device had not been calibrated.

4.7.2 Glitter and Gunpowder

We have used MAGLEV to measure the density of, and to separate, glitter, and gunpowder (Figure 4.9) [10]. Glitter is a sparkling material usually made from thin, flexible, and laminated layers of plastic and metal foil that are cut into tiny (approximately 100–700 μm in diameter) pieces of various shapes (e.g. hexagons, diamonds, and squares) [10]. Glitter is a common and abundant type of trace evidence that easily transfers through contact. Glitter is found in cosmetics, on gift cards, in decorations on clothing items, and in nail polish etc. [10, 71, 72]. Glitter is normally separated and characterized using techniques such as microscopy (e.g. light microscopy and scanning electron microscopy [SEM]), and spectroscopy (e.g. FTIR). Solubility tests in organic solvents and acids can be helpful when comparing a sample with authentic standards [10, 73]. The density of glitter particles found in cosmetics, such as nail polish, can be measured if the glitter is first separated from the solvents (e.g. butyl acetate and ethyl acetate) and the dissolved components (e.g. nitrocellulose) [74]. Separation of glitter from nail polish was accomplished by washing the particles with acetone on a filter paper before they were levitated in the MAGLEV device [10]. Biodegradable

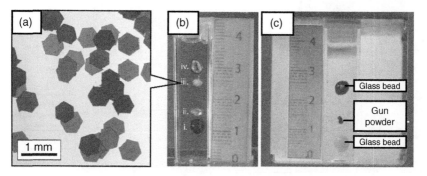

Figure 4.9 Separation of particles of glitter and gunpowder. (a) Light microscopy image of a mixture of glitter particles; the same particles that levitate in (b). (c) Separation and density determination of two types of glitter particles – chrome silver 1P (ii.) and mirror crystallina (iii.) – and glass-bead density standards (i and iv) The separation of the glitter particles in a solution of manganese(II) chloride (3.0 M) in a MAGLEV device (LWD). (b) The levitation and measurement of density for particles of gunpowder (Hercules Blue Dot) in a manganese(II) chloride solution (4 M) [10]. Source: Copyright 2012 Wiley. Modified and used with permission from John Wiley & Sons.

glitter, a substitute for glitter made from plastic is now commercially available (see Chapter 2 in this book). This type of glitter should be easily distinguishable by density.

When ignited, smokeless gunpowder produces, as the name implies, significantly less smoke than black powder on firing. The reduced amount of smoke is in part why smokeless gunpowder has become the standard propellant for bullets in handguns and rifles. Smokeless gunpowder consists of extruded granules that are shaped like ribbons, cylinders, disks, or balls [75]. The particles are usually made from gelatinized nitrocellulose with or without nitroglycerin, and stabilizers (e.g. diphenylamine).

In forensics, smokeless gunpowder is usually encountered in explosive devices [76] or as gunshot residue. A comparison between the density of the unknown samples and reference materials can be used to indicate the identity and composition of the unknown gunpowder. Other methods of analysis used to analyze gunpowder include high-performance liquid chromatography, ion chromatography, capillary electrophoresis, and FTIR [76]. We have reported the use of MAGLEV for separation of glitter and smokeless gunpowder from other particulate residues and pigments in mixtures of solids (Figure 4.9). Prolonged exposure to aqueous solutions of manganese(II) chloride did not affect the density of the materials [10].

4.7.3 Powdered Drugs, Polymorphs, and Enantiomers

We have demonstrated the use of MAGLEV for the analysis of illicit drugs (Figure 4.10) [8]. Producers of illicit drugs routinely combine powders of illicit drugs with adulterants and diluents. Examples of common illicit drugs include fentanyl, heroin, methamphetamine, and cocaine. Adulterants are semi-active compounds that are added to mimic a higher quality product (e.g. to modify taste) and include compounds such as acetaminophen, caffeine, lidocaine, and levamisole. Diluents are inactive compounds, e.g. lactose, dimethyl sulfone, which are added to dilute and produce a larger amount of product and to increase its profitability and safety [8, 77].

MAGLEV combines many features that are beneficial for the analysis of powdered drugs. For example, the method of separation of components based on their densities is cost-effective, portable, and operationally simple. Methods with these characteristics could be deployed at large scale in the field (because of their low cost) and used by professional forensic analysts with limited training. Field-based methods of analysis enable a more rapid chain of testing than analysis dependent on central laboratories and can provide analytical results within seconds to an hour. In comparison, due to backlogs in testing, it can take weeks to months (depending on the state) for state drug-testing laboratories to analyze samples [78]. MAGLEV, furthermore, solves a critical problem in field-based analysis of drugs,

Cocaine·HCl and levamisole·HCl

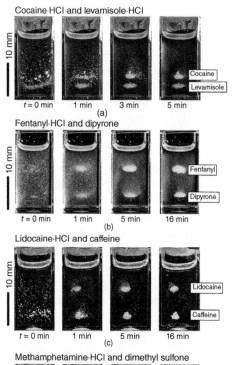

Fentanyl·HCl and dipyrone

Lidocaine·HCl and caffeine

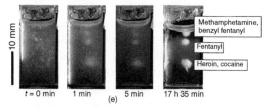

Methamphetamine·HCl and dimethyl sulfone

Methamphetamine·HCl, benzyl fentanyl·HCl, fentanyl·HCl, heroin·HCl, and cocaine·HCl

Figure 4.10 Time-lapse images of the separation of mixtures of powdered drugs, adulterants, and diluents using a HMFG MAGLEV device. The compounds levitate in a paramagnetic solution of Gd(DPM)$_3$TOPO (0.45 M) dissolved in a solvent mixture of 23 vol% hexane and 77 vol% tetrachloroethylene. Source: Reference [8] Copyright 2020 Wiley. Used with permission from John Wiley & Sons, Inc.

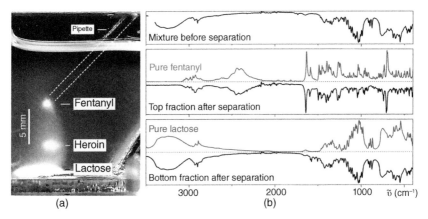

Figure 4.11 Separation via MAGLEV and identification by FTIR-ATR. (a) Image of the separation (30 minutes) of a mixture of (50 mg) of fentanyl·HCl (1.3 wt%), heroin·HCl (2.6 wt%), and α-lactose (96.1 wt%) using a HMFG MAGLEV device. The sample was separated in a paramagnetic solution of Gd(DPM)$_3$TOPO (0.45 M) dissolved in a solvent mixture of 23 vol% hexane and 77 vol% tetrachloroethylene. The dashed white lines delineate where the tip of the Pasteur pipette is located during the extraction of the fentanyl. (b) The FTIR-ATR spectra of the fentanyl- and lactose-rich fractions from the separation. Source: Reference [8] contains details. Copyright 2020 Wiley. Used with permission from John Wiley & Sons, Inc.

i.e. the separation and enrichment of dilute synthetic opioids (Figure 4.11a). This group of drugs, primarily fentanyl and its analogs, were estimated to be involved in 28 500 of the overdose-related fatalities in the United States in 2017 (the total number of fatalities due to overdoses of drugs was 72 000) [79, 80]. Fentanyl is one hundred times more potent than morphine and is therefore diluted to low levels (<10 wt%) in samples intended for street-level retail [79, 81]. The low levels prevent identification of these opioids in most samples with the commonly used field-based analytical methods, i.e. colorimetric spot tests, FTIR, and Raman spectroscopy [8]. These methods have limits of detection between 5 and 25 wt% and therefore cannot detect fentanyl in most seized samples [8, 82].

Our previous work shows that – by using MAGLEV as a separation and purification method, the limits of detection using FTIR-ATR spectroscopy for fentanyl are significantly lowered (Figure 4.11b) [8]. The average level of fentanyl in seized samples is 5.1 wt% [79]. In our study, fentanyl (1.3 wt%) could only be detected unambiguously with FTIR-ATR after separation with MAGLEV.

MAGLEV has previously been used to separate powdered mixtures of crystal polymorphs [11], enantiomers [24], and co-crystals in aqueous paramagnetic solutions [32]. These compounds had no solubility or low solubility in the aqueous solutions. These types of crystals are highly relevant to the analysis and profiling of seized illicit drugs.

4.7.4 Glass

Window glass is made by melting a mixture of materials such as sand, lime, soda, boric oxide, and alumina, and forming a transparent amorphous material. Pieces of crushed glass are common trace evidence found at crime scenes. For example, small shards of glass that are found on a suspect's clothes are often compared with crushed glass at the crime scene, such as an automobile accident.

The identity of the glass has historically been determined using methods such as refractive index, density, and elemental composition analysis [1, 83, 84]. Density and refractive index are fundamental properties of glass that are highly sensitive to variations in the chemical composition and the manufacturing process. There is strong correlation between density and the refractive index of glass [84]. The measurement of refractive index is, however, considered simpler and more convenient (the "immersion method") for small pieces of glass; therefore, measuring the density of glass for forensic purposes is not a common procedure [1]. The densities of small pieces of glass are usually determined using density-gradient columns or the float–sink method [1, 84, 85]. The development of MAGLEV in the last decade [4] as a method for simple and rapid determination of density could once again promote the use of density to identify the source of small pieces of glass. MAGLEV enables simple, fast, and low-cost measurement of density and separation of small pieces of glass (Figure 4.12).

MAGLEV can distinguish between and separate common types of glass, e.g. soda-lime ($\rho = 2.52\,\text{g/cm}^3$) and borosilicate ($\rho = 2.23\,\text{g/cm}^3$). MAGLEV can also separate glass from mixtures with other materials (Figure 4.12a, c), such as quartz sand ($\rho = 2.65\,\text{g/cm}^3$).

Mobile phones have screens made of glass, and are carried by almost everyone in today's society, making such glass an important source of trace evidence. MAGLEV can distinguish between Gorilla™ glass ($\rho = 2.43\,\text{g/cm}^3$ [86]), a common type of toughened glass found in mobile phones, and the glass from an iPhone SE ($\rho = 2.23\,\text{g/cm}^3$) based on density (Figure 4.12a, b).

Aluminum oxynitride (produced as ALON™ by the Surmet Corporation [87]) is a type of ceramic glass that is resistant to impacts by projectiles, such as bullets. ALON is used in windows and domes in, for example, armored trucks transporting valuables and in police and military vehicles [88]. These applications of ALON glass make it likely to be found in forensic investigations [89]. The density of ALON ($\rho = 3.96\,\text{g/cm}^3$ [87]) and air ($0.001\,\text{g/cm}^3$ [7]) can both be measured at the same time using a tilted MAGLEV device (Figure 4.12d). The cuvette in Figure 4.12d is sealed with a plastic sheet glued with five-minute epoxy to allow tilting without spillage of medium.

4.7.5 Polymers

Polymers comprise a vast array of products used throughout society, and these materials can play an important role in the investigation of a crime scene.

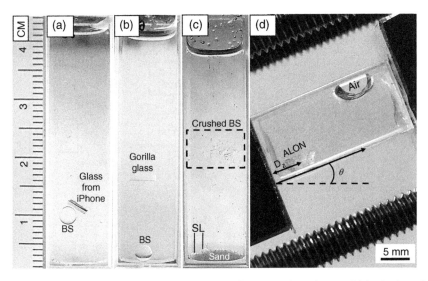

Figure 4.12 Levitation and separation of glass from granular mixtures. (a) Levitation of a borosilicate (BS) glass-bead (Ø = 3.18 mm) and a shard of glass from the screen from an iPhone SE in an aqueous solution of manganese(II) chloride (1.5 M) and zinc(II) bromide (16.21 M). (b) Levitation of a disc of Gorilla™ glass 5 (Ø = 5.00 mm, thickness = 1.00 mm) in an aqueous solution of manganese(II) chloride (1.5 M) and zinc(II) bromide (14.87 M). (c) Separation of crushed borosilicate glass from a mixture with sand and soda-lime (SL) glass beads (Ø = 1.00 mm) in an aqueous solution of manganese(II) chloride (0.73 M) and zinc(II) bromide (16.21 M). (a)–(c) Levitation in a LWD MAGLEV device. (d) MAGLEV of a shard of ALON bullet-resistant glass and an air bubble in an aqueous solution of manganese(II) chloride (4.00 M) inside a cuvette in a HMFG MAGLEV device.

Polymeric particles are frequently found at crime scenes in the form of adhesive, adhesive tape, waxes, fibers, flakes of paint, foam, ink on paper, wire insulation, buttons, and pieces of automotive trim from bumpers [65, 90–95]. Rubber and plastic pieces originating from automotive trim and bumper guards may have value in the investigation of hit-and-run automotive accidents. Blackledge reported the use of pyrolysis gas chromatography to compare fragments from rubber bumper guards. This article also discussed the use of a density-based method to distinguish rubber pieces from different sources based on differences in density. However, no density measurements were performed [92].

The shape, microstructure, color, and overall optical properties of plastic particles are characterized using polarized light microscopy or SEM. The composition of the particles is also determined using methods such as energy dispersive X-ray spectroscopy (in combination with SEM), high performance liquid chromatography, FTIR, and Raman spectroscopy [95–97]. Besides polymers, the plastic particles often contain fillers and pigments with unique chemical signatures, for example, carbon black, calcium carbonate, mica, or titanium dioxide that further differentiate the particles [94].

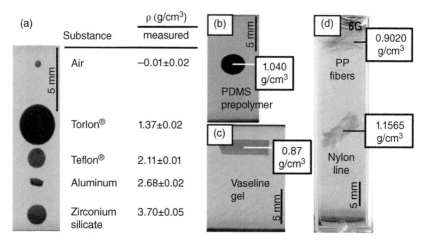

Figure 4.13 Density determinations of polymers, metals, and air. (a) Levitation of five objects in a solution of dysprosium(III) chloride (3.0 M) containing 0.01 (v/v%) Tween-20 [33]. (b) Levitation of a drop of liquid polydimethylsiloxane prepolymer mixed with powder of black graphite in an aqueous solution of manganese(II) chloride (0.5 M) [33]. (c) Levitation of Vaseline™ of cylindrical shape in an aqueous solution of manganese(II) chloride (3.0 M) and Tween-20 (0.1 v/v%). (a)–(c) Levitation in an Axial MAGLEV device. (d) Levitation of polypropylene fibers (PP) and a piece of Nylon line in an aqueous solution of 1.6 M manganese(II) chloride in a LWD MAGLEV device. Source: (a)–(c) is modified and reprinted with permission from (Ge, S. et al. [33]). Copyright (2018) American Chemical Society.

MAGLEV can separate and measure the density of polymers in their solid, paste, and liquid states in the laboratory and at the crime scene (Figure 4.13) [23, 28, 33, 34, 38, 60]. Figure 4.13d demonstrates separation of fibers made of polypropylene ($\rho = 0.92\,\text{g/cm}^3$; Tru-guard 642871 Diamond Braid Reflective Polypropylene Rope) and braided Nylon mason line ($\rho = 1.14\,\text{g/cm}^3$; T.W. Evans cordage Co.) using MAGLEV [98, 99].

Particles that are of spheroid shape and larger size ($\varnothing > 50\,\mu\text{M}$) separate rapidly (5 seconds to 30 minutes). The separation of fibrous particles is significantly slower (3–5 longer time of separation) compared to spheroid particles of similar mass because the fibrous shape causes an increase in viscous drag.

4.7.6 Hair and Dandruff

One of the most common types of trace evidence encountered at crime scenes is hair and dandruff. The determination of the origin of hair (from a human or animal) can indicate physical contact between a suspect, victim, and surfaces at a crime scene [100, 101]. Hair is a resilient material that remains stable over long periods of time, as opposed to blood or urine, which degrade rapidly [102].

In mammals, such as humans, hair grows as a thin fiber from a follicle that is embedded in the skin [103]. Dandruff is pieces of dry scale-like flakes which shed from skin, especially the base of the scalp [104–106].

Recovery of hair and dandruff is common from surfaces at crime scenes, such as clothing, bedding, and carpet [107]. The collection of these types of trace evidence is performed by picking, taping, scraping, shaking, or vacuuming (using a filter for collection) [102, 105, 106, 108]. Hair can contain significant amounts of morphological [109] and chemical information [102]. Both hair and dandruff trace evidence are potential sources of genetic information – a single strand of hair or a particle of dandruff will suffice in ideal cases [100, 104–106, 110]. Hair and dandruff commingled with other debris can be difficult to separate from complex mixtures. MAGLEV can in principle assist in the separation. Pieces of hair were separated successfully from a mixture with sand using MAGLEV (Figure 4.15a). MAGLEV can also be used to separate a mixture of dandruff and sand into two distinct fractions (Figure 4.15g). Most of the particles of dandruff had densities between 1.12 and 1.16 g/cm^3. DNA as well as protein profiles can be obtained from a single particle of dandruff [106, 111, 112]; this topic is further discussed in Chapter 8 of this book.

In the not too distant future, recovered dandruff particles could become trace evidence that could associate a violent sex offender with their deceased victim [101, 107]. All of us live in what might be considered our own personal cloud. Throughout the day and night, we breathe in and out of this cloud (dust, clothing lint, your own shed skin flakes). One's personal cloud may overlap with others (hence the need for social distancing during the COVID-19 pandemic). The particles in this cloud are varied and are affected by your location, but it is estimated that a person sheds around 50 million skin flakes a day, and breathes in around 700 000/day [113]. This is an ongoing dynamic process and the composition of particulate matter in the nasal passages and lungs is constantly changing. However, at death this dynamic process stops. The particulate makeup in the deceased will reflect his/her last few breaths. A deceased sexual assault victim who was strangled by his/her killer should have in their nasal passages skin flakes shed by their assailant [101, 107]. The recovery of one or more of these flakes that originated from the assailant, if successfully DNA typed, would provide highly relevant evidence. However, at present there are problems with attaining that goal. First, recovery of most flakes or other sources of DNA would be from the victim. If we assume a female victim and a male assailant, then the assailant's DNA will bear the Y-chromosome while the victim's will not. We need to develop an efficient means of recovering material (mucous) from the nasal passages, separating out the DNA containing fraction, and then find a way of separating out the Y-chromosome-containing DNA. Perhaps a daunting task but considering the past rate of improvements in DNA profiling, it should happen, and perhaps quite soon.

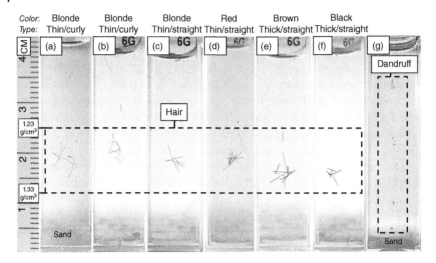

Figure 4.14 MAGLEV of cut pieces of human hair from the head and dandruff. (a) Separation of head hair (0.8 wt%) from a mixture with quartz sand, using MAGLEV. (b–f) Levitation of head hair of different color and type from five individuals. (g) Separation of dandruff (0.7 wt%, sourced from an older male, >65 years) from a mixture with quartz sand. Aqueous manganese(II) chloride solutions were used to levitate the hair (1.9 M) and the dandruff (1 M).

For this chapter, we measured the densities of pieces of hair from the head from five individuals with various types of hair (Figure 4.14b, c). The images were captured after five minutes of separation in a LWD MAGLEV device. The fibers of hair were washed in an aqueous solution of detergent and dried before levitation. The measured densities ranged between 1.23 and 1.33 g/cm^3 for the individuals. The results suggest that MAGLEV is suitable for separation of hair from other materials, such as sand ($\rho = 2.65$ g/cm^3). The measured densities for the samples of hair are consistent with earlier reports of density for human hair, i.e. 1.32 g/cm^3 for dry hair and 1.25 g/cm^3 for wet hair (in MAGLEV using polar medium the hair is, of course, wet) [114]. The densities of the hair from the different individuals are very similar. The samples of hair were from individuals with ages ranging between 28 and 40 years. The densities measured for different pieces of hair from the same individual varied over a similar range (Figure 4.14e). Density of hair is therefore not suitable as a method to distinguish between different individuals.

4.8 Instructions for the Construction of MAGLEV Devices

MAGLEV devices of the particular types we describe in this chapter are currently not commercially available; they are, however, simple to design, construct,

Table 4.3   Specifications for the NdFeB magnets used in MAGLEV devices.

Type of MAGLEV device	Strength of magnetic field (T)[a]	Dimensions of magnets (mm)	Shape of magnet/grade	References
LWD	0.38	$W \times L \times H : 50.8 \times 50.8 \times 25.4$	Rectangular cuboid/ N52 or N50	[5, 10, 30]
HMFG	0.51	$W \times L \times H : 25.4 \times 50.8 \times 25.4$	Rectangular cuboid/N52	[8, 23, 115]
Axial	0.33	$OD \times ID \times H : 76.2 \times 25.4 \times 25.4$	Flat ring/N45	[33]
Tilted	0.40	$W \times L \times H : 101.6 \times 101.6 \times 25.4$	Rectangular cuboid/N40	[7]

a)   Measured at the surface at the middle of the bottom magnet in the MAGLEV devices.

and use. In this chapter, we include technical drawings detailing recommended configurations of MAGLEV devices (Figure 4.15). The drawings can be used by the reader or a commissioned craftsman to facilitate the construction of the MAGLEV devices. The materials needed to build them are inexpensive ($50–300) and mostly associated with the magnets. The MAGLEV devices in Figures 4.3 and 4.15 contain magnets of various shape and magnetic field strength (Table 4.3). The magnets can be purchased from, e.g. appliedmagnets.com, kjmagnetics.com, or magnet4less.com.

The basic principle of construction is similar for all the MAGLEV devices described in this chapter. The devices have two like-poles-facing permanent magnets mechanically secured in two 3D-printed fixtures made of acrylonitrile–butadiene–styrene–plastic (ABS-plastic). We used computer-aided design software (Solidworks™) to design the fixtures for 3D printing. The files used for 3D printing are available on request. It is also possible to use subtractive tools such as a saw, lathe, or drill, to machine the fixtures out of materials such as plastics or aluminum, as long as the materials do not interact magnetically with magnets. The magnets can be further secured (if necessary) using adhesives, such as epoxy glue.

The two fixtures are held at a set distance from each other in the MAGLEV device using 4 threaded steel rods and 16 hex nuts; the position of the hex nuts can be adjusted along the rods to change the distance between the magnets. Steel cap nuts are not essential, but they may be placed at the end of the steel rods to prevent the MAGLEV device from scratching surfaces. We recommend the use of lock washers to hold hex nuts in position. The rods and hex nuts interact only weakly with the magnets and cause minimal disturbances to the magnetic field between the two magnets.

The paramagnetic solutions are corrosive to most metals, such as iron, and to the NdFeB magnets. We therefore use corrosion-resistant or super corrosion-resistant

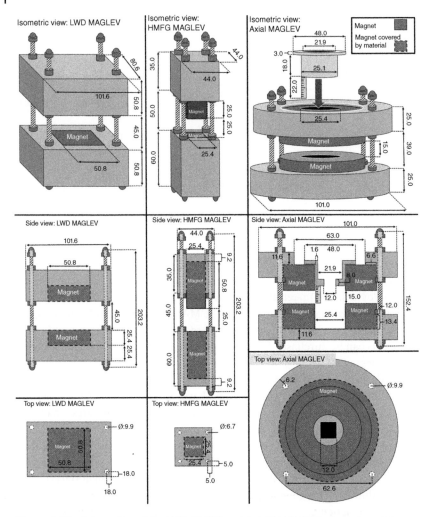

Figure 4.15 Schematics of the LWD, HMFG, and Axial MAGLEV devices viewed from different directions with dimensions specified in millimeter.

316 stainless steel for the parts of the MAGLEV devices, whenever possible, to reduce this problem. The use of nickel-plated magnets (this coating tends to be the standard coating on the magnets that are commercially available) reduces that rate of corrosion for the magnets [116, 117]. The threaded rods, hex nuts, and cap nuts we use to construct the LWD, Axial, and Tilted MAGLEV devices are of size and type 3/8″-24. The HMFG MAGLEV device uses threads of a smaller dimension (1/4″-20). We buy the rods and nuts from McMaster-Carr.

The MAGLEV devices described in this chapter use NdFeB permanent magnets (Table 4.3). These magnets have six characteristics making them attractive for use in MAGLEV devices: (i) commercial availability; (ii) stability and large magnetic field strength (e.g. up to ~1 T on the surfaces of block magnets [4]), (iii) low cost, (iv) ability to generate magnetic fields without using electricity, (v) resilience to demagnetization [4], and (vi) availability in many of sizes and shapes [116, 117].

4.9 Conclusion

MAGLEV has matured over the last one and a half decades to become a powerful method for separation of mixtures of particles and to make sensitive measurements of density. This chapter is the first publication to summarize the key practical lessons we have gained on this journey; broader reviews are also available [4, 16–18]. These instructions aim to lower the barriers for forensic analysts to get started with separations and density measurements using MAGLEV by providing simple guidelines and protocols. We also compiled relevant results from previous publications that use MAGLEV to separate common classes of trace evidence, for example, glitter, gunpowder, illicit drugs, plastics, metals, and minerals [4]. This work extends that list to include trace evidence such as polymeric fibers, new types of glass, bone, hair, and dandruff.

The low cost and small footprint of the MAGLEV devices could permit their use in larger numbers for analysis of trace evidence at crime scenes and in laboratories. The portability of the MAGLEV devices should help forensic investigators make rapid and better informed decisions at a crime scene – for example, investigators can decide quickly if an apartment needs a more thorough investigation should a few particles of illicit drugs be detected.

A drawback to MAGLEV devices is that they are only now becoming commercially available (LevitasBio, San Francisco, CA, makes one such device), and these initial commercial devices are intended for biological applications. Since the MAGLEV devices we describe can be constructed with relative ease using a 3D printer or a basic machine shop, we have included descriptions in Section 4.8 that detail the construction of the MAGLEV devices described in this book chapter.

The applications of MAGLEV described in this chapter constitute only a small part of its potential uses in forensics. A key area of improvement that will enable the use of MAGLEV at large scale in the field is to improve its user friendliness. Part of that goal is to develop a system for automated imaging and analysis of the levitating samples for quick and accurate determination of density. This development should take place in collaboration between developers of MAGLEV and law enforcement professionals to ensure that the needs of the latter are fulfilled. We are developing standardized protocols for the analysis of specific trace evidence using

MAGLEV and we hope eventually to have these protocols approved by forensic organizations, such as the Scientific Working Group for the Analysis of Seized Drugs [118].

References

1 Bottrell, M.C. (2009). Forensic glass comparison: background information used in data interpretation. *Forensic Sci. Commun.* 11: 2.

2 Rosensweig, R.E. (1966). Fluidmagnetic buoyancy. *AIAA J.* 4 (10): 1751–1758.

3 Khalafalla, S. (1976). Magnetic separation of the second kind: magnetogravimetric, magnetohydrostatic, and magnetohydrodynamic separations. *IEEE Trans. Magn.* 12 (5): 455–462.

4 Ge, S., Nemiroski, A., Mirica, K.A. et al. (2020). Magnetic levitation in chemistry, materials science, and biochemistry. *Angew. Chem. Int. Ed.* 59 (41): 17810–17855.

5 Mirica, K.A., Shevkoplyas, S.S., Phillips, S.T. et al. (2009). Measuring densities of solids and liquids using magnetic levitation: fundamentals. *J. Am. Chem. Soc.* 131 (29): 10049–10058.

6 Nemiroski, A., Kumar, A.A., Soh, S. et al. (2016). High-sensitivity measurement of density by magnetic levitation. *Anal. Chem.* 88 (5): 2666–2674.

7 Nemiroski, A., Soh, S., Kwok, S.W. et al. (2016). Tilted magnetic levitation enables measurement of the complete range of densities of materials with low magnetic permeability. *J. Am. Chem. Soc.* 138 (4): 1252–1257.

8 Abrahamsson, C., Nagarkar, A., Fink, M. et al. (2020). Analysis of powders containing illicit drugs using magnetic levitation. *Angew. Chem. Int. Ed.* 59 (2): 874–881.

9 Morrish, A.H. (2001). *The Physical Principles of Magnetism*, 31–77, 259–331, 486–538. Hoboken, NJ: IEEE Press.

10 Lockett, M.R., Mirica, K.A., Mace, C.R. et al. (2013). Analyzing forensic evidence based on density with magnetic levitation. *J. Forensic Sci.* 58 (1): 40–45.

11 Atkinson, M.B., Bwambok, D.K., Chen, J. et al. (2013). Using magnetic levitation to separate mixtures of crystal polymorphs. *Angew. Chem. Int. Ed.* 52 (39): 10208–10211.

12 Kahane, D. and Thornton, J. (1987). Estimation of the absolute density of glass following the sink/float technique. *J. Forensic Sci.* 32 (1): 87–92.

13 Fredricks, R.E. (1995). Density gradient columns made of water and sodium bromide solutions with isopropyl alcohol as a wetting agent. *J. Appl. Polym. Sci.* 57 (4): 509–511.

14 Stoney, D.A. and Stoney, P.L. (2015). Critical review of forensic trace evidence analysis and the need for a new approach. *Forensic Sci. Int.* 251: 159–170.

15 Fujii, K. (2006). Precision density measurements of solid materials by hydrostatic weighing. *Meas. Sci. Technol.* 17 (10): 2551.

16 Gao, Q.-H., Zhang, W.-M., Zou, H.-X. et al. (2019). Label-free manipulation via the magneto-Archimedes effect: fundamentals, methodology and applications. *Mater. Horiz.* 6: 1359–1379.

17 Yaman, S., Anil-Inevi, M., Ozcivici, E., and Tekin, H.C. (2018). Magnetic force-based microfluidic techniques for cellular and tissue bioengineering. *Front. Bioeng. Biotechnol.* 6: 192.

18 Turker, E. and Arslan-Yildiz, A. (2018). Recent advances in magnetic levitation: a biological approach from diagnostics to tissue engineering. *ACS Biomater Sci. Eng.* 4 (3): 787–799.

19 Castro, B., de Medeiros, M.S., Sadri, B., and Martinez, R.V. (2018). Portable and power-free serodiagnosis of Chagas disease using magnetic levitating microbeads. *Analyst* 143 (18): 4379–4386.

20 Shapiro, N.D., Mirica, K.A., Soh, S. et al. (2012). Measuring binding of protein to gel-bound ligands using magnetic levitation. *J. Am. Chem. Soc.* 134 (12): 5637–5646.

21 Shapiro, N.D., Soh, S., Mirica, K.A., and Whitesides, G.M. (2012). Magnetic levitation as a platform for competitive protein–ligand binding assays. *Anal. Chem.* 84 (14): 6166–6172.

22 Winkleman, A., Perez-Castillejos, R., Gudiksen, K.L. et al. (2007). Density-based diamagnetic separation: devices for detecting binding events and for collecting unlabeled diamagnetic particles in paramagnetic solutions. *Anal. Chem.* 79 (17): 6542–6550.

23 Ge, S., Semenov, S.N., Nagarkar, A.A. et al. (2017). Magnetic levitation to characterize the kinetics of free-radical polymerization. *J. Am. Chem. Soc.* 139 (51): 18688–18697.

24 Yang, X., Wong, S.Y., Bwambok, D.K. et al. (2014). Separation and enrichment of enantiopure from racemic compounds using magnetic levitation. *Chem. Commun.* 50 (56): 7548–7551.

25 Durmus, N.G., Tekin, H.C., Guven, S. et al. (2015). Magnetic levitation of single cells. *Proc. Natl. Acad. Sci. U.S.A.* 112 (28): E3661–E3668.

26 Whitesides, G.M., Sagué, A.L., Derda, R. et al. (2014). Density analysis of organisms by magnetic levitation. US2012/056655.

27 Ashkarran, A.A., Suslick, K.S., and Mahmoudi, M. (2020). Magnetically levitated plasma proteins. *Anal. Chem.* 92 (2): 1663–1668.

28 Hennek, J.W., Nemiroski, A., Subramaniam, A.B. et al. (2015). Using magnetic levitation for non-destructive quality control of plastic arts. *Adv. Mater.* 27 (9): 1587–1592.

29 Knowlton, S., Sencan, I., Aytar, Y. et al. (2015). Sickle cell detection using a smartphone. *Sci. Rep.* 5: 15022.

30 Mirica, K.A., Phillips, S.T., Mace, C.R., and Whitesides, G.M. (2010). Magnetic levitation in the analysis of foods and water. *J. Agric. Food Chem.* 58 (11): 6565–6569.

31 Aguilar-Arteaga, K., Rodriguez, J., and Barrado, E. (2010). Magnetic solids in analytical chemistry: a review. *Anal. Chim. Acta* 674 (2): 157–165.

32 Matheys, C., Tumanova, N., Leyssens, T., and Myerson, A.S. (2016). Magnetic levitation as a tool for separation: separating cocrystals from crystalline phases of individual compounds. *Cryst. Growth Des.* 16 (9): 5549–5553.

33 Ge, S. and Whitesides, G.M. (2018). "Axial" magnetic levitation using ring magnets enables simple density-based analysis, separation, and manipulation. *Anal. Chem.* 90 (20): 12239–12245.

34 Ge, S., Wang, Y., Deshler, N.J. et al. (2018). High-throughput density measurement using magnetic levitation. *J. Am. Chem. Soc.* 140 (24): 7510–7518.

35 Vince, J.J. and Sherlock, W.E. (2005). *Evidence Collection*. Burlington, MA: Jones & Bartlett Learning.

36 Shimoiizaka, J., Nakatsuka, K., Fujita, T., and Kounosu, A. (1980). Sink-float separators using permanent magnets and water based magnetic fluid. *IEEE Trans. Magn.* 16 (2): 368–371.

37 Fong, W. (1973). Value of glass as evidence. *J. Forensic Sci.* 18 (4): 398–404.

38 Bresee, R. (1980). Density gradient analysis of single polyester fibers. *J. Forensic Sci.* 25 (3): 564–570.

39 Lorefice, S. and Malengo, A. (2006). Calibration of hydrometers. *Meas. Sci. Technol.* 17 (10): 2560.

40 Martin, C., Herrman, T.J., Loughin, T., and Oentong, S. (1998). Micropycnometer measurement of single-kernel density of healthy, sprouted, and scab-damaged wheats. *Cereal Chem.* 75 (2): 177–180.

41 Haugh, C., Lien, R., Hanes, R., and Ashman, R. (1976). Physical properties of popcorn. *Trans. ASAE* 19 (1): 168–0171.

42 Peters, W.R. and Katz, R. (1962). Using a density gradient column to determine wheat density. *Cereal Chem.* 39: 487–493.

43 Moore, D.H. (1969). *Physical Chemical Techniques: Physical Techniques in Biological Research, Part B: Physical Chemical Techniques*, vol. 2, 285–313. Cambridge, MA: Academic Press.

44 Dell'Aquila, A. (2004). Cabbage, lentil, pepper and tomato seed germination monitored by an image analysis system. *Seed Sci. Technol.* 32 (1): 225–229.

45 Dell'Aquila, A. (2007). Towards new computer imaging techniques applied to seed quality testing and sorting. *Seed Sci. Technol.* 35 (3): 519–538.

46 Demilly, D., Ducournau, S., Wagner, M.-H., and Dürr, C. (2014). *Plant Image Analysis*, 147–164. Boca Raton, FL: CRC Press.

47 Dobraszczyk, B., Whitworth, M., Vincent, J., and Khan, A. (2002). Single kernel wheat hardness and fracture properties in relation to density and the modelling of fracture in wheat endosperm. *J. Cereal Sci.* 35 (3): 245–263.

48 Lay, D.C. (2003). *Linear Algebra and its Applications*. Boston, MA: Addison Wesley.

49 Potts, P.J. (1997). A glossary of terms and definitions used in analytical chemistry. *Geostand. Newsl.* 21 (1): 157–161.

50 Berensmeier, S. (2006). Magnetic particles for the separation and purification of nucleic acids. *Appl. Microbiol. Biotechnol.* 73 (3): 495–504.

51 Fitzpatrick, R.W., Raven, M.D., and Forrester, S.T. (2009). A systematic approach to soil forensics: criminal case studies involving transference from crime scene to forensic evidence. In: *Criminal and Environmental Soil Forensics* (ed. K. Ritz, L. Dawson and D. Miller), 105–127. Berlin: Springer-Verlag.

52 Strem Chemicals (2020). Tris(dipivaloylmethanato) gadolinium(III). https:// www.strem.com/catalog/v/93-3130/23/gallium_14405-43-7 (accessed 30 October 2020).

53 Santa Cruz Biotechnology (2020). Trioctylphosphine oxide. https://www.scbt .com/p/trioctylphosphine-oxide-78-50-2 (accessed 11 October 2020).

54 Alfa Aesar (2020). Acetylacetonate gadolinium(III). https://www.alfa.com/en/ catalog/013204 (accessed 30 October 2020).

55 Sigma-Aldrich (2020). Material safety data sheet: manganese(II) chloride tetrahydrate. https://www.sigmaaldrich.com/catalog/product/sigald/m3634 (accessed 28 September 2020).

56 Sigma-Aldrich (2020). Material safety data sheet: gadolinium(III) chloride hexahydrate. https://www.sigmaaldrich.com/catalog/product/aldrich/g7532 (accessed 28 September 2020).

57 Sigma-Aldrich (2020). Material safety data sheet: dysprosium(III) chloride hexahydrate. https://www.sigmaaldrich.com/catalog/substance/ dysprosiumiiichloridehexahydrate376951505952611 (accessed 28 September 2020).

58 Winkleman, A., Gudiksen, K.L., Ryan, D. et al. (2004). A magnetic trap for living cells suspended in a paramagnetic buffer. *Appl. Phys. Lett.* 85 (12): 2411–2413.

59 National Institute of Health (2020). ImageJ. https://imagej.nih.gov/ij/index .html (accessed 26 November 2020).

60 Xia, N., Zhao, P., Xie, J. et al. (2017). Density measurement for polymers by magneto-Archimedes levitation: simulation and experiments. *Polym. Test.* 63: 455–461.

61 Xie, J., Zhang, C., Gu, F. et al. (2019). An accurate and versatile density measurement device: magnetic levitation. *Sens. Actuators, B* 295: 204–214.

62 Zhao, P., Xie, J., Gu, F. et al. (2018). Separation of mixed waste plastics via magnetic levitation. *Waste Manag.* 76: 46–54.

63 White, T.D. and Folkens, P.A. (2005). *The Human Bone Manual.* Amsterdam, Netherlands: Elsevier.

64 Owsley, D.W., Mires, A.M., and Keith, M. (1985). Case involving differentiation of deer and human bone fragments. *J. Forensic Sci.* 30 (2): 572–578.

65 To, D. (2017). Forensic archaeology survey methods, scene documentation, excavation, and recovery methods. In: *Forensic Anthropology: A Comprehensive Introduction*, 2e (ed. N.R. Langley and M.T.A. Tersigni-Tarrant), 35–56. Boca Raton, FL: CRC Press.

66 Cattaneo, C., Porta, D., Gibelli, D., and Gamba, C. (2009). Histological determination of the human origin of bone fragments. *J. Forensic Sci.* 54 (3): 531–533.

67 Edson, S.M. and McMahon, T.P. (2016). Extraction of DNA from skeletal remains. In: *Forensic DNA Typing Protocols* (ed. W. Goodwin), 69–87. Berlin: Springer-Verlag.

68 Williams, P.A. and Saha, S. (1996). The electrical and dielectric properties of human bone tissue and their relationship with density and bone mineral content. *Ann. Biomed. Eng.* 24 (2): 222–233.

69 Blanton, P.L. and Biggs, N.L. (1968). Density of fresh and embalmed human compact and cancellous bone. *Am. J. Phys. Anthropol.* 29 (1): 39–44.

70 Evans, F.G. and Wood, J.L. (1976). Mechanical properties and density of bone in a case of severe endemic fluorosis. *Acta Orthop. Scand.* 47 (5): 489–495.

71 Rannazzisi, J.T. and Caverly, M.W. (2006). *Practitioner's Manual, U.S. Department of Justice.* Drug Enforcement Administration. https://www.brandeis.edu/ora/compliance/controlled-substances/dea-practitioner-manual.pdf.

72 Blackledge, R.D. (2007). GLITTER as forensic evidence. *Trace Evidence Symposium sponsored by the NIJ and the FBI*, Clearwater Beach, FL. http://projects.nfstc.org/trace/docs/final/Blackledge_Glitter.pdf (accessed 16 March 2022).

73 Gross, S., Igowsky, K., and Pangerl, E. (2010). Glitter as a source of trace evidence. *J. Am. Assoc. Trace Evid. Exam.* 1 (1): 62–72.

74 Chophi, R., Sharma, S., and Singh, R. (2020). Discrimination of nail polish using attenuated total reflectance infrared spectroscopy and chemometrics. *Aust. J. Forensic Sci.* 53: 1–12.

75 Morehead, W. (2007). Smokeless powders. In: *Forensic Analysis on the Cutting Edge: New Methods for Trace Evidence Analysis* (ed. R.D. Blackledge), 241–268. Hoboken, NJ: Wiley.

76 Heramb, R.M. and McCord, B.R. (2002). The manufacture of smokeless powders and their forensic analysis: a brief review. *Forensic Sci. Commun.* 4: 2.

77 Cole, C., Jones, L., McVeigh, J. et al. (2010). *CUT: A Guide to Adulterants, Bulking Agents and other Contaminants Found in Illicit Drugs*. Centre for Public Health, Liverpool John Moores University. www.cph.org.uk/wp-content/uploads/2012/08/cut-a-guide-to-the-adulterants-bulking-agents-and-other-contaminants-found-in-illicit-drugs.pdf (accessed 16 March 2022).

78 Albiges, M. (2019). The consequences of Virginia's drug testing backlog: cases delayed, defendants lingering in jail. *The Virginian-Pilot* (7 May).

79 Department of Justice, Drug Enforcement Administration (2018). *National Drug Threat Assessment*. U.S. Department of Justice, Drug Enforcement Administration. https://www.dea.gov/sites/default/files/2018-11/DIR-032-18%202018%20NDTA%20final%20low%20resolution.pdf (accessed 16 March 2022).

80 U.S. Department of Health and Human Services, Center for Disease Control and Prevention (2017). Annual surveillance report of drug-related risks and outcomes—United States. Surveillance special report 1. Published August 31, 2017. U.S. Department of Health and Human Services, Center for Disease Control and Prevention. https://www.cdc.gov/drugoverdose/pdf/pubs/2017-cdc-drug-surveillance-report.pdf (accessed 16 March 2022).

81 Statistical Bulletin (2017). *Price, Purity and Potency*. European Monitoring Centre for Drugs and Drug Addiction. http://www.emcdda.europa.eu/data/stats2017/ppp_en (accessed 16 March 2022).

82 Leary, P. (2020). *72st AAFS Annual Scientific Meeting*, Anaheim, CA (17–22 February 2020). Anaheim, CA: American Academy of Forensic Science.

83 Scientific Working Group for Materials Analysis United States of America. (2005). Glass refractive index determination. https://www.ojp.gov/ncjrs/virtual-library/abstracts/glass-refractive-index-determination (accessed 16 March 2022).

84 Stoney, D.A. and Thornton, J.I. (1985). The forensic significance of the correlation of density and refractive index in glass evidence. *Forensic Sci. Int.* 29 (3–4): 147–157.

85 Montero, S., Hobbs, A.L., French, T.A., and Almirall, J.R. (2003). Elemental analysis of glass fragments by ICP-MS as evidence of association: analysis of a case. *J. Forensic Sci.* 48 (5): 1101–1107.

86 Edmund Optics (2020). 5mm Diameter Uncoated, Gorilla Glass Window. https://www.edmundoptics.com/p/5mm-diameter-uncoated-gorilla-glassreg-window/26505 (accessed 19 August 2020).

87 Surmet (2020). ALON™ optical ceramic. http://www.surmet.com/docs/Product_sheet_ALON.pdf (accessed 19 August 2020).

88 Crouch, I., Franks, G., Tallon, C. et al. (2017). Glasses and ceramics. In: *The Science of Armour Materials* (ed. I. Couch), 331–393. Amsterdam, Netherlands: Elsevier.

89 Heard, B.J. (2009). *Wiley Encyclopedia of Forensic Science, Glass: Types and Substitutes*. Hoboken, NJ: Wiley.

90 Blackledge, R.D. (1980). Examination of automobile rubber bumper guards by synchronous excitation spectrofluorometry. *J. Forensic Sci.* 25 (3): 583–588.

91 Blackledge, R.D. (1981). Examination of automobile rubber bumper guards by attenuated total reflectance spectroscopy using a Fourier transform infrared spectrometer. *J. Forensic Sci.* 26 (3): 554–556.

92 Blackledge, R.D. (1981). Pyrolysis gas chromatography of automobile rubber bumper guard samples. *J. Forensic Sci.* 26 (3): 557–559.

93 Mistek, E., Fikiet, M.A., Khandasammy, S.R., and Lednev, I.K. (2018). Toward Locard's exchange principle: recent developments in forensic trace evidence analysis. *Anal. Chem.* 91 (1): 637–654.

94 Palenik, C.S. and Palenik, S. (2014). Seeing color: practical methods in pigment microscopy. *Microscope* 62 (2): 71–81.

95 Suzuki, E. and Gresham, W. (1987). Forensic science applications of diffuse reflectance infrared Fourier transform spectroscopy (DRIFTS): III. Direct analysis of polymeric foams. *J. Forensic Sci.* 32 (2): 377–395.

96 Kuptsov, A.H. (1994). Applications of Fourier transform Raman spectroscopy in forensic science. *J. Forensic Sci.* 39 (2): 305–318.

97 Vítek, P., Ali, E.M., Edwards, H.G. et al. (2012). Molecular and biomolecular spectroscopy, evaluation of portable Raman spectrometer with 1064 nm excitation for geological and forensic applications. *Spectrochim. Acta, Part A* 86: 320–327.

98 Ehrenstein, G.W. (2012). *Polymeric Materials: Structure, Properties, Applications*. Munich: Carl Hanser Verlag GmbH Co KG.

99 Haynes, W.M. (2014). *CRC Handbook of Chemistry and Physics*. Boca Raton, FL: CRC Press.

100 Almeida, M., Betancor, E., Fregel, R. et al. (2011). Genetics supplement series, efficient DNA extraction from hair shafts. *Forensic Sci. Int.* 3 (1): e319–e320.

101 Porter, L. (2013). Crime Magazine, Forensics: The "Head & Shoulders" Case, 15 July 2013.

102 Cooper, G.A., Kronstrand, R., and Kintz, P. (2012). Society of Hair Testing guidelines for drug testing in hair. *Forensic Sci. Int.* 218 (1–3): 20–24.

103 Harding, H. and Rogers, G. (1999). Physiology and growth of human hair. In: *Forensic Examination of hair*, 1e (ed. J. Robertson), 13–166. Boca Raton: CRC Press.

104 Herber, B. and Herold, K. (1998). DNA typing of human dandruff. *J. Forensic Sci.* 43 (3): 648–656.

105 Jian, H., Zhu, J., Wang, H. et al. (2017). Genetics supplement series, comparative study on methods of DNA genotyping between single piece of dandruff and EZ-tape. *Forensic Sci. Int.* 6: e244–e245.

106 Lorente, M., Entrala, C., Lorente, J.A. et al. (1998). Dandruff as a potential source of DNA in forensic casework. *J. Forensic Sci.* 43 (4): 901–902.

107 Blackledge, R. (2014). Dander, death, & DNA. *The CAC News* (4th quarter) 2014, 14–15.

108 Hodges, C., Hendra, P., Willis, H., and Farley, T. (1989). Fourier transform Raman spectroscopy of illicit drugs. *J. Raman Spectrosc.* 20 (11): 745–749.

109 Deedrick, D.W. and Koch, S.L. (2004). Microscopy of hair part 1: a practical guide and manual for human hairs. *Forensic Sci. Commun.* 6: 1.

110 Linch, C.A., Whiting, D.A., and Holland, M.M. (2001). Human hair histogenesis for the mitochondrial DNA forensic scientist. *J. Forensic Sci.* 46 (4): 844–853.

111 Parker, G.J., Leppert, T., Anex, D.S. et al. (2016). Demonstration of protein-based human identification using the hair shaft proteome. *PLoS One* 11 (9): e0160653.

112 Schneider, H., Sommerer, T., Rand, S., and Wiegand, P. (2011). Hot flakes in cold cases. *Int. J. Leg. Med.* 125 (4): 543–548.

113 Holmes, H. (2001). *The Secret Life of Dust: From the Cosmos to the Kitchen Counter the Big Consequences of Little Things*. Hoboken, NJ: Wiley.

114 Bouillon, C. and Wilkinson, J.D. (2005). *The Science of Hair Care*, 2. Boca Raton, FL: Taylor & Francis.

115 Lynch, M.J., Suyama, J., and Guyette, F.X. (2018). Scene safety and force protection in the era of ultra-potent opioids. *Prehosp. Emerg. Care* 22 (2): 157–162.

116 KJMagnetics (2020). https://www.kjmagnetics.com (accessed 28 May 2020).

117 Applied Magnets (2020). https://appliedmagnets.com (accessed 28 May 2020).

118 Scientific Working Group for the Analysis of Seized Drugs (2020). http://swgdrug.org (accessed 3 June 2020).

5

Forensic Applications of Gas Chromatography – Vacuum Ultraviolet Spectroscopy Paired with Mass Spectrometry

Ryan Schonert

VUV Analytics, Inc., Cedar Park, TX, USA

5.1 Introduction

Vacuum ultraviolet (VUV) spectroscopy is a tool for spectral analysis that has recently been configured into a benchtop detector for gas chromatography (GC). Like mass spectrometry (MS), it is a universal detector, as it can detect almost all compounds eluting from a GC column; additionally, it can use a library of reference spectra to identify each compound using its unique VUV spectral profile. Although MS is one of the most common GC detection methods used in forensic laboratories, hyphenating MS with VUV improves the analysis by providing VUV spectral confirmation of MS identifications. The strengths of VUV spectroscopy often cover the weaknesses of MS and vice-versa, reducing the number of incorrect identifications made by GC/MS or GC/VUV alone. In this chapter, the theory of VUV and forensic applications for GC/VUV/MS are discussed.

5.2 Background of Mass Spectrometry

After being developed in the mid-twentieth century, GC quickly became an invaluable tool for sample analysis in the laboratory. GC exploded in popularity throughout several industries, including petroleum [1], pharmaceuticals [2], environmental [3], and foods [4]. The flame ionization detector (FID) was developed in the 1950s as one of the first GC detectors. At the time, it was considered one of the most reliable and sensitive detectors available, and it remains the detector of choice for many applications to this day [5, 6].

Leading Edge Techniques in Forensic Trace Evidence Analysis: More New Trace Analysis Methods, First Edition. Edited by Robert D. Blackledge.

Despite their value, detectors such as the FID only provide information on a compound's retention in the stationary phase, which does not provide any spectral or structural information that can be used for compound identification. To overcome this limitation, research groups began coupling MS to GC in the late 1950s. MS added a second dimension of identification to sample analysis, providing information on both compound retention and compound structure. Although the first GC/MS systems used time-of-flight mass spectrometry (TOF-MS), several other variants of MS were developed soon thereafter, and GC/MS systems eventually became commonplace in the lab [7–9]. Additionally, due to its popularity, considerable effort was given to produce mass spectral reference databases, such as the NIST Mass Spectral Library, which has been developed over many decades since the late twentieth century [10]. Because of its utility and large userbase, GC/MS is considered by many to be the gold standard for complex sample analysis.

MS detection techniques all follow the same general principles. Compounds of interest, such as those eluting from a GC column, are ionized, and the resulting ions are detected and distinguished according to their mass-to-charge ratio (m/z). In practice, mass spectrometers can vary according to traits such as their ionization technique or type of mass analyzer [11, 12]. For example, the common benchtop mass spectrometer found in forensic science laboratories is the single-quadrupole mass spectrometer using electron ionization (also known as electron impact ionization). Compounds eluting from the GC cross a beam of electrons released from a heated filament, which cause the compounds to ionize and break apart, or fragment. Because compounds fragment in reproducible ways, fragmentation patterns obtained from the mass spectrometer can be used to reconstruct the structure of the original compound [13, 14] (Figure 5.1).

Like many other disciplines, forensic science has come to depend on GC/MS for several analyses. For example, the analysis of seized drugs is described in ASTM

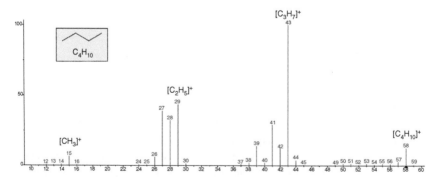

Figure 5.1 Mass spectrum of butane and selected mass fragments. Source: Adapted from NIST [15].

E2329. There are three basic categories of analytical techniques described in this method; Category A, B, and C. Category A techniques have the highest discriminating power and offer the most information; Category B and C techniques have progressively less discriminating power but can still offer useful information to an investigation. Because MS offers detailed structural information, it is placed in Category A. GC only provides information on retention time, which is not as discriminating, so it is placed in Category B. However, analyzing a sample with GC/MS provides both Category A and Category B results with a single injection [16, 17]. Other areas within forensic science, such as toxicology, fire debris analysis, and trace evidence analysis, also make use of GC/MS as an analytical scheme.

5.3 Background of Vacuum Ultraviolet Spectroscopy

Since MS, there have been few new GC detectors introduced to the market. However, a benchtop VUV detector has been recently developed. Like MS, VUV spectrometers can detect compounds eluting from a GC column and provide spectral identification of each compound. Unlike MS, however, VUV spectroscopy identifies compounds by their unique absorption profiles in the 125–240 nm range [5, 18]. The VUV detector has several advantages over other detection techniques, and it shows considerable promise as an ideal complement to MS.

Since the twentieth century, VUV absorption data has been collected from synchrotron radiation facilities. While these facilities can gather detailed and useful VUV absorption data, the facilities themselves need to be quite large to accommodate the equipment necessary for production of synchrotron radiation, so they are not practical for GC work [19]. GC-amenable far UV detectors (FUVDs) were later developed, but these detectors suffer from limited capabilities at lower wavelengths [18, 20–25]. One study [25] described the use of a GC-UV system that measured absorbance in the 178–330 nm range. While this wavelength range is useful for some compounds, many other compounds do not absorb well in this range. For example, saturated hydrocarbons such as hexane absorb very little past 160 nm [26–28]. Recently, a GC-amenable VUV detector capable of measuring wavelengths down to 125 nm has been developed [18, 29, 30]. Although new to the market, this detector has already demonstrated utility in a variety of fields, including the petroleum industry [29–33], environmental analysis [34, 35], and flavors/fragrances [36], among others [37–41].

Figure 5.2 shows a diagram of the benchtop VUV detector developed by VUV Analytics, Inc., which can operate on all GCs (Figure 5.3). At the beginning of the GC run, a dark noise reading is taken to ensure no stray UV light is reaching the detector. Then, a reference spectrum of the deuterium lamp is captured, after which the detector collects VUV absorption data at an adjustable frequency for

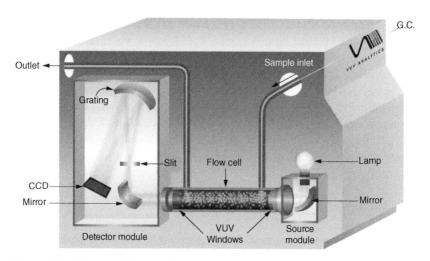

Figure 5.2 Diagram of the VUV Spectrometer developed by VUV Analytics, Inc. Currently, there are two models on the market: the VGA-100 and VGA-101. The VGA-100 scans the 125–240 nm range and has a maximum temperature setting of 300 °C, while the VGA-101 scans the 125–430 nm range and has a maximum temperature setting of 430 °C.

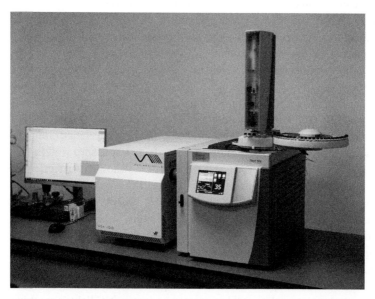

Figure 5.3 VUV Analytics VGA-100 spectrometer connected to a Thermo Scientific Trace 1310 gas chromatograph.

the duration of the run. As compounds elute from the GC column, they travel through a heated transfer line to a heated flow cell along with a flow of makeup carrier gas, which can be adjusted to alter the compounds' residence time in the flow cell. Once in the flow cell, the compounds absorb some amount of radiation over the VUV range, while the unabsorbed radiation is captured and recorded by a back-thinned charged-coupled device (CCD). The measured absorbance data is subtracted from the reference spectrum taken at the beginning of the run, yielding a VUV absorbance spectrum. VUV spectra are collected at a frequency of up to 75 Hz, and the average absorption over 125–240 nm is plotted for each time point to generate a chromatogram. Additionally, spectral filters can be applied to further distinguish eluting peaks and improve quantitation by using only regions where the compounds of interest are absorbing [18].

Photons in the 125–240 nm VUV wavelength region are absorbed by a molecule's valence electrons, causing electronic transitions from the highest occupied molecular orbital (HOMO) to the lowest unoccupied molecular orbital (LUMO). High energy, low wavelength photons cause transitions from sigma bonding (σ) or non-bonding (n) orbitals to sigma antibonding (σ*) orbitals; low energy, high wavelength photons cause transitions from pi bonding (π) or n orbitals to pi antibonding (π*) orbitals [30, 42]. Therefore, aside from spectrally invisible carrier gases such as He or H_2, almost all molecules can be detected by VUV spectroscopy, making it a universal detection technique. Additionally, because each molecule has a unique gas-phase absorption cross section, the resulting VUV absorption profiles are unique, allowing for compound identification through spectral library matching [18]. Figures 5.4–5.6 show example VUV spectra.

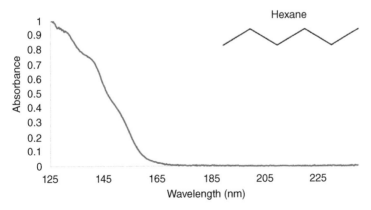

Figure 5.4 VUV absorption spectrum of hexane. Because hexane has only sigma (σ) bonds, absorption is observed in the lower wavelength/higher energy region of the VUV spectrum.

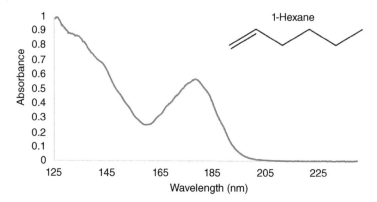

Figure 5.5   VUV absorption spectrum of 1-hexene. Because a pi (π) bond is present, absorption is observed at higher wavelengths relative to linear alkanes such as hexane.

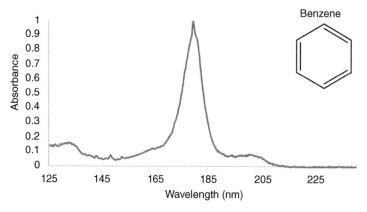

Figure 5.6   VUV absorption spectrum of benzene. Compounds containing more pi (π) bonds will absorb more at higher wavelengths.

If any absorbance spectrum is measured that does not match the library spectrum of a single compound, it is possible that another compound is coeluting and the measured spectrum is a combination of two or more compounds. In such cases, these coelutions can be distinguished using a process called time interval deconvolution (TID). TID is performed by dividing the chromatogram into equal time intervals, usually 0.02 minutes each. At each time interval, the measured absorbance spectrum is matched against the VUV spectral library. If the measured spectrum does not match the library spectrum of a single compound, the library spectra of two or more compounds may be combined to best match the measured spectrum. Like other light spectroscopy techniques, VUV spectroscopy follows the Beer–Lambert Law, and thus a simple linear combination of multiple library spectra can be used to fit a measured spectrum. These single or multi-analyte

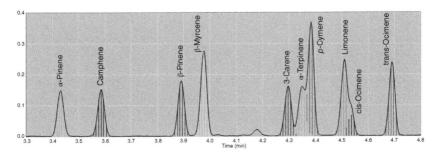

Figure 5.7 TID analysis of a terpenes mixture. The chromatogram is divided into time intervals (usually 0.02 minutes), represented by the colored bars. The measured spectrum at each interval is compared to the spectral library. Compounds are identified using spectral and RI information, even for coelutions such as the coelutions from 4.32 to 4.40 minutes or 4.48 to 4.56 minutes.

matches are made at each time interval throughout the duration of the run, and they can be done quickly and automatically. Multiple runs can be analyzed in sequence, providing a simple and effective automated data analysis. Figure 5.7 shows an example of TID [30, 43–46].

Using TID can mitigate the need for baseline peak resolution, which can be difficult to achieve with complex samples. For example, in the fuels industry, gasolines are often characterized by GC/FID. Standard methods using FID, such as detailed hydrocarbon analysis (DHA), must chromatographically separate each individual peak for identification and quantification. While useful, this method has an enormous run time of 174 minutes and requires a cryogenic oven start to separate each compound of interest [47]. Instead, coelutions in a GC/VUV run can be resolved using TID, as shown in Figures 5.8 and 5.9, allowing for the same gasoline samples to be analyzed in approximately 30 minutes [43, 44].

Like MS, VUV spectroscopy allows for both detection and identification of eluting compounds. However, VUV spectroscopy has two distinct advantages over MS:

1) *No vacuum required*

   Mass spectrometers operate under vacuum, which introduces limitations to GC method development. For example, high GC carrier gas flow rates could overwhelm the MS turbo pump, so GC/MS methods have a limited range of flow rates available.

   On the other hand, VUV spectroscopy is an atmospheric detection technique. Originally, radiation in the low UV range was collected in evacuated chambers,

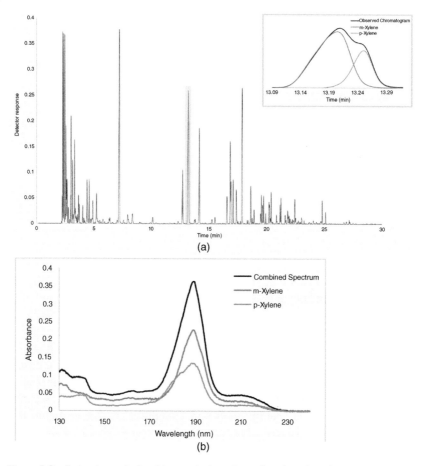

Figure 5.8 Example analysis with a coelution of *m*-xylene/*p*-xylene in a gasoline sample (a) and spectral overlay of the coelution at a single time interval (b). Despite eluting very closely to each other, each species can be individually identified and quantified.

giving it the name VUV radiation, because oxygen and water from the air would dominate the spectral measurements at atmospheric pressure [19]. However, the optics and detector housings of the GC/VUV instrument are kept under positive pressure of an inert gas that either absorbs poorly (N_2) or not at all (He), keeping oxygen and water from interfering. Furthermore, GC/VUV subtracts the background reference spectrum from the data gathered during each run, so any minimal absorption from a gas like N_2 will not affect results [18]. The lack of vacuum allows for a wider range of flow rates and makes the VUV detector easier to maintain.

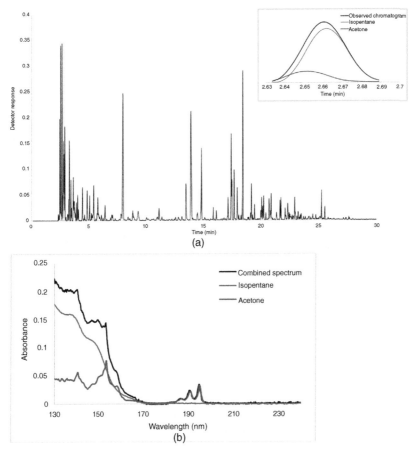

Figure 5.9 Example analysis with a coelution of acetone/isopentane in a gasoline sample (a) and spectral overlay of the coelution at a single time interval (b). Even though the small acetone peak is hidden under the large isopentane peak, it can be identified and quantified.

2) *Isomer Distinction*

While MS can easily identify many compounds, identification may be difficult when distinguishing between closely related compounds. For example, Figure 5.10 shows the mass spectra of m-xylene, p-xylene, and o-xylene. These structural isomers have mass spectra that are difficult to distinguish. However, their VUV spectra are clearly distinguishable, as shown in Figure 5.11. The only isomers that cannot be distinguished by VUV are enantiomers, as they have the same absorption cross section.

Finally, like other light spectroscopy techniques, VUV absorption is linearly proportional to concentration via the Beer–Lambert Law, allowing for simple

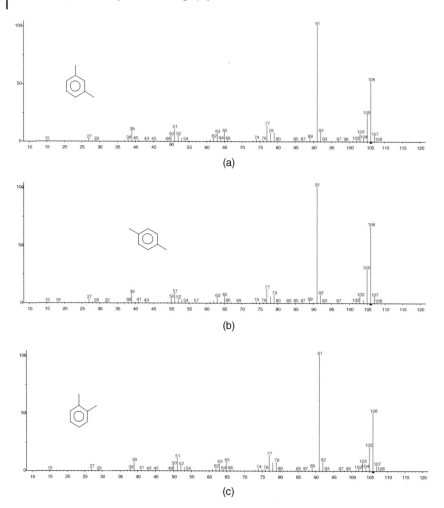

Figure 5.10 Mass spectra of *m*-xylene (a), *p*-xylene (b), and *o*-xylene (c), adapted from NIST [15]. Because the compounds are closely related in structure, their mass spectra are difficult to distinguish.

quantitation of eluting compounds [30]. Quantitation is achieved for individual compounds using their RRF, a measure of its absorbance relative to methane [43].

5.4 Combining GC/VUV and GC/MS

Although GC/VUV is a promising technique, it is still an incomplete technique from a forensics perspective. One disadvantage is the small spectral library. The

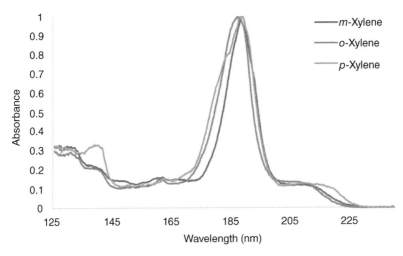

Figure 5.11 VUV absorption spectra of *m*-xylene, *o*-xylene, and *p*-xylene. Despite having very similar mass spectra, the VUV absorption spectra are distinguishable even to the eye.

NIST mass spectral library, for example, has hundreds of thousands of entries amassed over the past 40 years; comparatively, the VUV spectral library currently contains under 2000 compounds. Additionally, GC/MS tends to have better sensitivity, as it can achieve a limit of detection (LOD) on the fg level [48], while the LOD of GC/VUV is on the pg level [18].

However, by combining VUV and MS, each detector can compensate for the other detector's weaknesses. For example, consider the following hypothetical situations:

- Samples containing compounds not in the VUV spectral library could be identified by MS. That information could then be used to make a new entry for the VUV spectral library.
- Samples containing coeluting compounds not deconvolved by MS could be individually quantified using VUV.
- Samples containing closely related isomers not distinguishable by MS could be identified using VUV.
- Trace level compounds not detectable by VUV could be detected by MS.

In addition, gathering additional information about each sample is always desirable for forensics investigations. For example, while a GC/MS analysis satisfies the requirements of ASTM E2329 by providing Category A (MS) and Category B (GC) data, adding VUV to complement the GC/MS results would provide additional confidence in the results.

VUV and MS can be combined into a single GC/VUV/MS setup. Because VUV spectroscopy is non-destructive, it is possible to run the detectors in-series

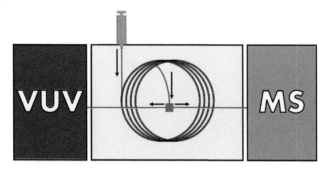

Figure 5.12 Basic diagram of a GC/VUV/MS configuration using a column splitter. Using this setup, both VUV and MS data can be collected with a single injection.

[49, 50], although there are significant technical issues associated with connecting an atmospheric detector to a vacuum detector [51]. Instead, it is recommended to use a column splitter, as diagrammed below in Figure 5.12.

The following sections detail five selected forensics applications using GC/MS and GC/VUV.

5.5   Analysis of Fentanyl Analogues

Fentanyl is a synthetic opioid developed by Paul Janssen and his team at Janssen Pharmaceutica in 1959 and introduced to the pharmaceutical market in the 1960s as an intravenous analgesic [51–54]. Because it is more lipid-soluble than morphine and can more easily penetrate the blood–brain barrier, it was observed to be approximately 100x more potent than morphine [53]. Due to its effectiveness, demand for fentanyl grew exponentially, which encouraged the development of fentanyl analogues such as sufentanil, alfentanil, remifentanil, and carfentanil [54]. In particular, carfentanil is one of the most potent drugs available today; it is 100x more potent than fentanyl, making it 10 000x more potent than morphine [55].

Increased access to fentanyl and its derivatives has sharply increased the number of opioid overdoses, leading the United States to label fentanyl and its derivatives as controlled substances [51, 54]. According to the Controlled Substances Act, a compound may be scheduled as a controlled substance if it is "substantially similar" in structure or pharmacological effect to a compound already listed as a schedule 1 or 2 controlled substance [56, 57]. However, because "substantially similar" is a vague term, demonstrating the similarity between a controlled substance and an analogue can be difficult; in fact, several court cases have hinged on the interpretation of the Controlled Substances Act [58–60]. Therefore, when

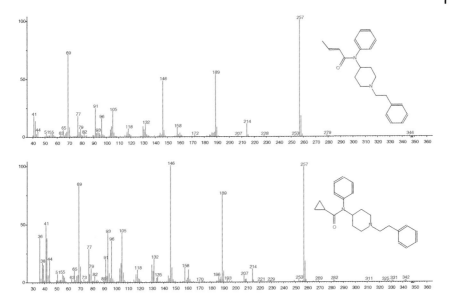

Figure 5.13 Structures and mass spectra of crotonyl fentanyl (top) and cyclopropyl fentanyl (bottom). Because these compounds have very similar structures, their mass spectral profiles share many of the same major peaks [61, 62].

analyzing fentanyl analogues, it is imperative to gather as much information as possible so differences and similarities between suspected drugs of abuse can be accurately determined.

Fentanyl analogues can be screened using GC/MS, but structurally similar fentanyl analogues may have very similar mass spectral profiles. For example, the mass spectra of cyclopropyl fentanyl and crotonyl fentanyl are shown in Figure 5.13. However, VUV can be used to accurately distinguish controlled substance isomers, as shown in Figure 5.14 [51, 63].

To demonstrate the combined analytical power of MS and VUV, Buchalter et al. performed a study to characterize 24 fentanyl analogues using tandem cold electron ionization MS (cold EI MS) and VUV. Cold EI enhances the survival of molecular ions during ionization by cooling molecules emerging from the GC [51]. The full experimental details can be found in the publication by Buchalter et al. Figure 5.15 shows the structure of each fentanyl analogue chosen for this study.

Even using cold EI MS, many of the analogues were indistinguishable by their mass spectra. While most of the compounds could be baseline resolved, some compounds coeluted, preventing identification using retention time and mass spectral analysis alone. However, using a tandem cold EI MS/VUV approach allowed for proper identification of all 24 compounds (Figures 5.16 and 5.17). Additionally, quantification was achieved using VUV with excellent determination coefficients

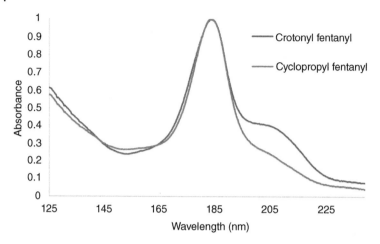

Figure 5.14 VUV absorbance spectra of crotonyl fentanyl and cyclopropyl fentanyl. Despite being similar in structure, their VUV profiles are easily distinguishable.

($R^2 \geq 0.9991$) over a concentration range of 1.56–200 μg/ml for most compounds without the need for deuterated standards; Table 5.1 lists the concentration ranges quantified and limits of detection determined for each compound [51].

Real world drug samples received by forensic laboratories are impure and cut with adulterants and/or diluents. To demonstrate the applicability of this method to real world samples, a simulated sample containing fentanyl, heroin, diphenhydramine, lidocaine, caffeine, acetylcodeine, O6-monoacetylmorphine, morphine, and papaverine was analyzed. Even in the presence of impurities, all compounds were properly identified [51]. With a comprehensive spectral library, VUV can greatly supplement abused substances analysis in the lab [64].

5.6 Analysis of Smokeless Powders

Although gunpowder was first developed in ancient China, Europeans began using it in guns around the 1400s and 1500s. Gunpowder generally refers to black powder, which is a mixture of potassium nitrate, charcoal, and sulfur. Many centuries later, a new type of powder, called smokeless powder, was developed. Smokeless powders could achieve significantly higher bullet speeds, leading to firearms with a longer range and more stopping power. Since then, smokeless powders have become the gold standard for use in firearms [65–67].

Certain types of ammunition, such as handgun or rifle ammunition, include cartridge cases with a charge of smokeless powder, primer, and a bullet (Figure 5.18). Once fired, the primer and smokeless powder are spent, propelling the bullet

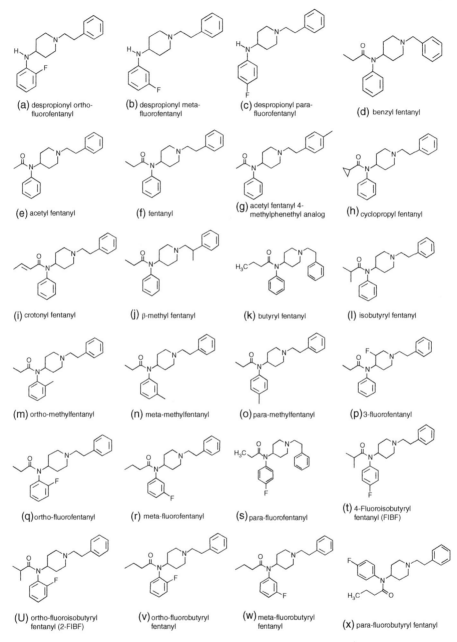

(a) despropionyl ortho-fluorofentanyl

(b) despropionyl meta-fluorofentanyl

(c) despropionyl para-fluorofentanyl

(d) benzyl fentanyl

(e) acetyl fentanyl

(f) fentanyl

(g) acetyl fentanyl 4-methylphenethyl analog

(h) cyclopropyl fentanyl

(i) crotonyl fentanyl

(j) β-methyl fentanyl

(k) butyryl fentanyl

(l) isobutyryl fentanyl

(m) ortho-methylfentanyl

(n) meta-methylfentanyl

(o) para-methylfentanyl

(p) 3-fluorofentanyl

(q) ortho-fluorofentanyl

(r) meta-fluorofentanyl

(s) para-fluorofentanyl

(t) 4-Fluoroisobutyryl fentanyl (FIBF)

(U) ortho-fluoroisobutyryl fentanyl (2-FIBF)

(V) ortho-fluorobutyryl fentanyl

(W) meta-fluorobutyryl fentanyl

(X) para-fluorobutyryl fentanyl

Figure 5.15 Twenty-four fentanyl analogues characterized by Buchalter et al. using cold EI MS and VUV [51]. This group includes sets of positional isomers having molecular masses of 298 (a–c), 322 (d–e), 336 (f–g), 348 (h–i), 350 (j–o), 354 (p–s), and 368 (t–x) [51].

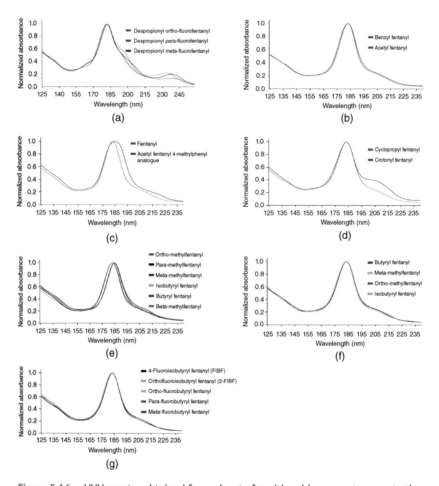

Figure 5.16 VUV spectra obtained for each set of positional isomers at concentrations of 50 μg/ml from 125 to 240 nm. The spectra for (a) were smoothed using a 24-point average on Microsoft Excel, while the spectra for (b–g) were smoothed with a polynomial order of 3 and nR and nL 24 in DataView (VUV Analytics, Inc.) [51].

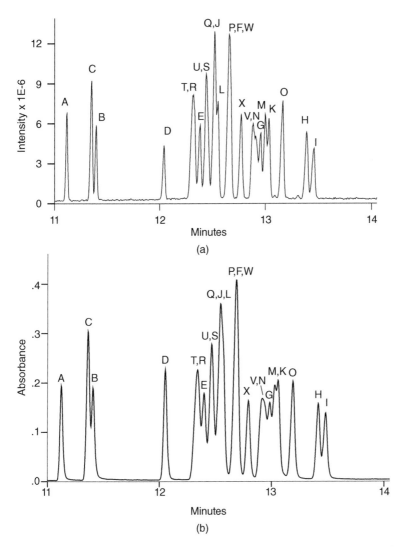

Figure 5.17 Simultaneous cold EI MS (a) and VUV (b) detection for a mixture of the 24 fentanyl analogues. Peak identifications can be found in Figure 5.15. Although many of these compounds coelute, VUV's deconvolution can be used to distinguish each compound [51].

toward the target. Afterward, the empty cartridge case is ejected from the firearm [68]. Because these cartridges cases are often made of brass, experienced firearm owners can save money by refilling these cartridge cases rather than buying new ammunition [69]. To this end, smokeless powders can be purchased in bulk at most sporting goods stores.

Table 5.1 Determination coefficients and limits of detection for each fentanyl analogue studied.

Compound	Concentration range (μg/ml)	Determination coefficient (R^2)	VUV LOD (ng/ml)	Cold Ei-MS LOD (ng/ml)
Despropionyl *ortho*-fluorofentanyl	1.56–200	0.9996	334	585
Despropionyl *meta*-fluorofentanyl	1.56–200	0.9999	390	936
Despropionyl *para*-fluorofentanyl	1.56–200	0.9999	260	669
Benzyl fentanyl	1.56–200	0.9999	292	468
Acetyl fentanyl	1.56–200	1.0000	260	585
Fentanyl	1.56–200	1.0000	260	936
Acetyl fentanyl 4-methylphenethyl analogue	1.56–200	0.9996	390	936
Cyclopropyl fentanyl	1.56–200	0.9995	292	936
Crotonyl fentanyl	1.56–200	0.9999	390	936
Beta-methyl fentanyl	1.56–200	0.9992	292	520
Butyryl fentanyl	1.56–200	0.9979	390	585
Isobutyryl fentanyl	1.56–200	0.9920	260	334
Ortho-methylfentanyl	1.56–200	0.9993	334	936
Meta-methylfentanyl	1.56–200	0.9999	334	936
Para-methylfentanyl	1.56–200	0.9995	292	520
3-Fluorofentanyl	1.56–200	0.9967	292	585
Ortho-fluorofentanyl	1.56–200	0.9999	260	585
Meta-fluorofentanyl	1.56–200	1.0000	260	780
Para-fluorofentanyl	3.12–200	0.9996	585	1040
4-Fluoroisobutyryl fentanyl	1.56–200	0.9999	292	790
Ortho-fluoroisobutyryl fentanyl	1.56–200	0.9998	292	585
Ortho-Fluorobutyryl fentanyl	1.56–200	0.9991	390	585
Meta-Fluorobutyryl fentanyl	1.56–200	0.9999	260	936
Para-Fluorobutyryl fentanyl	1.56–200	0.9965	334	520

Determination coefficients were calculated by analyzing each sample using VUV. Limits of detection are based on three times signal-to-noise [51].

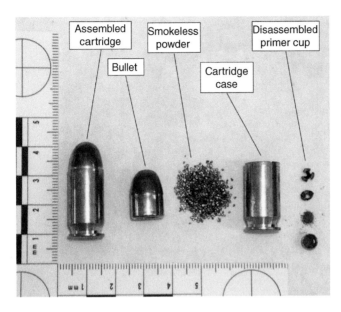

Figure 5.18 A fully assembled cartridge compared to a disassembled cartridge (Winchester 230 grain 0.45 ACP).

Despite its legitimate purpose, commercially available smokeless powders can be used in improvised explosive devices (IEDs). Out of 289 bombing incidents in 2018, 131 incidents involved the use of an IED, and many of these IEDs use smokeless powders as the main charge [70]. However, because IED explosions do not consume all of the smokeless powder particles, unburnt smokeless powder residue can be collected from a crime scene and analyzed in the lab [69].

Smokeless powders consist primarily of nitrocellulose (NC), also called guncotton, and various additives. Additives may include deterrents, stabilizers, suppressants, plasticizers; wear reduction additives, and taggants [71–74]. Double base powders also use nitroglycerin (NG) as another propellant in addition to NC [69, 75]. These additives can be extracted from the NC matrix and analyzed in the lab using techniques such as GC/MS, liquid chromatography–mass spectrometry (LC–MS), Fourier-transform infrared spectroscopy (FTIR), or capillary electrophoresis (CE). While smokeless powder can be fully dissolved in solvents like acetone, it is advantageous to inhibit NC dissolution during extraction, as NC can contaminate laboratory instruments [67, 69, 75–78]. With an optimized extraction method, a database of additive profiles can be developed and used for identification of unknown smokeless powder residues, such as the Smokeless Powders Database developed by the National Center for Forensic Sciences. [69, 79].

Figure 5.19 A deconstructed IED using smokeless powder. Image provided by Dr. Peter Diaczuk, John Jay College of Criminal Justice, CUNY.

As a proof of concept, four smokeless powders containing various additive profiles were purchased and analyzed by VUV Analytics, Inc. An extraction procedure adapted from Reardon and MacCrehan (2001) was used to isolate additives from the nitrocellulose matrix, and the extracts were analyzed by GC/VUV. Additives were successfully identified and characterized by their unique spectral features, as shown in Figures 5.19–5.22. Similar work has been demonstrated by Cruse and Goodpaster (2019) in which explosive compounds and additives were characterized by GC/VUV [80].

By combining GC/VUV and GC/MS, investigators can more accurately identify compounds in unburned particles and post-blast debris, allowing for additional confidence in smokeless powder identification.

5.7 Analysis of Lipstick

Evidence of lipstick use dates back as far back as 3500 BCE, where crushed red rocks mixed with white lead were used as lip colorings [81]. Modern lipsticks have a more sophisticated design: wax is used as a base to provide structure, various organic dyes and inorganic pigments provide the desired color, and various other additives can be used to improve smell, appearance, or shelf life of the product [82]. Samples of lipstick and other cosmetic products can be valuable pieces of

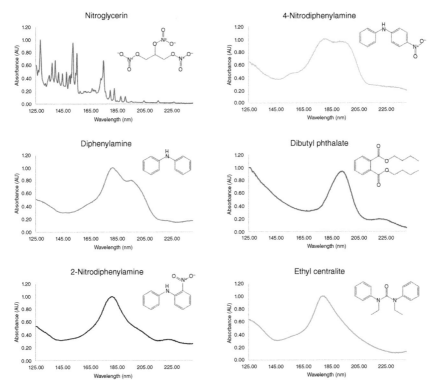

Figure 5.20 VUV spectra and structures of additives found in smokeless powders. Note that the spectrum for nitroglycerin contains several sharp absorbance features caused by the decomposition of nitrogcerin [71]. The decomposition products are quite small and have several sharp absorption features not seen in other, larger molecules [18].

evidence at a crime scene; even if DNA is not found, the lipstick itself may be useful to investigators (Figure 5.23).

On a basic level, lipsticks can be discriminated by color and texture; additionally, lipsticks may contain components in the wax matrix, such as glitter particles, that can be characterized microscopically [83]. Lipsticks have also been analyzed by methods such as thin-layer chromatography (TLC), pyrolysis GC/MS, HPLC, Raman spectroscopy, and FTIR [82, 84–90]. Although pyrolysis is useful for breaking apart large molecules such as polymers, a standard GC/MS analysis of a lipstick extract can yield a profile usable for distinguishing lipsticks [91, 92]. Many components of lipsticks are too nonvolatile for GC analysis, but some GC-amenable compounds can be extracted and analyzed. Like with GC/MS, these compounds can be identified by GC/VUV.

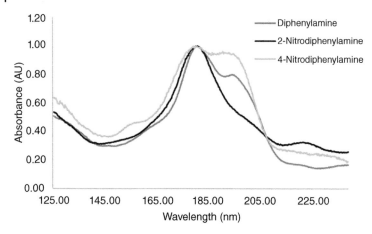

Figure 5.21 VUV spectra of diphenylamine, 2-nitrodiphenylamine, and 4-nitrodiphenylamine. Despite having similar structures, each of these additives has unique spectral features which can be exploited for identification and quantification.

An in-house experiment was performed to demonstrate the utility of GC/VUV for lipstick characterization. Samples of seven lipsticks were extracted using 90 : 10 hexane: dichloromethane for one hour, filtered through a 0.45 µm syringe filter, and analyzed by GC/VUV and GC/MS on a 30 m 100% PDMS column. Each lipstick's extract profile was distinguishable from the others, but there were several compounds in each profile that couldn't be fully identified by MS or VUV alone. Consider the following profile in Figure 5.24.

When examining the major peak at 8.83 minutes, the top hit from MS is 1-eicosanol, but it only has a 5% probability of correct identification. Searching the VUV library, the VUV spectrum most closely matches behenic acid methyl ester. However, this compound has a retention index of 2511, but it elutes before tetracosane (C24), which has a retention index of 2400 [15]. However, both the mass spectrum and VUV spectrum can be used to classify the compound. The mass spectrum has large peaks that differ by 14, indicating the presence of a linear alkane functional group. This is confirmed by VUV, as linear alkanes absorb strongly with the same spectral shape around 125–160 nm [30]. However, since the mass spectral peaks do not exactly match those of a linear alkane, it is likely that another functional group is present, and because the VUV spectrum is matching well with other fatty acid methyl esters (FAMEs), the peak is likely a FAME or another similar compound.

VUV can also take advantage of spectral filters to quickly deduce the types of compounds present in a sample. These lipstick samples were analyzed using a VGA-101 with a 125–430 nm range. The chromatogram obtained using the 125–430 nm filter is shown in Figure 5.25. However, more information can be obtained by using spectral filters specific to certain types of compounds;

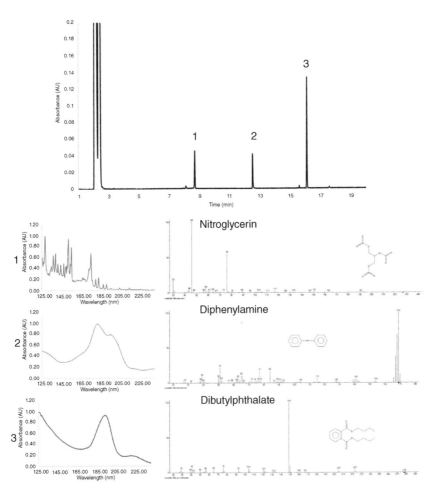

Figure 5.22 Example analysis of a smokeless powder extract. Each additive compound can be identified by both VUV and MS.

Figure 5.23 Simulated evidence containing traces of a cosmetic product. While cosmetic product samples may contain DNA evidence, the chemical profile of the product may also be useful to investigators. Image provided by Dr. Peter Diaczuk, John Jay College of Criminal Justice, CUNY.

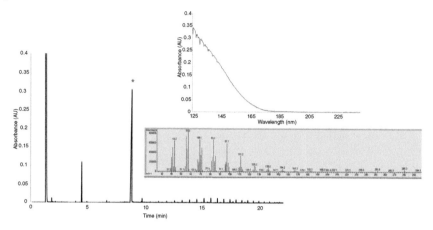

Figure 5.24 Extract profile of one lipstick sample, along with the mass spectrum and VUV spectrum of the major peak at 8.83 minutes. Neither the mass spectrum nor VUV spectrum gives good identifications. However, using information from both sources, it is possible to deduce some of the compound's properties.

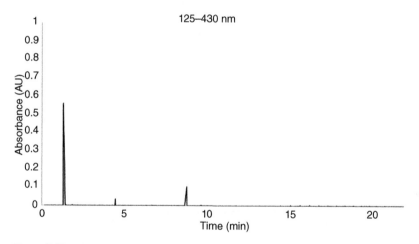

Figure 5.25 Extract profile of one lipstick sample at 125–430 nm. This filter averages absorbance over the whole wavelength range, so if compounds only absorb in a certain portion of the range, only major peaks will be visible.

for example, saturated compounds absorb strongly around 125–160 nm, while compounds containing pi bonds absorb strongly in the 170–240 nm range. Applying the 125–160 and 170–205 nm filters yields the chromatograms listed in Figures 5.26–5.28.

Even if cosmetic samples like lipstick may prove challenging for MS or VUV alone, using both approaches allows for a much more informative analysis.

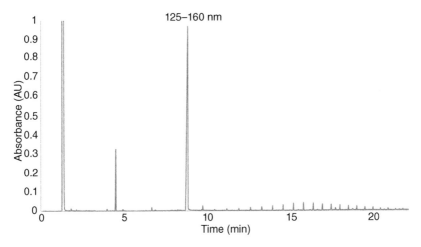

Figure 5.26 Extract profile of one lipstick sample at 125–160 nm. Using this filter, many more peaks appear. These compounds absorb strongly from 125 to 160 nm but not at higher wavelengths, indicating compounds that are saturated.

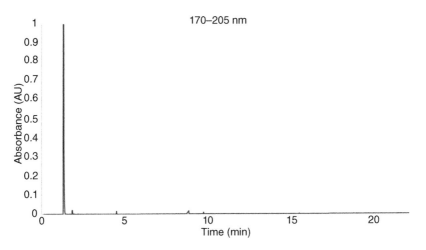

Figure 5.27 Extract profile of one lipstick sample at 170–205 nm. Using this filter, there are fewer peaks visible. The peaks that are visible are also visible at 125–160 nm, but many of them are smaller, indicating that these compounds absorb more strongly at 125–160 nm. Thus, these compounds likely contain few, if any, pi bonds.

5.8 Analysis of Blood Alcohol Content and Inhalants

In 2016, around 10,500 people died in alcohol-impaired driving crashes, accounting for 28% of traffic-related deaths in the United States, and an estimated 29

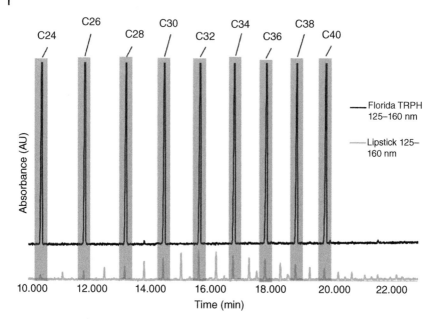

Figure 5.28 Extract profile of one lipstick sample at 125–160 nm compared to a linear alkanes standard. Given the MS and spectral information available for this lipstick, it can be deduced that it contains petroleum-based wax, distinguishing it from other lipsticks that are made with beeswax, for example.

people each day die in motor vehicle crashes that involve an alcohol-impaired driver [93]. Alcohol is a depressant that impairs brain function, affecting the judgment and muscle coordination necessary for safe driving. It is absorbed through the walls of the stomach and small intestines into the bloodstream. In small quantities, alcohol can be safely metabolized in the liver without noticeable effects on a person's behavior. However, when consumed in excess, the concentration of alcohol in the blood may significantly increase faster than the liver can metabolize it, resulting in an elevated blood alcohol content (BAC, also called blood alcohol concentration) that impairs driving ability. Most states have designated 0.08 g/dl of ethanol in the bloodstream as the legal BAC limit while driving; however, concentrations as low as 0.02 g/dl have demonstrable effects on brain function [94–96].

For screening purposes, a person's BAC can be measured with a breathalyzer; for more comprehensive testing, a person's blood or urine may be analyzed using techniques such as GC-FID, GC/MS, HPLC, chemical assays, and enzymatic assays [97]. Of the various options available, a traditional choice is a simultaneous dual-column GC-FID analysis. Two different columns, such as the Restek Rtx®-BAC Plus 1 and -BAC Plus 2 columns, have slightly different stationary phases and thus different retention times for ethanol which can be used for

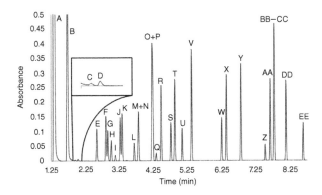

A	Oxygen	I	Acetonitrole	Q	2-Butanol	Y	Tetrachloroethane
B	Water	J	Dichloromethane	R	Chloroform	Z	Unknown
C	Acetaldehyde	K	t-Butanol	S	Isobutyl alcohol	AA	Ethylbenzene
D	Methanol	L	n-Hexane	T	Benzene	BB	m-Xylene
E	Ethanol	M	n-Propanol	U	Fluorobenzene	CC	p-Xylene
F	Propionaldehyde	N	Halothane	V	Trichloroethane	DD	o-Xylene
G	Acetone	O	Ethyl acetate	W	Isoamyl alcohol	EE	4-Bromofluorobenzene
H	2-Propanol	P	2-Butanone	X	Toluene		

Figure 5.29 Example chromatogram from the sample spiked with 22 μl of the standard solution.

identification [98, 99]. However, as with controlled substances, techniques such as MS and VUV provide more discriminatory information than retention times alone [16, 17, 100]. Additionally, a setup using these techniques can also be applied to abused inhalants in blood [101]. Inhalants are compounds that may be inhaled, either accidentally or intentionally, from commercially available products such as gasoline, refrigerants, solvents, or adhesives. Inhalant compounds include aliphatic, aromatic, halogenated, and oxygenated compounds [98].

BAC and inhalant analysis can be achieved using VUV and MS using headspace sample introduction. To demonstrate this analysis, an experiment to measure BAC and inhalant compounds was performed at VUV Analytics, Inc. The full experimental details can be found in the publication by Diekmann et al. [102].

All compounds in the standard were identified using VUV spectroscopy in under nine minutes, as shown in Figure 5.29. Three sets of coelutions occurred – n-propanol/halothane, ethyl acetate/2-butanone, and m-xylene/p-xylene – but each was spectrally deconvolved and quantified.

Ethanol was successfully quantified over a range 0.009–0.495% with a determination coefficient of 0.997 as shown in Table 5.2, indicating that BAC concentrations around the legal limit of 0.08% can be accurately determined with this method. Additionally, most BAC and nonaromatic inhalant compounds also had determination coefficients over 0.99, while the heavier aromatic inhalant

Table 5.2 Quantification results of ethanol and BAC/inhalant compounds.

Compound	Retention time (minutes)	Concentration range (mg/dl)	Limit of detection (mg/dl)	R^2
Methanol	2.00	38–99	8.8	0.981
Ethanol	2.58	9–495	3.1	0.997
Propionaldehyde	2.84	2.4–99	0.63	0.998
Acetone	2.90	2.4–99	0.34	0.996
2-Propnaol	2.99	2.4–99	0.54	0.998
Acetonitrile	3.11	11–99	5.5	0.998
Dichloromethane	3.25	2.4–99	0.63	0.994
t-Butanol	3.32	2.4–99	0.70	0.998
n-Hexane	3.66	2.4–99	2.8	0.822
n-Propanol	3.80	2.4–99	1.3	0.997
Halothane	3.80	2.4–99	1.0	0.983
Ethyl acetate	4.20	2.4–99	0.44	0.998
2-Butanone	4.20	2.4–99	0.36	0.998
2-Butanol	4.31	11–99	4.5	0.998
Chloroform	4.44	2.4–99	0.89	0.991
Isobutyl alcohol	4.74	2.4–99	0.37	0.998
Benzene	4.85	0.5–24	0.10	0.987
Trichloroethene	5.34	0.5–24	0.12	0.981
Isoamyl alcohol	6.22	2.4–99	0.44	0.997
Toluene	6.35	0.5–24	0.098	0.982
Tetrachloroethene	6.78	0.5–24	0.11	0.975
Ethylbenzene	7.64	0.5–24	0.11	0.977
m-Xylene	7.74	0.5–24	0.12	0.972
p-Xylene	7.74	0.5–24	0.11	0.979
o-Xylene	8.10	0.5–24	0.10	0.979

Ethanol and many lighter BAC/inhalant compounds had determination coefficients above 0.99, indicating highly linear data. Even though the heavier inhalant compounds did not have determination coefficients above 0.99, they have detection limits down to the low mg/dl level, which is still useful for screening of blood samples.

compounds had determination coefficients under 0.99, indicating that only some BAC and inhalant compounds can be quantified with this method. It is suspected that the heaver compounds did not partition to the headspace as efficiently as the more volatile compounds, making quantification using headspace sample introduction difficult. However, other work has shown similar results for some heavier compounds. Additionally, because inhalant compounds are not controlled substances, many labs only identify these compounds in samples rather than quantify them [101, 103–105]. To this end, headspace GC/VUV has proven to be a viable alternative to traditional methods, as it provides spectral identification of BAC and inhalant compounds. Although it was not paired with MS in this study, using this method with a tandem GC/VUV/MS approach would provide further confirmation of the BAC and inhalant compounds, giving more useful data than what could be achieved with a traditional dual-column GC-FID analysis.

5.9 Analysis of Fire Debris Samples

From the burning of the Library of Alexandria to the Great Chicago Fire, destructive fires have plagued civilization since ancient times. For most of history, eyewitnesses were some of the only sources of information available to determine the cause of fires. In the twentieth century, investigators began to examine fire scenes more closely; they could smell traces of ignitable liquids in ashes placed in glass jars, and others could sometimes identify them using taste by chewing on a piece of bread pressed on burned material. Modern investigations generally rely on gas chromatographic analysis for fire investigations [106].

A common tool used in fire investigations is fire debris analysis. As the name implies, debris from the scene of a fire can be collected and brought to the lab to extract as much information as possible. Despite the fire's destructive nature and comparably intense conditions of fire extinguishment, traces of ignitable liquids may remain in debris at the scene, called fire debris [106]. Once collected, ignitable liquid traces can be extracted from the debris and characterized. Methods of extraction and sampling include solvent extraction, passive or direct headspace sampling, and solid-phase microextraction (SPME); analysis methods include GC-FID, GC/MS, and GCxGC/MS [106–113].

Currently, pattern matching of chromatograms is the most common way to characterize fire debris samples. Additionally, if a detector such as MS is used, certain peaks may be identified to provide additional information [106, 109]. However, adding VUV to fire debris analysis is especially advantageous: while VUV spectral profiles are able to identify specific compounds like MS, it is also possible to classify compounds by their spectral shape even if they are not in the VUV library. For example, linear/branched alkanes (paraffins/isoparaffins),

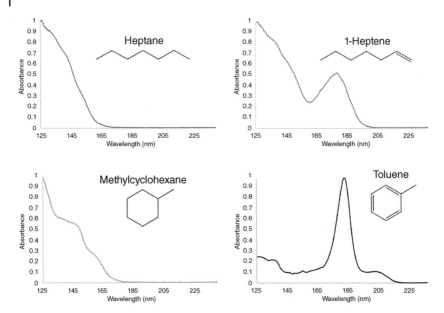

Figure 5.30 Spectral shapes of linear/branched alkanes (heptane), alkenes (1-heptene), cycloalkanes (methylcyclohexane), and aromatics (toluene). While each compound's spectrum is unique, compounds in the same class will have similarly shaped spectra. For example, the spectra in Figures 5.4–5.6 (hexane, 1-hexene, and benzene) share the same general spectral shapes as heptane, 1-heptene, and toluene, respectively.

alkenes (olefins), cycloalkanes (naphthenes), and aromatics can be classified by their spectral shapes, as shown in Figures 5.30 and 5.31 [30].

In an experiment performed by VUV Analytics, Inc., 12 different ignitable liquids – including six gasolines, three diesels, paint thinner, charcoal lighter fluid, and mineral spirits – were characterized by GC/VUV under a single method. Neat samples of each ignitable liquid were analyzed by GC/VUV, and unique chromatograms for each were obtained. Like with MS, it is possible to use these chromatograms to distinguish ignitable liquids, and it is possible to identify compounds in each sample using VUV, as shown in Figure 5.32. Fire debris analysis using GC/VUV has great potential, as fuels analysis is a strong and well-established application of VUV [30].

Although fuel matrices can be incredibly complex, VUV's ability to classify compounds by spectral shape provides additional information to investigators. To demonstrate this potential, one gasoline sample was used to partially burn a piece of wood, and the resulting debris was extracted according to ASTM E1412-16 [112]. The extract was analyzed by GC/VUV and characterized using the PIONA classification described in ASTM D8071-19 [14] (Figure 5.33). With enough characterization of weathered samples of ignitable liquids, it may be possible to

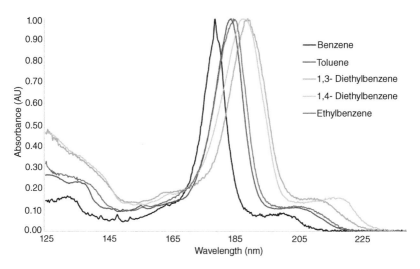

Figure 5.31 Spectral shapes of five aromatic compounds. While each compound maintains the same general aromatic spectral shape, each spectrum is distinguishable.

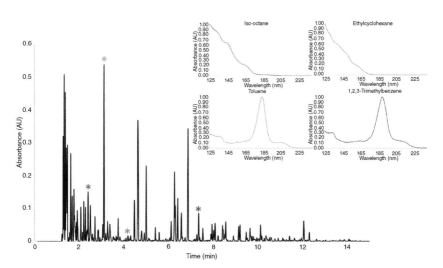

Figure 5.32 Profile of a gasoline sample analyzed by GC/VUV with selected peaks highlighted. Like with MS, compounds can be matched against a spectral library for identification.

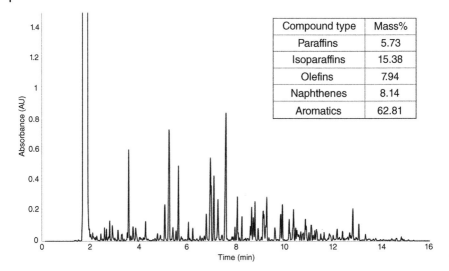

Figure 5.33 Characterization of trace fuel extracted from a piece of fire debris, including the mass% content of paraffins (linear alkanes), isoparaffins (branched alkanes), olefins (alkenes), naphthenes (cycloalkanes), and aromatics.

use a similar classification scheme to analyze samples from crime scenes, which has been demonstrated by Pina et al. [114].

5.10 Conclusion

Separately, both VUV and MS are powerful detection techniques that have value in a forensic laboratory, but the combination of VUV and MS can be greater than the sum of its parts. By using a column splitter to divert flow to both detectors simultaneously, more information can be obtained from a single injection. Because VUV and MS can uniquely identify compounds eluting from a GC, a tandem GC/VUV/MS approach provides VUV confirmation on top of MS confirmation, allowing greater confidence in results presented in court. Additionally, since VUV and MS often cover each other's weaknesses, GC/VUV/MS can be used for more types of analysis than either VUV or MS could accomplish alone.

References

1 Bartle, K.D. and Myers, P. (2002). History of gas chromatography. *Trends Anal. Chem.* 21: 547–557.

2 Brochmann-Hanssen, E. (1962). Gas chromatography and its application to pharmaceutical analysis. *J. Pharm. Sci.* 51 (11): 1017–1031.

3 Wilmshurst, J.R. (1964). Gas-chromatographic analysis of polynuclear arenes. *J. Chromatogr.* 17: 50–59.

4 Creveling, R.K. and Jennings, W.J. (1970). Volatile components of Bartlett pear: higher boiling esters. *J. Agric. Food. Chem.* 18 (1): 19–24.

5 Schug, K.A., McNair, H.M., and Hinshaw, J.V. (2015). GC detectors: from thermal conductivity to vacuum ultraviolet absorption. *LC-GC Eur.* 28 (1): 45–50.

6 Ettre, L.S. (2002). The invention, development, and triumph of the flame ionization detector. *LC-GC North Am.* 20 (1): 48–60.

7 Wiley, W.C. (1956). Bendix time-of-flight mass spectrometer. *Sci.* 124 (3226): 817–820.

8 Gohlke, R.S. (1959). Time-of-flight mass spectrometry and gas-liquid partition chromatography. *Anal. Chem.* 31 (4): 535–541.

9 Hinshaw, J.V. (2018). A compendium of GC detection, past and present. *LC-GC North Am.* 36 (3): 178–182.

10 Heller, S.R. (1999). The history of the NIST/EPA/NIH mass spectral database. *Today Chem. Work* 8 (2): 45–46, 49–50.

11 Harris, D.C. (2010). *Quantitative Chemical Analysis: Mass Spectrometry*, 8e. New York: W.H. Freeman and Co.

12 Sneddon, J., Masuram, S., and Richert, J.C. (2007). Gas chromatography-mass spectrometry – basic principles, instrumentation and selected applications for detection of organic compounds. *Anal. Lett.* 40 (6): 1003–1012.

13 Hübschmann, H. (2009). *Handbook of GC/MS: Fundamentals and Applications: Fundamentals*, 2e. Weinheim: Wiley-VCH.

14 Taylor, T. (2012). Electron ionization for GC/MS. *LCGC North Am.* 30 (4): 358.

15 Wallace, W.E. (2019). Mass spectra. In: *NIST Chemistry WebBook*, NIST Standard Reference Database Number 69 (ed. P.J. Linstrom and W.G. Mallard). Gaithersburg, MD: National Institute of Standards and Technology.

16 ASTM International (2017). *ASTM E2329–17, Standard Practice for Identification of Seized Drugs*. West Conshohocken, PA: ASTM International.

17 SWGDRUG (2016). Scientific Working Group for the Analysis of Seized Drugs (SWGDRUG). *Recommendations, Version 7.1*.

18 Schug, K.A., Sawicki, I., Carlton, D.D. Jr. et al. (2014). Vacuum ultraviolet detector for gas chromatography. *Anal. Chem.* 86 (16): 8329–8335.

19 Winick, H. (1987). Synchrotron radiation. *Sci. Am.* 257 (5): 88–101.

20 Middletich, B.S., Sung, N., Zlatkis, A., and Settembre, G. (1987). Trace analysis of volatile polar organics by direct aqueous injection gas chromatography. *Chromatography* 23 (4): 273–278.

21 Duffy, M., Driscoll, J.N., and Pappas, S. (1988). Capillary gas chromatographic analysis with the far-UV absorbance detector. *J. Chromatogr.* 441 (1): 63–71.

22 Lagesson, V., Lagesson-Andrasko, L., Andrasko, J., and Baco, F. (2000). Identification of compounds and specific functional groups in the wavelength region 168-330 nm using gas chromatography with UV detection. *J. Chromatogr. A* 867 (1): 187–206.

23 Hatzinikolaou, D.G., Lagesson, V., Stavridou, A.J. et al. (2006). Analysis of the gas phase of cigarette smoke by gas chromatography coupled with UV-diode Array detection. *Anal. Chem.* 78 (13): 4905–4516.

24 Nilsson, A., Lagesson, V., Bornehag, C. et al. (2005). Quantitative determination of volatile organic compounds in indoor dust using gas chromatography-UV spectrometry. *Environ. Int.* 31 (8): 1141–1148.

25 Andrasko, J., Lagesson-Andrasko, L., Dahlén, J., and Jonsson, B. (2017). Analysis of explosives by GC-UV. *J. Forensic Sci.* 62 (4): 1022–1027.

26 Au, J.W., Cooper, G., Burton, G.R. et al. (1993). The valence shell photoabsorption of the linear alkanes, C_nH_{2n+2} (n = 1-8): absolute oscillator strengths (7-220 eV). *Chem. Phys.* 173 (2): 209–239.

27 Lombos, B.A., Sauvageau, P., and Sandorfy, C. (1967). The electronic spectra of n-alkanes. *J. Mol. Spectrosc.* 24 (1): 253–269.

28 Keller-Rudek, H., Moortgat, G.K., Sander, R., and Sörenson, R. (2013). The MPI-Mainz UV/Vis spectral atlas of gaseous molecules of atmospheric interest. *Earth Syst. Sci. Data* 5 (2): 365–373.

29 Bai, L., Smuts, J., Walsh, P. et al. (2015). Permanent gas analysis using gas chromatography with vacuum ultraviolet detection. *J. Chromatogr. A* 1388: 244–250.

30 Hodgson, A., Cochran, J., Diekmann, J., and Schonert, R. (2019). Gas chromatography-vacuum ultraviolet spectroscopy: a versatile tool for analysis of gasoline and jet fuels. *Chromatogr. Today* 18–23.

31 Johnson, P. (2016). Novel detector dramatically simplifies PIONA analysis. *Petroleum Industry News.* https://www.petro- http://online.com/news/analytical-instrumentation/11/vuv-analytics-inc/novel-detector-dramatically-simplifies-piona-analysis/37824 (accessed 16 March 2022).

32 Schonert, R., Wispinski, J.D., and Cochran, J. (2019). Fast analysis of non-traditional gasoline additives with gas chromatography-vacuum ultraviolet spectroscopy. *Chromatogr. Today* 8–13.

33 Bai, L., Smuts, J., Schenk, J. et al. (2018). Comparison of GC/VUV, GC-FID, and comprehensive two-dimensional GC/MS for the characterization of weathered and unweathered diesel fuels. *Fuel* 214: 521–527.

34 Qui, C., Cochran, J., Smuts, J. et al. (2017). Gas chromatography-vacuum ultraviolet detection for classification and speciation of polychlorinated biphenyls in industrial mixtures. *J. Chromatogr. A* 1490: 191–200.

35 Fan, H., Smuts, J., Walsh, P. et al. (2015). Gas chromatography-vacuum ultraviolet spectroscopy for multiclass pesticide identification. *J. Chromatogr. A* 1389: 120–127.

36 Hodgson, A. and Cochran, J. (2019). Vacuum ultraviolet spectroscopy as a new tool for GC analysis of terpenes in flavors and fragrances. *J. AOAC Int.* 102 (2): 655–658.

37 de Koning, S. and Ram, L. (2016). Improved analysis of FAMEs using gas chromatography with vacuum ultraviolet detection (GC/VUV). *The Column* 3 (12): 25–29.

38 Santos, I.C., Smuts, J., Crawford, M.J. et al. (2019). Large-volume injection gas chromatography-vacuum ultraviolet spectroscopy for the qualitative and quantitative analysis of fatty acids in blood plasma. *Anal. Chim. Acta* 1053: 169–177.

39 Shear-Laude, L. (2017). Gas chromatography-vacuum ultraviolet absorbance spectroscopy for the quantitative determination of trace and bulk water in organic solvents. *The Column* 13 (2): 9–13.

40 Santos, I.C., Smuts, J., Choi, W. et al. (2018). Analysis of bacterial FAMEs using gas chromatography-vacuum ultraviolet spectroscopy for the identification and discrimination of bacteria. *Talanta* 182: 536–543.

41 Zoccali, M., Schug, K.A., Walsh, P. et al. (2017). Flow-modulated comprehensive two-dimensional gas chromatography combined with a vacuum ultraviolet detector for the analysis of complex mixtures. *J. Chromatogr. A* 1497: 135–143.

42 Dunkle, M.N., Pijcke, P., Winniford, B., and Bellos, G. (2019). Quantification of the composition of liquid hydrocarbon streams: comparing the GC/VUV to DHA and GCxGC. *J. Chromatogr. A* 1587: 239–246.

43 ASTM D8071-19 (2017). *Standard Test Method for Determination of Hydrocarbon Group Types and Select Hydrocarbon and Oxygenate Compounds in Automotive Spark-Ignition Engine Fuel Using Gas Chromatography with Vacuum Ultraviolet Absorption Spectroscopy Detection (GC/VUV)*. West Conshohocken, PA: ASTM International.

44 Schenk, J., Mao, J.X., Smuts, J. et al. (2016). Analysis and deconvolution of Dimethylnaphthalene isomers using gas chromatography vacuum ultraviolet spectroscopy and theoretical computations. *Anal. Chim. Acta* 945: 1–8.

45 Johnson, P., Scussel, M., Patzelt, B. et al. (2017). VUV PIONA+™ Improved Accuracy of Hydrocarbon Reporting in Gasoline. *Petro Industry News*. https://www.petro-online.com/article/measurement-and-testing/14/vuv-analytics-inc/pvuv-pionasupsup-improves-accuracy-of-hydrocarbon-reporting-in-gasolinenbspp/2255 (accessed 16 March 2022).

46 Samokhin, A. (2018). Spectral skewing in gas chromatography-mass spectrometry: misconceptions and realities. *J. Chromatogr. A* 1576: 113–119.

47 ASTM D6730-01 (2016). *Standard Test Method for Determination of Individual Components in Spark Ignition Engine Fuels by 100–Metre Capillary (with Precolumn) High-Resolution Gas Chromatography*. West Conshohocken, PA: ASTM International.

48 Fialkov, A.B., Steiner, U., Lehotay, S., and Amirav, A. (2007). Sensitivity and noise in GC/MS: achieving low limits of detection for difficult analytes. *Int. J. Mass spectrom.* 260 (1): 31–48.

49 Anthony, I.G.M., Brantley, M.R., Gaw, C.A. et al. (2018). Vacuum ultraviolet spectroscopy and mass spectrometry: a tandem detection approach for improved identification of gas chromatography-eluting compounds. *Anal. Chem.* 90 (7): 4878–4885.

50 Anthony, I.G.M., Brantley, M.R., Floyd, A.R. et al. (2018). Improving accuracy and confidence of chemical identification by gas chromatography/vacuum ultraviolet spectroscopy-mass spectrometry: parallel gas chromatography, vacuum ultraviolet, and mass spectrometry library searches. *Anal. Chem.* 90 (20): 12307–12313.

51 Buchalter, S., Marginean, I., Yohannan, J., and Lurie, I.S. (2019). Gas chromatograph with tandem cold electron ionization mass spectrometric detection and vacuum ultraviolet detection for the comprehensive analysis of fentanyl analogues. *J. Chromatogr. A* 1596: 183–193.

52 Fuster, D. and Muga, R. (2018). The opioid crisis. *Med. Clin.* 151 (12): 487–488.

53 Stanley, T.H. (2014). The fentanyl story. *J. Pain* 15 (12): 1215–1226.

54 Armenian, P., Vo, K.T., Barr-Walker, J., and Lynch, K.L. (2018). Fentanyl, fentanyl analogs and novel synthetic opioids: a comprehensive review. *Neuropharmacology* 134: 121–132.

55 George, A.V., Lu, J.J., Pisano, M.V. et al. (2010). Carfentanil – an ultra potent opioid. *Am. J. Emerg. Med.* 28 (4): 530–532.

56 United States Government Printing Office (2016). United States Code Title 21, Chapter 13, Subchapter I, Part A, Section 802. United States Government Printing Office.

57 Drug Enforcement Agency (2017). *Drugs of Abuse*. U.S. Department of Justice.

58 United States Court of Appeals (2005). *United States v. Brown, 415 F.3d 1257*. United States Court of Appeals, 11th Circuit – Alabama.

59 United States Court of Appeals (2004). *United States v. Roberts, 363 F.3d 118*. United States Court of Appeals, 2nd Circuit – New York.

60 United States District Court (1992). *United States v. Forbes, 806 F. Supp. 232*. United States District Court – District of Colorado.

61 Cayman Chemical. (2018). Crotonyl Fentayl. *SWGDRUG Mass Spectral Library*. SWGDRUG. http://www.swgdrug.org/ms.htm (accessed 16 March 2022).

62 Drug Enforcement Administration (2018). Cyclopropylfentanyl. *SWGDRUG Mass Spectral Library*. SWGDRUG. http://www.swgdrug.org/ms.htm (accessed 16 March 2022).

63 Lurie, I.S., Tremeau-Cayel, L., and Rowe, W.F. (2017). Recent advances in comprehensive chromatographic analysis of emerging drugs. *LC-GC North Am.* 35 (12): 878–883.

64 Kranenburg, R.F., García-Cicourel, A.R., Kukurin, C. et al. (2019). Forensic Science International, "Distinguishing Drug Isomers in the Forensic Laboratory: GC/VUV in Addition to GC/MS for Orthogonal Selectivity and the Use of Library Match Scores as a New Source of Information. 302,109900.

65 Deng, Y. (2011). *The Four Major Inventions: Ancient Chinese Inventions*. Cambridge: Cambridge University Press.

66 Koller, L.R. (1959). *Work and War: The Fireside Book of Guns*. New York: Simon & Schuster.

67 Dahl, D.B. and Lott, P.F. (1987). Determination of black and smokeless powder residues in firearms and improvised explosive devices. *Microchem. J.* 35 (1): 40–50.

68 Robinson, C.S. (1943). *Ignition of Propellant Powders: The Thermodynamics of Firearms*. New York, NY: McGraw Hill.

69 Moorehead, W. (2007). Characterization of smokeless powders. In: *Forensic Analysis on the Cutting Edge: New Methods for Trace Evidence Analysis*, Chapter 10 (ed. R.D. Blackledge), 241–268. Hoboken, NJ: Wiley.

70 United States Bomb Data Center (2018). *Explosives Incident Report (EIR): 2018*. United States Department of Justice. https://www.atf.gov/file/128106/download (accessed 16 March 2022).

71 Davis, T.L. (1943). *Smokeless Powders: The Chemistry of Powder and Explosives*. Hollywood, CA: Angriff Press.

72 Heramb, R.M. and McCord, B.R. (2002). The manufacture of smokeless powders and their forensic analysis: a brief review. *Forensic Sci. Commun.* 4 (2): 1–4.

73 Ward, J.R. and Brosseau, T.L. (1980). The reduction of barrel erosion by wear-reducing additives. *Wear* 60 (1): 145–155.

74 National Research Council (1998). *Black and Smokeless Powders: Technologies for Finding Bombs and the Bomb Makers: Identification*. Washington, DC: National Research Council, National Academy Press.

75 MacCrehan, W.A., Reardon, M.R., and Duewer, D.L. (2002). A quantitative comparison of smokeless powder measurements. *J. Forensic Sci.* 47 (6): 1283–1387.

76 Hopper, K.G. and McCord, B.R. (2005). A comparison of smokeless powders and mixtures by capillary zone electrophoresis. *J. Forensic Sci.* 50 (2): 307–315.

77 MacCrehan, W.A., Reardon, M.R., and Duewer, D.L. (2002). Associating gunpowder and residues from commercial ammunition using compositional analysis. *J. Forensic Sci.* 47 (2): 260–266.

78 Reardon, M.R. and MacCrehan, W.A. (2001). Developing a quantitative extraction technique for determining the organic additives in smokeless handgun powder. *J. Forensic Sci.* 46 (4): 802–807.

79 National Center for Forensic Sciences, University of Central Florida, SWGFEX (2006). *Smokeless Powders Database*. National Institute of Justice. http://www.ilrc.ucf.edu/powders (accessed 16 March 2022).

80 Cruse, C.A. and Goodpaster, J.V. (2019). Generating highly specific spectra and identifying thermal composition products via gas chromatography / vacuum ultraviolet spectroscopy (GC/VUV): application to nitrate ester explosives. *Talanta* 195: 580–596.

81 Shaffer, S.E. (2007). Reading our lips: the history of lipstick regulation in western seats of power. *Food Drug Law. J.* 62 (1): 165–225.

82 Wong, J.X.W., Sauzier, G., and Lewis, S.W. (2019). Forensic discrimination of lipsticks using visible and attenuated total reflectance infrared spectroscopy. *Forensic Sci. Int.* 298: 88–96.

83 Grieve, M.C. (1987). Glitter particles – an unusual source of trace evidence? *Forensic Sci. Soc.* 27: 405–412.

84 Barker, A.M.L. and Clarke, P.D.B. (1972). Examination of small quantities of lipstick. *J. Forensic Sci. Soc.* 12 (3): 449–451.

85 Andrasko, J. (1981). Forensic analysis of lipsticks. *Forensic Sci. Int.* 17 (3): 235–251.

86 Jasuja, O.P. and Singh, R. (2005). Thin-layer chromatographic analysis of liquid lipsticks. *J. Forensic Ident.* 55 (1): 28–35.

87 Bardner, P., Bertino, M.F., Weimer, R., and Hazelrigg, E. (2013). Analysis of lipsticks using Raman spectroscopy. *Forensic Sci. Int.* 232 (1): 67–72.

88 Zellner, M. and Quarino, L. (2009). Differentiation of twenty-one glitter lip-glosses by pyrolysis gas chromatography/mass spectroscopy. *J. Forensic Sci.* 54 (5): 1022–1028.

89 Salahioglu, F., Went, M.J., and Gibson, S.J. (2013). Application of Raman spectroscopy for the differentiation of lipstick traces. *Anal. Methods* 5 (20): 5392–5401.

90 Sharma, V., Bharti, A., and Kumar, R. (2019). On the spectroscopic investigation of lipstick stains: forensic trace evidence. *Spectrochim. Acta, Part A* 215: 48–57.

91 Roda, G., Arnoldi, S., Casagni, E. et al. (2019). Determination of polycyclic aromatic hydrocarbons in lipstick by gas-chromatography coupled to mass spectrometry: a case history. *J. Pharm. Biomed. Anal.* 165: 386–392.

92 Gładysz, M., Król, M., Własiuk, P. et al. (2018). Development and evaluation of semi-destructive, ultrasound assisted extraction method followed by gas chromatography coupled to mass spectrometry enabling discrimination of red lipstick samples. *J. Chromatogr. A* 1577: 92–100.

93 Centers for Disease Control and Prevention (2019). *Impaired Driving: Get the Facts. Centers for Disease Control and Prevention.* Centers for Disease Control and Prevention. https://www.cdc.gov/motorvehiclesafety/impaired_driving/impaired-drv_factsheet.html (accessed 16 March 2022).

94 National Highway Traffic Safety Administration (n.d.). *Drunk Driving.* United States Department of Transportation. https://www.nhtsa.gov/risky-driving/drunk-driving (accessed 16 March 2022).

95 National Institutes of Health (2007). *NIH Curriculum Supplement Series, Information about Alcohol.* National Institutes of Health. https://www.ncbi.nlm.nih.gov/books/NBK20360 (accessed 16 March 2022).

96 Governor's Highway Safety Administration (2019). *Alcohol Impaired Driving. Governor's Highway Safety Administration.* https://www.ghsa.org/state-laws/issues/alcohol%20impaired%20driving (accessed 16 March 2022).

97 Taglario, F., Lubli, G., Ghielmi, S. et al. (1992). Chromatographic methods for blood alcohol determination. *J. Chromatogr.* 580: 161–190.

98 Levine, B., Caplan, Y.H., and Jones, A.W. (2013). *Principles of Forensic Toxicology.* Washington, DC: AACC Press.

99 Boswell, H.A. and Dorman, F.L. (2015). Uncertainty of blood alcohol concentration (BAC) results as related to instrumental conditions: optimization and robustness of BAC analysis headspace parameters. *Chromatography* 2 (4): 791–708.

100 Tiscione, N.B., Alford, I., Yeatman, D.T., and Shan, X. (2011). Ethanol analysis by headspace gas chromatogrpahy with simultaneous Fame ionization and mass spectrometry detection. *J. Anal. Toxicol.* 35 (7): 501–511.

101 Wasfi, I.A., Al-Awadhi, A.H., Al-Hatali, Z.N. et al. (2004). Rapid and sensitive static headspace gas chromatography-mass spectrometry method for the analysis of ethanol and abused inhalants in blood. *J. Chromatogr. B* 799: 331–336.

102 Diekmann, J.A. III, Cochran, J., Hodgson, J.A., and Smuts, D.J. (2019). Quantitation and identification of ethanol and inhalant compounds in whole blood using static headspace gas chromatography vacuum ultraviolet spectroscopy. *J. Chromatogr. A* 1611: 460607.

103 McCarver-May, D.G. and Durisin, L. (1997). An accurate, automated, simultaneous gas chromatographic headspace measurement of whole blood ethanol and acetaldehyde for human in vivo studies. *J. Anal. Toxicol.* 21 (2): 134–141.

104 Broussard, L.A. (2000). The role of the laboratory in detecting inhalant abuse. *Clin. Lab. Sci.* 13 (4): 205–209.

105 Baydala, L. (2010). Inhalant abuse. *Pediat. Child. Health.* 15 (7): 443–448.

106 Stauffer, E., Dolan, J.A., and Newman, R. (2008). *Fire Debris Analysis.* Amsterdam, Netherlands: Elsevier.

107 Nowicki, J. (1991). Analysis of fire debris samples by gas chromatography/mass spectrometry (GC/MS): case studies. *J. Forensic Sci.* 36 (5): 1536–1550.

108 Abel, R.J., Zadora, G., Sandercock, P.M.L., and Harynuk, J.J. (2018). Modern instrumental limits of identification of ignitable liquids in forensic fire debris analysis. *Separat.* 5 (4): 58.

109 ASTM E1618-14 (2014). *Standard Test Method for* Ignitable Liquid Residues in Extracts from Fire Debris Samples by Gas *Chromatography-Mass Spectrometry.* West Conshohocken, PA: ASTM International, ASTM International.

110 ASTM E1386-15 (2015). *Standard Practice for Separation of Ignitable Liquid Residues from Fire Debris Samples by Solvent Extraction.* West Conshohocken, PA: ASTM International.

111 ASTM E1388-17 (2017). *Standard Practice for Static Headspace Sampling of Vapors from Fire Debris Samples.* ASTM International, West Conshohocken, PA.

112 ASTM E1412-16 (2016). *Standard Practice for Separation of Ignitable Liquid Residues from Fire Debris Samples by Passive Headspace Concentration with Activated Charcoal.* West Conshohocken, PA: ASTM International.

113 ASTM E2154-15a (2015). *Standard Practice for Separation and Concentration of Ignitable Liquid Residues from Fire Debris Samples by Passive Headspace Concentration with Solid Phase Microextraction (SPME).* West Conshohocken, PA: ASTM International.

114 Pina, S., Diekmann, J., Steen, T. et al. (2017). Vacuum ultraviolet (VUV) absorption spectroscopy: a novel method in the forensic analysis of fire debris evidence? *Presented at Gulf Coast Conference (GCC),* 2017.

6

Surface Acoustic Wave Nebulization-Mass Spectrometry

Roselina Medico[1] and Alina Astefanei[2]

[1] olam food ingredients (ofi) Department of Plant Science, Koog aan de Zaan, The Netherlands
[2] Van't Hoff Institute for Molecular Sciences, Faculty of Science, Analytical Chemistry Group, University of Amsterdam, Amsterdam, The Netherlands

6.1 Theory and Instrumentation

Currently, in designing lab equipment for chemistry and biology procedures on chip devices there is a trend of reducing the size to as small as possible. Different types of miniature devices especially designed for faster and cheaper analysis compared to the available conventional technologies are being introduced. The interest in portable devices is constantly increasing for many different relevant applications, including cell biology, forensic analysis, DNA, and protein analysis [1–3]. Currently, "lab on a chip" devices are mature and advanced technologies of different microfluidic systems.

Among these, ultrasonic nebulizers, also called atomizers, are very important due to their extensive applications such as drug delivery, mass spectrometry, humidity control, spray pyrolysis, nanoparticle synthesis, coatings, etc. [3–14]. Within this category, SAWN is a recent technology mostly used as ultra-fast microfluidics for atomization (i.e. drug delivery) and as an ionization technique in mass spectrometry [15–18].

It was shown that using SAWN, there is almost negligible denaturation of proteins or biomolecules during nebulization [7, 18, 19] and little damage on the viability, proliferation, and differentiation of primary osteoblast-like cells [20]. Another extremely attractive feature of SAW devices is their availability to quickly synthesize nanoparticles and to encapsulate therapeutic drugs within polymeric nanoparticles and pulmonary delivery systems using a one-step simple procedure [5, 16, 21, 22].

SAWN utilizes a nanometer-order amplitude acoustic wave that propagates along and near the surface of a single-crystal piezoelectric substrate actuation to

Leading Edge Techniques in Forensic Trace Evidence Analysis: More New Trace Analysis Methods, First Edition. Edited by Robert D. Blackledge.

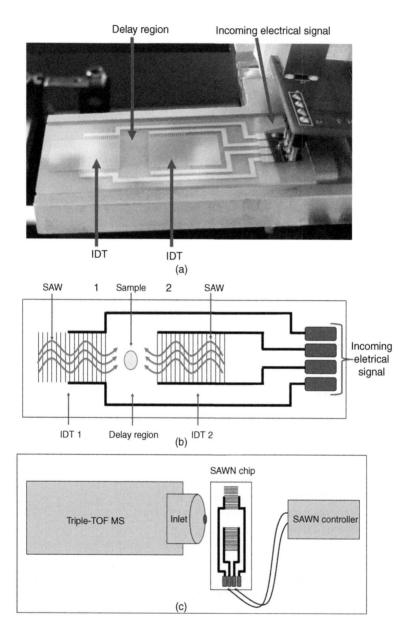

Figure 6.1 (a) Schematic of the commercially available SAWN chip (LiNbO$_3$), (b) the nebulization process, and (c) set up the SAWN device in front of the MS inlet. Source: Medico [24].

produce multiple-charged ions [12, 23]. It consists of two interdigital transducers (IDTs) which can be made of a chrome adhesion layer and gold, (Figure 6.1a) [24] with a piezoelectric substrate such as quartz, lithium tantalite (LiTaO$_3$), zinc oxide (ZnO), and lithium niobate (LiNbO$_3$). The commercially available version uses a lithium niobite substrate and it is operated at 9.6 MHz in dual mode (Figure 6.1) [25]. As can be seen in Figure 6.1, between the two IDTs there is a space (delay zone) where the liquid sample is placed for nebulization. A sinusoidal electrical signal is then applied to the chip through a radiofrequency (RF) connector. This signal is transformed through the piezoelectric effect produced by the IDT, into an acoustic wave that propagates through the chip. The IDTs are located on both extremities of the chip, which allows it to generate recirculation that induces the capillary waves on the surface of the droplet. The capillary waves subsequently destabilize, which produces a vertical nebulization (Figure 6.2). An acoustic streaming is created in the droplet [27], and plumes of droplets with ionized molecules are generated. By this process, the sample droplet is introduced into the inlet of the MS (Figure 6.3) where it is further desolvated and analyzed [12, 23, 28, 29]. The nebulization is an extremely fast process of only few seconds. The smaller the droplet size, the faster the nebulization.

Many aspects of the nebulization process are still poorly understood, but they are most likely linked to the high voltages that are present on the chip [26]. The limited understanding is to a great extent related to the extremely complicated

Figure 6.2 The nebulization of a droplet on the SAWN chip taken with a five figure µs flash light. Two droplet size regimes can be observed as indicated by the red arrows. Source: Kooij et al. [26]/Springer Nature/CC BY 4.0.

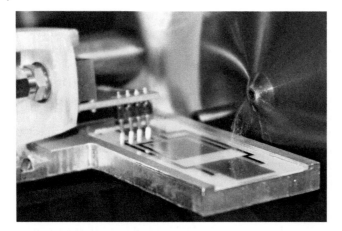

Figure 6.3 The nebulization of a droplet on a SAWN chip and the subsequent introduction into a mass spectrometer.

dynamics and to the speed at which this process takes place. Even less studied, and hence understood, is the size distribution of the formed aerosols. The droplet size distribution plays an important role for applications such as mass spectrometry where the presence of large droplets can lead to loss of sensitivity and if not properly desolvated may lead to vacuum fluctuations.

Recently, we have reported the first systematic study of droplet size distributions for SAWN and two other types of nebulizer technologies, working at different frequencies [26]. The nebulization mechanisms can be very different and the droplet sizes can be both narrowly or broadly distributed. Interestingly, we observed that the differences relate to the discrepancies in the wavelengths of the waves that contribute to the droplet formation. Interestingly, despite the order-of-magnitude differences in the parameters of different nebulizers, the median droplet size changes with the capillary wavelength, with a proportionality constant that does not depend significantly on the type of nebulizer [26]. Once the RF is applied to the ITDs, the edges of the liquid droplet flatten out, creating a thin fluid layer. Next, a jet of small droplets is formed from large droplets upon their break-up [26] (see Figure 6.4).

Laser diffraction measurements have shown that by SAWN both micron-sized droplets and also much larger ones (approximately 50 μm) are generated (see Figure 6.5). As illustrated in Figure 6.5, the big droplets are narrowly distributed whereas the smallest droplets show a broader size distribution. The fluid viscosity has an impact on the size distribution of the produced aerosols. The mean droplet size in the dominant droplet fraction decreases with increasing viscosity [30].

It has been shown that SAWN-MS provides a uniquely soft ionization mechanism, with less fragmentation than in electrospray ionization (ESI), as it was

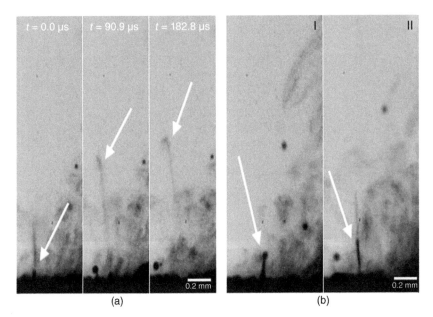

(a)   (b)

Figure 6.4 High-speed microscopic images from nebulizing a water droplet on a SAWN chip. (a) Parametrically driven capillary waves superposed on much larger waves set by the wavelength of the chip. The acceleration of these waves induces the sudden release of small droplets, visible here as jets and indicated by arrows. The three images are sequential, with time steps of 90.91 µs (11 000 fps); (b) The breakup of droplets is governed by the statics of the interference of waves, which can lead to the production of large amplitudes (ligaments) that subsequently break up to form droplets. The two peaks shown in (b) are two extreme cases; most big droplets are created by waves of smaller amplitude.

reported for the analysis of lipids [7]. This feature reduces the uncertainty about whether the observed degradation products were already present in the sample or may have been formed during the analysis. Additionally, this technique allows micro-sampling, requires a simple sample preparation and the results can be obtained within one minute. Due to its simple setup of spraying the sample directly into the MS from the surface of the chip, neither capillaries/tubing that can be cross-contaminated/clogged nor syringes are involved. The much larger ones (approximately 50 µm) are generated (see Figure 6.5). As illustrated in Figure 6.5, the big droplets are narrowly distributed whereas the smallest droplets show a broader size distribution. The fluid viscosity has an impact on the size distribution of the produced aerosols. The mean droplet size in the dominant droplet fraction decreases with increasing viscosity [30].

When used in combination with mass spectrometry, it has been shown that SAWN provides a uniquely soft ionization mechanism, with less fragmentation

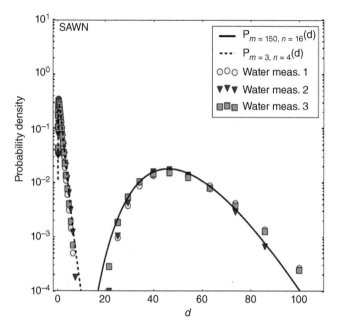

Figure 6.5 Droplet size distribution of the SAWN including the larger droplets. Three measurements are shown along with the fit (solid line). Source: Kooij et al. [26]/American Physical Society/CC BY 4.0.

than in ESI, as it was reported for the analysis of lipids [7]. This feature reduces the uncertainty about whether the observed degradation products were already present in the sample or may have been formed during the analysis. Additionally, this technique allows micro-sampling, requires a simple sample preparation, and the results can be obtained within one minute. Due to its simple setup of spraying the sample directly into the MS from the surface of the chip, neither capillaries/tubing that can be cross-contaminated/clogged, and nor syringes are involved. The chip is efficiently and quickly cleaned between runs, and there is less carryover compared to conventional atmospheric pressure ionization techniques such as ESI. Regarding the ions generated by using SAWN, they are similar to ESI-derived ones including doubly charged analyte ions.

While still in its infancy, SAWN-MS has been proven to work efficiently in the study of complex samples including blood [31], fermented food products [32], organic explosives in pre- and post-explosion case samples, [3] dyed wool samples [8], and oil paints [33]. Ho et al. [31] have successfully used the SAWN-MS technology in combination with a cheap and simple paper-based sample delivery system (Figure 6.6), an ion trap mass spectrometer for the analysis of trace amounts (nM levels) of drugs (i.e. caffeine and 5-fluorouracil) in human whole blood and plasma

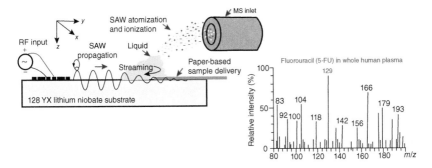

Figure 6.6 Paper-based SAWN–MS interface for direct analysis of blood samples. Source: Ho et al. [31]/American Chemical Society.

and also of heavy metals in tap water. In this setup, the surface acoustic waves draw the liquid sample from a reservoir through a paper wick onto the substrate followed by the nebulization of the sample and the generation of droplets with ions that are analyzed by the MS. The reported method requires minimal to no sample pretreatment. The high ionic strength and high viscosity of complex samples that can be challenging for conventional ionization techniques such as electrospray ionization, does not represent a problem for this new technology. Trace level compounds, below 300 pg for tested drugs in blood samples and in the order of 1 nM for 13 heavy metals in tap water, could be directly detected. Although the detection limits values obtained for the heavy metals are not as low as the ones that can be obtained by ICP-MS, the proposed SAWN-MS setup has the advantage of being easy to use, portable, and enables the analysis of high ionic strength samples.

6.2 Analysis of Complex Samples

When it comes to forensic investigations on incidents involving energetic materials, there is a need for fast screening and detection methods for highly sensitive chemical analysis. The technique should be sensitive and accurate to avoid false negative and false positive results which may have severe consequences. Ideally, these techniques should be portable for on-scene analysis. We have recently reported the first study on the potential of SAWN-MS for the analysis of organic explosives (TATP, HMTD, PETN, ETN, NG, RDX, HMX, TNT, and tetryl) [3]. Excellent sensitivity was achieved for nitrate-based organic explosives in negative ionization mode (Figure 6.7). The studied peroxides (TATP and HMDT) analyzed in positive mode, did not show dominant adduct peaks but extensive fragmentation was observed for these (Figure 6.8). When applying the proposed SAWN-MS method to the analysis of pre- and post-explosion case samples, the technique

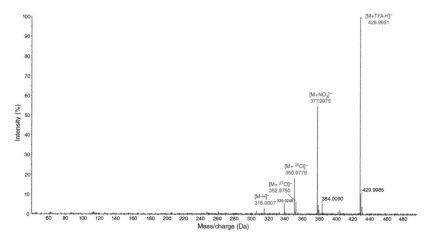

Figure 6.7 SAWN-MS spectrum of 1 µg/ml PETN diluted in MeOH/H$_2$O (70 : 30 v/v%) + 0.1 vol% CHCl$_3$, analyzed in negative ionization mode. Peaks labeled in red were identified as adducts of PETN.

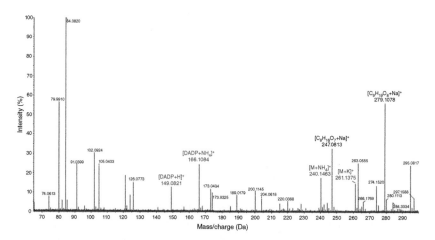

Figure 6.8 SAWN-MS spectrum of 30 µg/ml TATP in isopropanol/water (85 : 15 v/v%) with 15 vol% NH$_4$OH in aqueous phase, analyzed in positive ionization mode. Peaks labeled in black were identified as characteristic fragments, while those in red were identified as adducts of TATP and DADP, the analyte ion and degradation product, respectively.

was able to accurately identify nitrate-based organic explosives (Figure 6.9) and the answers were obtained in a significantly reduced analysis time and minimal sample treatment when compared to conventionally used laboratory analysis methods. Limits of detection (LODs) down to 1 ng/l were found for

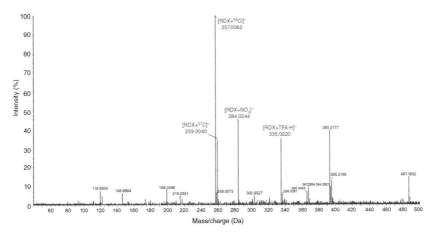

Figure 6.9 SAWN-MS analysis of forensic case extract 11 (post-explosion swab) showing the presence of RDX.

nitro-explosives. Significantly higher LODs (15–30 ppm) were determined for TATP and HMTD, which indicated that the sensitivity of this ambient technique may not always be sufficient to detect traces of organic explosives in case samples. Nevertheless, this technique has a simple setup, it is easy to operate, fast, cheap, and has potential to be built within portable mass spectrometers (external gas flows, heating, or high voltages are not required).

6.2.1 Single Fibers Having Synthetic Organic Dyes and Other Trace Evidence Examples

The identification and elucidation of dyes can be particularly complex. Factors such as limited sample size, the variance between samples, and unknown state of degradation are all issues that can easily affect the quality of the results obtained from the chosen analytical technique. The ideal analysis of these items involves the use of non-destructive techniques that can be applied in situ, which in some cases, is feasible. The ultimate choice when it comes to selecting the type of instrument to use for the analysis of dyes on fiber depends on the aim of the experiment. Is the goal to screen, confirm, or identify? How much sample is available for analysis? Additionally, for an analytical technique to be suitable for the study of dyes, it must adhere to the following parameters: minimally invasive or non-destructive, fast, multi-elemental, sensitive, versatile, and universal. Techniques based on spectroscopy allow the item to be analyzed both in situ and without the need for sample extraction. However, because these techniques do not always provide in-depth information about the dyes themselves (i.e. degradation products) and the surrounding environment (e.g. fibers and binding agents) can interfere with the results. Therefore, there is a need for a micro-destructive technique that can overcome these challenges.

The number of reported studies related to dyes analysis in forensic cases is much lower when compared to numerous articles published on cultural heritage research. Most of the literature is focused on the improvement of the current analytical techniques used in the forensic lab rather than exploiting new and upcoming analytical techniques. In a survey taken in 2001, 121 forensic experts from the United States and Europe were asked to rate which analytical technique was most commonly used for the analysis of dyed cotton [34]. The top three techniques at that time were microscopy, microspectrophotometry, and TLC. For instance, 38% of the forensic experts stated that if no microspectrophotometer were available, they would then use TLC for color examination [34]. Little has changed since 2001 regarding the choice of the techniques for analysis, which remains highly dependent on the instrument availability, with the exception of the inclusion of LC methods as part of the dye extraction process. Figure 6.10

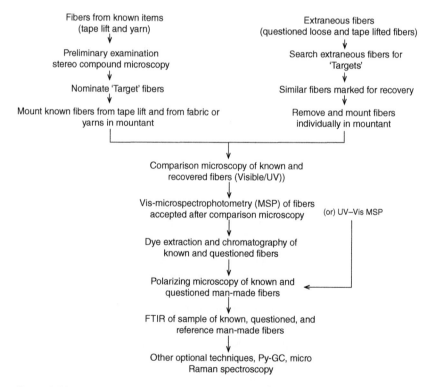

Figure 6.10 Flow diagram for fiber examination. Source: Based on Robertson and Roux [35].

displays the suggested analytical workflow for the analysis of dyes on fiber, as suggested by the book Forensic Examination of Fibers [35]. However, not all cases will require all the analytical techniques outlined in the flowchart.

Due to the nature of samples and its evidential value in forensic cases, the guidelines for selecting a proper analytical technique are strict. Approval of new techniques in the field of forensic chemistry involves rigorous testing to prove the robustness, reliability, and repeatability of the selected technique. According to the Frye Standards [36] (Frye v. the United States, 293 F. 1013 [D.C. Cir. 1923]), an expert's opinion is only acceptable "if the scientific technique on which the opinion is based is "generally acceptable" as reliable in the relevant scientific community." Hence, the following parameters must have applied: (i) the technique can be and has been tested and validated, (ii) the technique must be both peer-reviewed and published, (iii) known or potential error rates have been established, (iv) the existence and maintenance of standards controlling its operation, and (v) it must have the general approval of the pertinent scientific community. Additional to the issue of cost, the general acceptance and low volume of cases are other determining factors that limit the implementation of new techniques. The number of cases involving fibers and dye analysis is relatively small in day-to-day basis in forensic laboratories [16]. Based on these points, it can be inferred that it will be a long time before the use of ambient MS techniques become part of the routine analysis in a forensic lab. At least until more rigorous testing has been performed for all the ambient techniques, including all the possible dye classes. The use of these techniques can significantly decrease the time spent on analysis of the evidence and, thereby providing information to the court in a quick, efficient, and reliable manner, which in some cases can be extremely crucial.

Very little information can be found in the literature regarding the analysis of modern synthetic organic pigments by ambient MS techniques, only the works of Wyplosz [37] with laser desorption ionization-mass spectrometry (LDI-MS) and Astefanei et al. [8] with SAWN-MS referenced them. Regarding the SAWN-MS study, the authors have successfully applied the technique for the analysis of synthetic dyes (e.g. basic violet 3, acid violet 7, basic blue 26, acid orange 6, acid red 88, acid red 44, and acid blue 74) on wool fibers (e.g. 1–2 mm) [8]. While the extraction process of the synthetic dyes from wool samples took no longer than two minutes [38], it does require some previous knowledge of the fiber type in order to select the suitable solvent for extraction. The wool samples (a few mm of fibers) were dissolved in 100 μl MeOH-H_2O and vortexed for two minutes. A total of 10 μl aliquot was placed on the SAWN chip, where it was subsequently nebulized and introduced into the MS to produce a spectrum in an acquisition time up to one minute. This work demonstrated that with the use of this ionization method, complex dye mixtures, such as that of wool dyed in acid violet 7 ($C_{20}H_{16}N_4Na_2O_9S_2$) (Figure 6.11), can be elucidated, and the dye

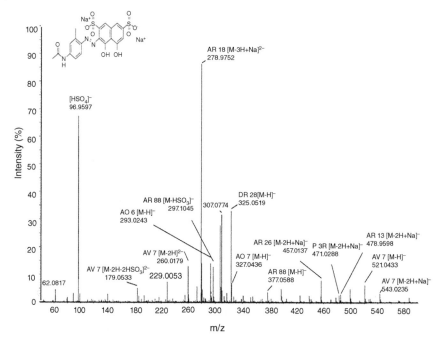

Figure 6.11 SAWN-MS spectrum of wool fiber dyed with Acid Violet 7. Source: Based on Bang et al. [6].

mixtures and degradation products identified. In addition, in combination with high-resolution mass spectrometry, trace level peaks of the demethylated products with a 2-ppm accuracy were detected, demonstrating its capabilities of detecting dye degradation products (Figure 6.12). Moreover, LOQ's as low as 0.001 pg were obtained for basic dyes and between 0.0005 and 0.5 pg for acidic dyes.

6.2.2 The Case of Denim Fibers

The identification of dyes can be particularly difficult given the quantity and the state of the sample. Especially challenging is distinguishing between samples that seem to originate from the same dye source yet are produced in different regions of the world. Such is the case of the indigo dye found in denim fibers. In the context of forensic cases, the analysis of denim is omitted in routine laboratory work due to the lack of distinguishing factors between the different brands of denim. Original denim is characterized as a twill cotton fabric with intertwined white weft and indigo (Vat Blue 1) warp threads [39]. Cotton, its main fiber component, has a heterogeneous composition that creates intrasample variation that does not allow for individualization from different sources [40]. Indigo or Vat Blue 1, the

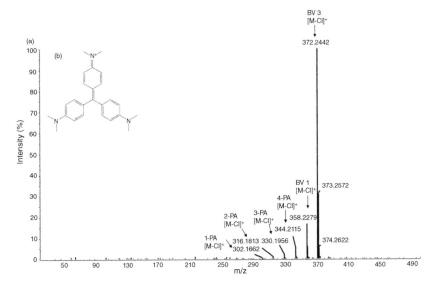

Figure 6.12 Analysis of a wool fiber dyed with Basic Violet 3 ($C_{25}H_{30}ClN_3$) and the degradation products observed. (a) SAWN-MS full scan spectrum and (b) SAWN-MS/MS spectrum. Source: Based on Bang et al. [6].

original dye used for denim, has been shown to display little to no discrimination power when analyzed via its absorbance spectrum [40]. Hence, denim evidence has been disregarded in forensic cases. However, in the past 20 years, there has been an increase in the reported variety of denim fibers and dyes [41]. Yet, no studies have been found describing the potential of these new dyes [42], the new indigo reduction techniques [42–45] and the new denim dyeing mechanism [42, 46, 47] which have the potential to increase the evidential values of denim in forensic applications.

The denim composition has changed significantly in the last 30 years. A study utilizing 108 denim products found in the United States but originating from other countries, found that the amount of denim solely containing cotton was about 30%, while the other 70% contained a mixture of cotton with either spandex, polyester, and ramie [41]. Moreover, denim commonly known to be dyed with indigo or its synthetic version, Vat Blue 1 (CI 73000), has also had some changes in the past couple of decades [48]. Recently, it was determined that only 8% of denim contain Vat Blue 1 only, while 85% contained Vat Blue 1 plus some additives, and the remaining 7% contained other unspecified dyes [41]. The variation in dye mixture arose from the use of other fibers found in denim since vat dyes do not bind in the same manner for all types of fiber. Nevertheless, the analysis of denim is still excluded from forensic analysis to this day.

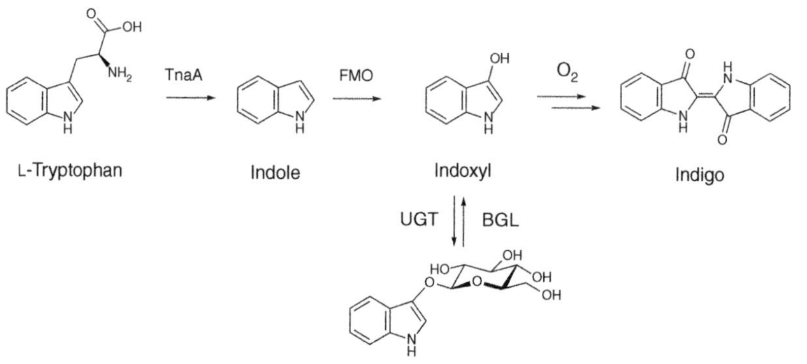

Figure 6.13 The proposed mechanism for the production of Indigo from *E. coli* for the dyeing of cotton fibers. Source: Hsu et al. [42].

6.2.3 From Natural to Synthetic Indigo for Denim Dyeing

In addition to the changes seen to the denim fiber and dye composition over the last 30 years, new changes have also been occurring in the denim dye process itself. In the past decade, much research has been done for the development of sustainable dyes. Indigo, or Vat Blue 1, is an insoluble dye pigment without affinity to cotton fiber in its natural state. For vat dyes to be applied to cotton fibers, they must first be reduced from its keto structure to its leuco structure, its water-soluble version, with the aid of a reducing agent (e.g. sodium dithionate). During the dyeing process, the leuco form saturates the cotton fiber, after which the leuco structure is oxidized back to its keto structure, where it remains trapped [43]. Due to the morphology of the cotton fibers, the dyeing process may have to be repeated several times to achieve the desired hue. Additional processes may take place to alter the physical appearance of the denim fabric. This includes finishing effects, which is intended to produce lighter or "faded" colors. This can occur via the washing of the denim material with pumice stones, oxidative bleaching, or through the use of enzymes (e.g. cellulase) [39].

The synthesis of synthetic indigo involves the use of toxic compounds such as aniline, formaldehyde, hydrogen cyanide, sodamide, and strong bases [42]. In addition, sodium dithionite, the reducing agent used in the dyeing process, decomposes into sulfate and sulfite, both of which are known to corrode equipment, pipes, and end up in wastewater treatment facilities [43]. Moreover, the bleaching process to create the "faded look" of denim also contribute to the corrosion of machinery and to the exposure of chemicals to the workers, such as potassium permanganate and sodium hypochlorite [39]. Altogether, the process of creating denim is considered one of the most polluting processes in the textile industry.

To counteract or stop the use of these toxic compounds and reduce the amount of water consumed, greener ways of dyeing denim have been developed. Some of the methods include the synthetic production of indigo from bacteria [42] and the reuse of indigo obtained via ultrafiltration from wastewater [49]. Novel ways of reducing indigo have also been developed, such as the use of reducing sugars [42, 44], bacterial reduction [45], catalytic hydrogenation, and electrochemical reductions [43]. Moreover, the use of enzymes [39] and CO_2 laser [50] have also been proposed as a more sustainable process for the washing of denim. Additionally, methods involving the use of spray [42], foam [46], and waterless [47] dyeing mechanism have also been applied for reducing the amount of water in the denim dyeing process.

The work of Hsu et al. [42] demonstrates an alternative indigo dyeing process that utilizes a new source for the production of indigo with a more environmentally reducing agent (Figure 6.13). In this study, indigo was produced from the *E. coli* bacteria by transforming the L-tryptophan protein to an indole with the use of the TnaA enzyme. Afterward, the production of indole, the flavin-dependent monooxygenase (FMO) gene, is added, which in turn produces Indoxyl molecules. The prereduction process begins by first protecting the hydroxyl group of the indoxyl molecules through the use of glucosyltransferase, or PtUGT1. At this point, the protected indican is ready to be used and can be stored for an extended period. Application of the prereduced indican to the fiber occurs through a spray mechanism (the bacteria broth), after which the crystallization of dye occurs by applying a layer of β-glucosidases solution to the fiber. The β-glucosidases deprotect the hydroxyl groups, which returns the molecule into the Indoxyl form that, when exposed to air, oxidizes immediately into its insoluble form, indigo. The process of spraying the prereduced indican solution followed by the β-glucosidases is repeated for at least three cycles. Denim dyed via this method produces a much lighter color when compared to the normal dyeing process (Figure 6.14). This difference in color is attributed to the delicate nature of the enzyme, which requires a neutral pH bath rather than the typical pH of 11. Additionally, the use of β-glucosidases is, at the moment, more expensive than the standard reducing agent. However, with the production of a higher quality enzyme hydrolysis become faster, and a higher pH can be achieved. From a sustainable point of view, this process manages to reduce the need for indican purification, as well as the need to use hazardous chemicals for the production of synthetic indigo. Additionally, the use of the standard reducing agent is also eliminated, all while reducing the overall consumption of water.

The reuse of indigo obtained from wastewater has also been proposed as another way to produce more or less greener denim [49]. In the standard production of denim dyeing, around 15% of indigo is lost in the rinsing process and ends up in the wastewater. Due to the insolubility of the dye in water, ultrafiltration module

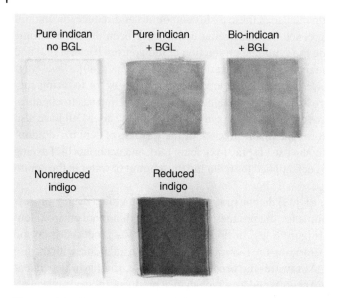

Figure 6.14 Comparison of cotton fibers dyed with a different type of Indigo dyes. Source: Hsu et al. [42]/Springer Nature.

membranes (e.g. U-1b and ZW-1), can be used to collect and reuse any left-over dye. A pilot study of each membrane resulted in 96% and 99% removal of the dye when U-1b and ZW-1 membrane were utilized, respectively. Additionally, with the use of the U-1b membrane concentrates up to 3 g/l was obtained, while with the ZW-1 membrane up to 20 g/l concentrate was obtained. For a dye bath, a concentrate at least 20 g/l is needed, hence for their semi-industrial experiment, both membranes were utilized and were able to achieve 98% dye removal with the same amount concentrate amount. With the obtained 20 g/l concentrate, they were then able to dye the cotton fabric following the unspecified protocol of Tejidos Royo. The dye results obtained had similar chroma and hue values. However, the reused dyes had higher intensity values signifying that the cotton fabric dyed with reused indigo where much darker in color. This darker color is attributed to the excess dispersant found in the wastewater, which is also present in the reused indigo. This work demonstrated the capability of reusing indigo dyeing from wastewater, which reduced the cost of production with the additional benefit of being able to reuse the effluent for further dyeing.

In the work of Shin et al. [44], banana peels were utilized to produce a biodegradable reducing agent for indigo dyeing. Banana peels were considered as a suitable replacement due to its antioxidant capabilities and its high sugar content. It was hypothesized that its antioxidant capabilities would prevent the leuco-indigo from reoxidizing. The banana peel contains galactopyranose, a cyclic isomer of galactose, that is known for its ability to reduce indigo. After the extraction of the peel in water, the powder form was added to a synthetic indigo bath with a pH of 11. Then the ramie fabric was placed in a bath heated to 60 °C for 20 minutes. Afterward, the fabric was exposed to air to recrystallize the dye molecule. It was subsequently rinsed, neutralized, then rinsed again, and finally dried. The results showed that with the use of the banana peel extract, the duration of the reduction was longer and more stable than with sodium dithionate but yielded a less intense color. Nevertheless, it was also possible to reuse the dyebath for up to 18 days at 3.0% (o.w.b.) of the extract concentration proving the capabilities of reducing sugars as a greener alternative to the use sodium dithionate as a reducing agent.

6.2.3.1 MS and Denim Analysis

From the synthesis of indigo to the changes in the denim dyeing process, sustainability is changing the denim production. These new methods for dyeing denim fiber reintroduce the potential for the characterization and individualization of denim in the legal field. With the developments of indigo dyes from new sources (e.g. bacteria) [42] and the use of dye mixtures, the evidential value of denim can become significant. Moreover, the use of sulfur dyes [51], reactive dyes [52], and other indigo dye derivatives [48] have also been proposed as a relatively cheaper alternative, hence providing an even more distinguishable chemical profile. The methods mentioned above all produce different shades of blue that are visibly distinguishable even though they originate from the same indigo molecule. Additionally, the broad range of reducing agents and the potential residues left after the completion of the dyeing process can also further enhance the evidential value of these fibers. These novel techniques provide manufacturers the chance to adopt new strategies for the development of dyes. Given that not all the methods cost the same or are as efficient, if and when these methods are available to the market, they will create a new generation of denim that will be distinguishable from one another based on their fiber (e.g. pure cotton, cotton, and polyester) and dye composition (e.g. dye source and dyeing technique).

Although previous studies have shown that it was not possible to distinguish denim by the use of spectroscopic methods [41, 42], this may soon change with the use of more sensitive and selective analytical techniques such as mass spectrometry. Studies done with MS-based techniques have shown that it is possible to distinguish between natural and synthetic dyes and even between plant and animal origin. The extraction of modern synthetic dyes is more straightforward than those of natural origins, which indicates that ambient MS-based methods could be easily applied for the analysis of both synthetic and natural dyes. The use of mass spectrometry allows for the characterization of the dye as a standard and on denim. Indigo dyes produced from new sources can be identified by studying their degradation products (e.g. isatin and isatoic anhydride), additives, intermediates products or precursors (e.g. indican), and other possible impurities found on the sample, much like it has been done for indigo in cultural heritage cases [37, 53–56]. They can be used with the ultimate goal of creating a dye database that can be applied in real-world cases to increase the evidential value of denim.

In addition, different techniques can identify different molecules within the same sample hence providing a broader overview of the dye. Ambient MS techniques such as atmospheric solid analysis probe (ASAP), direct analysis in real-time (DART), laser desorption analysis (LDI), and liquid chromatography (LC) are already being used for the identification and elucidation of natural and synthetic indigo dyes in fiber (e.g. cotton, linen, silk, and wool) [37, 53–57]. For instance, ASAP-MS, the least sensitive of the techniques, can be initially used to screen for the presence of indigo and its molecular peak at m/z 263.2 [54]. Afterward, another more sensitive technique such as LC, DART, or LDI, in combination with the use of an MS/MS or HRMS, can be applied to confirm the presence, detecting the degradation products and byproducts to deduce the source of the dye [37, 55].

The use of complementary techniques aid in the understanding and characterization of the present dyes. Ambient MS techniques, can provide quick screening results often without the need to extract the dye from the fiber or with fast and simple extraction procedures [54, 57], while LC–MS can provide separation for complex mixture using low sample volumes [55]. Together, they can provide enough detailed information to individualize denims and their dyes and increase the potential of denims as a valuable source of evidence in forensic cases.

Advances in the miniaturization of MS instruments can lead to further developments of ambient MS techniques. This allows these technologies to become more versatile, economical, and widespread. One of the most significant selling points is its ability to bring the "lab" to wherever it is needed. For instance, this would mean bringing the portable MS to a crime scene. Miniaturized MS, in combination with ambient ionization techniques such as SAWN-MS, can simplify workload by removing the need for sample preparation. Importantly, these types of instruments could allow for a nonexpert to perform the analysis.

References

1 Wang, J. (2002). Portable electrochemical systems. *Trends. Anal. Chem.* 21: 226–232. https://doi.org/10.1016/S0165-9936(02)00402-8.

2 Verpoorte, E. (2002). Microfluidic chips for clinical and forensic analysis. *Electrophoresis* 23: 677–712. https://doi.org/10.1002/1522-2683(200203)23:5 3.0.CO;2-8.

3 Pintabona, L., Astefanei, A., Corthals, G.L., and van Asten, A.C. (2019). Utilizing surface acoustic wave nebulization (SAWN) for the rapid and sensitive ambient ionization mass spectrometric analysis of organic explosives. *J. Am. Soc. Mass. Spectrom.* 30 (12): 2655–2669.

4 Huang, Y., Heron, S.R., Clark, A.M. et al. (2016). Surface acoustic wave nebulization device with dual interdigitated transducers improves SAWN-MS performance. *J. Mass Spectrom.* 424–429. https://doi.org/10.1002/jms.3766.

5 Alvarez, M., Friend, J., and Yeo, L.Y. (2008). Rapid generation of protein aerosols and nanoparticles via surface acoustic wave atomization. *Nanotechnology* 19: 455103. https://doi.org/10.1088/0957-4484/19/45/455103.

6 Bang, J.H., Helmich, R.J., and Suslick, K.S. (2008). Nanostructured ZnS:Ni^{2+} photocatalysts prepared by ultrasonic spray pyrolysis. *Adv. Mater.* 20: 2599–2603. https://doi.org/10.1002/adma.200703188.

7 Yoon, S.H., Huang, Y., Edgar, J.S. et al. (2012). Surface acoustic wave nebulization facilitating lipid mass spectrometric analysis. *Anal. Chem.* 84: 6530–6537.

8 Astefanei, A., van Bommel, M., and Corthals, G.L. (2017). Surface acoustic wave nebulisation mass spectrometry for the fast and highly sensitive characterisation of synthetic dyes in textile samples. *J. Am. Soc. Mass. Spectrom.* 28: 2108–2116.

9 Länge, K., Rapp, B.E., and Rapp, M. (2008). Surface acoustic wave biosensors: a review. *Anal. Bioanal.Chem.* 391: 1509–1519. https://doi.org/10.1007/s00216-008-1911-5.

10 Tveen-Jensen, K., Gesellchen, F., Wilson, R. et al. (2015). Interfacing low-energy SAW nebulization with liquid chromatography-mass spectrometry for the analysis of biological samples. *Sci. Rep.* 5: 9736.

11 Thalhammer, S. and Wixforth, A. (2013). Surface acoustic wave actuated lab-on-chip system for single cell analysis. *Biosens. Bioelectron.* 4: 1–7. https://doi.org/10.4172/2155-6210.100013.

12 Heron, S.R., Wilson, R., Shaffer, S.A. et al. (2010). Surface acoustic wave nebulization of peptides as a microfluidic interface for mass spectrometry. *Anal. Chem.* 82: 3985–3989. https://doi.org/10.1021/ac100372c.

13 Rocha-Gaso, M.I., March-Iborra, C., Montoya-Baides, Á., and Arnau-Vives, A. (2009). Surface generated acoustic wave biosensors for the detection of pathogens: a review. *Sensors* 9: 5740–5769. https://doi.org/10.3390/s9095740.

14 Jin, H., Zhou, J., He, X. et al. (2013). Flexible surface acoustic wave resonators built on disposable plastic film for electronics and lab-on-a-chip applications. *Sci. Rep.* 3: 2140. https://doi.org/10.1038/srep02140.

15 Qi, A., Friend, J.R., Yeo, L.Y. et al. (2009). Miniature inhalation therapy platform using surface acoustic wave microfluidic atomization. *Lab Chip* 9: 2184–2193. https://doi.org/10.1039/b903575c.

16 Yeo, L.Y., Friend, J.R., McIntosh, M.P. et al. (2010). Ultrasonic nebulization platforms for pulmonary drug delivery. *Expert. Opin. Drug. Deliv.* 7: 663–679. https://doi.org/10.1517/17425247.2010.485608.

17 Yeo, L.Y. and Friend, J.R. (2009). Ultrafast microfluidics using surface acoustic waves. *Biomicrofluidics* 3: 12002. https://doi.org/10.1063/1.3056040.

18 Huang, Y., Yoon, S.H., Heron, S.R. et al. (2012). Surface acoustic wave nebulization produces ions with lower internal energy than electrospray ionization. *J. Am. Soc. Mass. Spectrom.* 23: 1062–1070. https://doi.org/10.1007/s13361-012-0352-8.

19 Qi, A., Yeo, L., Friend, J., and Ho, J. (2010). The extraction of liquid, protein molecules and yeast cells from paper through surface acoustic wave atomization. *Lab Chip* 10: 470–476. https://doi.org/10.1039/b915833b.

20 Li, H., Friend, J., Yeo, L. et al. (2009). Effect of surface acoustic waves on the viability, proliferation and differentiation of primary osteoblast-like cells. *Biomicrofluidics* 3: 1–11. https://doi.org/10.1063/1.3194282.

21 Friend, J.R., Yeo, L.Y., Arifin, D.R., and Mechler, A. (2008). Evaporative self-assembly assisted synthesis of polymeric nanoparticles by surface acoustic wave atomization. *Nanotechnology* 19: 145301. https://doi.org/10.1088/0957-4484/19/14/145301.

22 Alvarez, M., Yeo, L.Y., Friend, J.R., and Jamriska, M. (2009). Rapid production of protein-loaded biodegradable microparticles using surface acoustic waves. *Biomicrofluidics* 3: 14102. https://doi.org/10.1063/1.3055282.

23 Qi, A., Yeo, L.Y., Friend, J.R. et al. (2014). https://doi.org/10.1063/1.2953537). Interfacial destabilization and atomization driven by surface acoustic waves. *Phys. Fluids* 20: 074103.

24 Medico, R.F. (2019). *The Potentials of SAWN-MS for the Analysis of Polyphenols in Oenological Samples.* University of Amsterdam.

25 Zhang, X., Wang, F.Y., and Li, L. (2007). Optimal selection of piezoelectric substrates and crystal cuts for SAW-based pressure and temperature sensors. *IEEE Trans. Ultrason. Ferroelectr. Freq. Control* 54: 1207–1215. https://doi.org/10.1109/TUFFC.2007.374.

26 Kooij, S., Astefanei, A., Corthals, G.L., and Bonn, D. (2019). Size distributions of droplets produced by ultrasonic nebulizers. *Sci. Rep.* 9: 1–8. https://doi.org/10.1038/s41598-019-42599-8.

27 Smith, J.P. and Hinson-Smith, V. (2006). The new era of SAW devices. Commercial SAW sensors move beyond military and security applications. *Anal. Chem.* 78: 3505–3507. https://doi.org/10.1021/ac0694122.

28 Blamey, J., Yeo, L.Y., and Friend, J.R. (2013). Microscale capillary wave turbulence excited by high frequency vibration. *Langmuir* 29: 3835–3845. https://doi.org/10.1021/la304608a.

29 Tengfei, Z., Chaohui, W., Dong, N. et al. (2014). Exploitation of surface acoustic waves to drive nanoparticle concentration within an electrification-dependent droplet. *RSC Adv.* 4: 46502–46507. https://doi.org/10.1039/c4ra07090a.

30 Winkler, A., Bergelt, P., Hillemann, L., and Menzel, S. (2016). Influence of viscosity in fluid atomization with surface acoustic waves. *Open J. Acoust.* 06: 23–33. https://doi.org/10.4236/oja.2016.63003.

31 Ho, J., Tan, M.K., Go, D.B. et al. (2011). Paper-based microfluidic surface acoustic wave sample delivery and ionization source for rapid and sensitive ambient mass spectrometry. *Anal. Chem.* 83: 3260–3266.

32 Oyler, B.L., Goodlett, D.R., Schneider, T. et al. (2018). Rapid food product analysis by surface acoustic wave nebulization coupled mass spectrometry. *Food Anal. Methods* 11: 2447–2454. https://doi.org/10.1007/s12161-018-1232-z.

33 Baij, L., Astefanei, A., Hermans, J. et al. (2019). Solvent-mediated extraction of fatty acids in bilayer oil paint models: a comparative analysis of solvent application methods. *Herit. Sci.* 7: 1–8. https://doi.org/10.1186/s40494-019-0273-y.

34 Wiggins, K.G. (2003). The European Fibres Group (EFG) 1993-2002: "understanding and improving the evidential value of fibres". *Anal. Bioanal.Chem.* 376: 1172–1177. https://doi.org/10.1007/s00216-003-1997-8.

35 Robertson, J. and Roux, C. (2018). Trace fibre evidence. In: *Forensic Exam. Fibres*, 3e (ed. K. Robertson, J. Roux and C. Wiggins), 119–122.

36 Frye v. United States - 293 F. 1013 (D.C. Cir. 1923).

37 Wyplosz, N. (2003). *Laser Desorption Mass Spectrometric Studies of Artists' Organic Pigments*. University of Amsterdam.

38 Kramell, A.E., Brachmann, A.O., Kluge, R. et al. (2017). Fast direct detection of natural dyes in historic and prehistoric textiles by flowprobe™-ESI-HRMS. *RSC Adv.* 7: 12990–12997. https://doi.org/10.1039/c6ra27842f.

39 Patra, A.K., Madhu, A., and Bala, N. (2018). Enzyme washing of indigo and sulphur dyed denim. *Fash. Text.* https://doi.org/10.1186/s40691-017-0126-9.

40 Grieve, M.C., Dunlop, J., and Haddock, P. (1988). An assessment of the value of blue, red, and black cotton Fibers as target Fibers in forensic science investigations. *J. Forensic. Sci.* 33: 12577J. https://doi.org/10.1520/jfs12577j.

41 Tinnin, A. (2015). *Analysis of Dyes in Blue Denim by Microspectrophotemetry*. Marshall University.

42 Hsu, T.M., Welner, D.H., Russ, Z.N. et al. (2018). Employing a biochemical protecting group for a sustainable indigo dyeing strategy. *Nat. Chem. Biol.* 14: 256–261. https://doi.org/10.1038/nchembio.2552.

43 Blackburn, R.S., Bechtold, T., and John, P. (2009). The development of indigo reduction methods and pre-reduced indigo products. *Color. Technol.* 125: 193–207. https://doi.org/10.1111/j.1478-4408.2009.00197.x.

44 Shin, Y., Choi, M., and Il Yoo, D. (2013). Utilization of fruit by-products for organic reducing agent in indigo dyeing. *Fibers Polym.* 14: 2027–2031. https://doi.org/10.1007/s12221-013-2027-x.

45 Nicholson, S.K. and John, P. (2004). Bacterial indigo reduction. *Biocatal. Biotransform.* 22: 397–400. https://doi.org/10.1080/10242420400024490.

46 Sarwar, N., Mohsin, M., Bhatti, A.A. et al. (2017). Development of water and energy efficient environment friendly easy care finishing by foam coating on stretch denim fabric. *J. Cleaner Prod.* 154: 159–166. https://doi.org/10.1016/j.jclepro.2017.03.171.

47 Rashid, M.A., Hoque, M.S., and Hossain, M.J. (2020). Developing a new hydrose wash technique for treating denim fabric. *J. Inst. Eng. Ser. E.* https://doi.org/10.1007/s40034-020-00161-6.

48 Grieve, M.C., Biermann, T.W., and Schaub, K. (2006). The use of indigo derivatives to dye denim material. *Sci. Just. Forensic Sci. Soc.* 46: 15–24. https://doi.org/10.1016/S1355-0306(06)71563-5.

49 Buscio, V., Crespi, M., and Gutiérrez-Bouzán, C. (2015). Sustainable dyeing of denim using indigo dye recovered with polyvinylidene difluoride ultrafiltration membranes. *J. Cleaner Prod.* 91: 201–207. https://doi.org/10.1016/j.jclepro.2014.12.016.

50 Kan, C.W. (2014). Colour fading effect of indigo-dyed cotton denim fabric by CO_2 laser. *Fibers Polym.* 15: 426–429. https://doi.org/10.1007/s12221-014-0426-2.

51 Božič, M. and Kokol, V. (2008). Ecological alternatives to the reduction and oxidation processes in dyeing with vat and sulphur dyes. *Dyes Pigm.* 76: 299–309. https://doi.org/10.1016/j.dyepig.2006.05.041.

52 Rosa, J.M., Tambourgi, E.B., Santana, J.C.C. et al. (2014). Development of colors with sustainability: a comparative study between dyeing of cotton with reactive and vat dyestuffs. *Text. Res. J.* 84: 1009–1017. https://doi.org/10.1177/0040517513517962.

53 Humphrey, P.I. and Humphrey, P.I. (2017). The use of mass spectrometry to differentiate blue dyes from indigo and woad. *Mc. Nier. Sch. Res. J.* 10: 73–87.

54 Kramell, A., Porbeck, F., Kluge, R. et al. (2015). A fast and reliable detection of indigo in historic and prehistoric textile samples. *J. Mass Spectrom.* 50: 1039–1043. https://doi.org/10.1002/jms.3625.

55 Witkoš, K., Lech, K., and Jarosz, M. (2015). Identification of degradation products of indigoids by tandem mass spectrometry. *J. Mass Spectrom.* 50: 1245–1251. https://doi.org/10.1002/jms.3641.

56 Thile, T. (2008). *The Discrimination of Blue Cotton Fibers by High Pressure Liquid Chromatography Mass Spectrometry*. University of California.

57 Selvius Deroo, C. and Armitage, R.A. (2011). Direct identification of dyes in textiles by direct analysis in real time-time of flight mass spectrometry. *Anal. Chem.* 83: 6924–6928.

7

Elemental Imaging of Forensic Traces with Macro- and Micro-XRF

Alwin Knijnenberg[1], Annelies van Loon[2], Joris Dik[3], and Arian van Asten[4, 5]

[1]*Netherlands Forensic Institute, P.O. Box 24044, The Hague 2490 AA, The Netherlands*
[2]*Department of Conservation & Science, Rijksmuseum, P.O. Box 74888, 1070 DN Amsterdam, The Netherlands*
[3]*Department of Materials Science and Engineering, Delft University of Technology, P.O. Box 5, 2600 AA Delft, The Netherlands*
[4]*Van 't Hoff Institute for Molecular Sciences, University of Amsterdam, P.O. Box 94157, 1090 GD Amsterdam, The Netherlands*
[5]*Co van Ledden Hulsebosch Center (CLHC), Amsterdam Center for Forensic Science and Medicine, University of Amsterdam, Van 't Hoff Institute for Molecular Sciences, P.O. Box 94157, 1090 GD Amsterdam, The Netherlands*

7.1 Introduction

Chemical imaging can generally be described as spatially resolved chemical analysis to provide insight in the distribution of compounds in two (i.e. a plane) or three (i.e. a volume) dimensions. Using imaging approaches on a completely homogenous sample is typically useless unless the specific aim is to demonstrate the high degree of homogeneity or to visualize the entire sample using the special contrast offered by the chemical analysis. Important information with respect to material properties can be provided when the analytical chemist is able to reveal the distribution of constituents in space. Such insight is typically lost when a relatively large sample needs to be processed for chemical analysis resulting in the determination of the average composition without any indication on potential spatial variations and inhomogeneities. However, materials with the same average composition but a different spatial arrangement of its constituents can behave and function very differently. Spatial features play an important role in, for instance, construction materials, electronics, home and personal care products and food stuffs. Through its general definition chemical imaging encompasses a broad and diverse range of analytical techniques. Such techniques, however, must share two features: (i) the ability to chemically detect, differentiate, characterize and/or identify compounds and (ii) to do so at multiple locations

Leading Edge Techniques in Forensic Trace Evidence Analysis: More New Trace Analysis Methods, First Edition. Edited by Robert D. Blackledge.
© 2023 John Wiley & Sons, Inc. Published 2023 by John Wiley & Sons, Inc.

within a given sample or object of interest. Analytical chemistry instrumentation capable of chemical imaging must therefore be able to *"probe"* or *"interrogate"* a small area or volume of the sample. Ideally, this sampling process is noninvasive enabling additional investigations as the material remains virtually unaltered. However, this is not a prerequisite and many of the mass spectrometric imaging techniques (e.g. matrix-assisted laser desorption ionization or MALDI, desorption electrospray ionization or DESI, time-of-flight secondary ion mass spectrometry or TOF-SIMS and Laser Ablation – Inductively Coupled Plasma – Mass Spectrometry or LA-ICP-MS) are to variable extent invasive. Either by altering the sample surface as part of the probing process (e.g. LA-ICP-MS), requiring surface modification for the analysis (MALDI) or exposing the sample to high vacuum (TOF-SIMS). Alternatively, spectroscopic techniques based on the interrogation of a sample with electromagnetic radiation are virtually noninvasive as long as the energy of the photons, the intensity of the beam (number of photons per time unit in a given area) and the associated exposure time are not too high. Spectroscopic based imaging can be conducted within a broad range of the electromagnetic spectrum including techniques such as X-ray transmission imaging (as used in security and medical setting, e.g. computed tomography or CT scanning), X-ray fluorescence or XRF imaging, UV–vis hyperspectral imaging or HSI, near infrared (NIR) and infrared (IR) imaging. Millimeter wave scanners introduced in the last decade in many airports use electromagnetic microwaves to produce whole body images while radio waves with wavelengths in the *m-km* range are used in RADAR (radio detection and ranging) systems in air traffic control and in a military setting. Based on the wavelength employed, different features can be investigated, including elemental composition (XRF imaging) and electronic (HSI, fluorescence imaging) and vibrational states (NIR, IR, and Raman) of organic compounds. An interesting feature of spectroscopic imaging is the possibility of irradiating/illuminating the entire sample surface at once and to make use of optics and detection technologies to perform the spatially resolved characterization. An approach that is very similar to human vision and is therefore often used with HSI. A third family of imaging techniques uses particle beams to interrogate samples and object surfaces. This includes electron microscopy with energy-dispersive X-ray spectroscopy or scanning electron microscopy (SEM)-EDX which is used in a similar fashion as XRF to obtain spatial information on the elemental composition. With proton-induced X-ray emission or PIXE a proton beam is used to create the element specific photon emissions to investigate elemental spatial distributions. TOF-SIMS also uses a particle beam to interrogate sample surfaces but in this case the analysis of the resulting secondary ions is performed with a time-of-flight mass spectrometer.

An essential characteristic of every imaging technique is its resolution. Resolution determines the detail with which the spatial variations in chemical

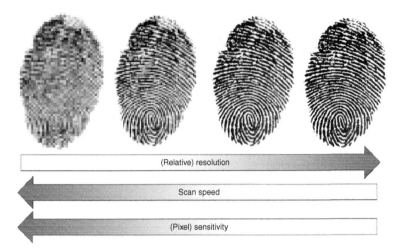

Figure 7.1 The relation between resolution, scan speed, and sensitivity in chemical imaging (Fingermark picture was obtained from Pixabay and serves for illustration only.)

composition can be reconstructed in the chemical image. This does not only depend on the technique but also on the dimensions of the object of interest and this could be reflected in a relative resolution measure. The smaller the interrogated sample or its spatial details the more demanding the instrumental requirements become to create sufficient resolution in the chemical image. In the end, the minimal dimensions of the beam interrogating the surface or the maximum magnification enabled by the optics will determine the smallest dimensions that can still be imaged with sufficient detail. Although (relative) resolution in chemical imaging instrumentation can often be optimized by adjusting method parameters this typically comes at the cost of reduced sensitivity, increased scan times or both. This intrinsic inverse relation between resolution on the one hand and sensitivity and speed on the other is illustrated in Figure 7.1.

Interestingly, when considering spectroscopy and mass spectrometry-based chemical imaging techniques, resolution has multiple dimensions. In addition to spatial resolution, method performance is also determined by the spectral resolution. The spectral resolution in spectroscopy and the mass resolution in mass spectrometry affects the selectivity and often also the sensitivity of the spatially resolved chemical analysis. This determines with what detail chemical information can be provided. As an example, isobaric compounds can typically not be distinguished at unit mass resolution whereas individual distributions can be created when using a high-resolution mass spectrometer capable of measuring the small mass difference due to the deviating elemental compositions.

When dealing with forensic traces such as fingermarks, gunshot residues (GSRs), glass fragments, paint deposits, fibers, human and animal hairs and minute stains of biological fluids, employing chemical imaging techniques serves three distinct purposes, (i) trace detection, (ii) trace characterization, and (iii) capturing trace position and distribution. The added value of these three applications in criminal investigations will be briefly discussed next. At the crime scene and in forensic laboratories, crime scene officers, and forensic experts search for forensic trace evidence that could provide crucial information to reconstruct events. Special light sources and optical filters are used and/or contrast agents are applied to visualize latent traces. This visualization step is crucial to prevent vital evidence from going unnoticed. For some substrates the use of regular techniques is less effective. The use of fluorescence to detect human semen is for instance hampered on fabric that provides strong background fluorescence either because of the fabric type or the laundry detergent used. Similarly, the use of optical techniques can be difficult when dealing with dark colored substrates that are strong absorbers and thus minimize selective reflection of incident light. In these circumstances the use of chemical imaging techniques can provide novel contrast enabling the visualization of latent traces based on their chemical composition. In addition to visualization the spatially resolved chemical analysis also provides information on the nature of the forensic trace. This allows for further chemical characterization that could lead to classification (what kind of trace are we dealing with) and in some cases even more detailed information such as the time of deposition (e.g. the use of hyperspectral imaging to estimate the age of blood stains on the basis of blood discoloration over time) or activities that led to the stains (e.g. the detection of explosives traces in a fingermark). Such activity level information can be very important for forensic reconstruction and capturing the entire trace pattern through large area chemical imaging can also be valuable in this respect. By situating forensic traces in space and studying their position relative to each other and other objects, forensic experts can make inferences regarding the associated activities that resulted in these patterns. This type of investigation is for instance conducted in forensic bloodstain pattern analysis or BPA. However, when the substrates and surfaces do not facilitate detailed photographic recording, chemical imaging could still provide the visual information that is needed for the forensic interpretation.

In this chapter, the use of micro- and macro-XRF imaging of forensic microtraces is discussed. The opportunities available through the unique *"elemental contrast"* offered by XRF will be demonstrated for important forensic trace evidence material including gunshot residues or GSR, traces of human biological fluids such as blood, sperm, saliva and urine and stains of personal care products and cosmetics. The possibility of detecting and visualizing concealed forensic stains in a virtually noninvasive manner by exploiting X-ray surface penetration will be illustrated. This chapter will conclude with an outlook on future trends with respect to the use of XRF imaging in forensic and crime scene investigations. The results presented by the authors are based on a recent scientific publication in Nature's Scientific Reports [1] complemented with supplemental information material of this paper and new imaging data on personal care products and cosmetics. The chapter has not been written as an extensive literature review and includes a very limited number of literature references. Instead, the authors aim to provide the readers with new insights and ideas regarding the use of elemental imaging in forensic science. For related scientific literature the readers are referred to the Scientific Reports paper.

7.2 XRF Imaging Methods and Instrumentation

The two-dimensional scanning of large, flat surfaces using XRF has first been developed within the field of cultural heritage science, in particular for studying paintings [2]. The introduction of mobile XRF scanners creates the possibility to perform XRF imaging on paintings directly in situ, at the museum or conservation laboratory [3, 4]. The elemental distribution maps generated can identify and visualize the pigments in the paint and reveal underlayers and changes in composition. Macro-XRF has become an important diagnostic technique, not only in the examination of paintings but also of other works of art, and more recently it has also been applied in forensic studies.

The Bruker M6 Jetstream, a commercial scanner, was used in the studies described in this chapter. It allows objects to be scanned vertically or horizontally. The scanner is positioned at 1–2 cm working distance to the object (no actual contact is made). The measure head includes a rhodium-target microfocus X-ray tube (max. HV 50 kV, max. power 30 W), positioned perpendicular to the sample, and a poly-capillary optic with flexible spot sizes of 100 to approximately 500 µm depending on the working distance between sample and tube. The fluorescence spectrum is acquired using one or two 60 mm^2 XFlash silicon drift detectors (SDD) positioned at a 45° angle (energy resolution <145 eV for Mn-Kα). The measure head travels on an X,Y-stage that has a resolution of 10 µm and maximum travel speed of 100 mm/s. An area of maximum 80 × 60 cm is scanned in one session that typically lasts several hours, using a step size of 700 µm and acquisition time of 70 ms/step. Objects of larger dimensions need to be scanned in sections that are then assembled ("stitched"). The acquired data cubes can be processed – fitted and stitched – using PyMca [5] and Datamuncher [6], both open-access software packages, to produce the elemental distribution maps. Macro-XRF can detect all elements with a mass equal to or higher than of silicon. A picture of an M-6 setup to scan a black cotton T-shirt with forensic GSR traces as created in this study at the NFI is shown in Figure 7.2.

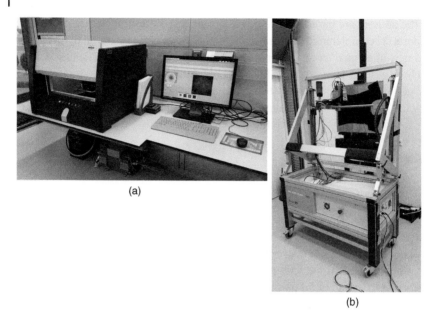

(a)

(b)

Figure 7.2 The M4 micro-XRF scanner from Bruker at the NFI (a) and the M6 MA-XRF scanner from Bruker at the Rijksmuseum "Ateliergebouw" scanning a black cotton T-shirt with GSR traces (b).

7.3 Elemental Imaging of Gun Shot Residues

GSR analysis typically involves the use of a scanning electron microscope combined with energy dispersive spectroscopy to locate and analyze potential GSR particles on small aluminum stubs that were used as a sampling tool. Combined with a visual inspection and the use of chemographic methods such as the sodium rhodizonate and modified Griess tests one can obtain information on the presence and distribution of GSRs. Although this type of investigation provides important forensic evidence, more detailed information is needed for the reconstruction of complex shooting incidents. Instead of sampling large areas and investigating such samples, in situ scanning of trace patterns would be preferable in order to preserve the trace patterns while obtaining direct information on the presence and distribution of GSR.

In this section, we show that the present generation of macro-XRF scanners has suitable characteristics (i.e. elemental sensitivity and spatial resolution) to provide such scans and thus generate detailed information which can be used for determining proper sampling strategies and for interpretation of results on activity level [7]. According to the definition in the Best Practice Manual for Chemographic Methods in Gunshot Residue Analysis, GSR analysis consists of the visualization, and identification of GSR deposited on various types of objects and persons [8]. In this paragraph, we focus on the investigation of targeted materials that often contain high concentrations of GSR-traces, especially compared to samples taken from a suspected shooter where single micron-sized particles can play an important role.

Bullet holes and their corresponding trace patterns provide essential information for the forensic reconstruction of shooting incidents. Typically, their analysis provides information on the nature of the bullet hole (entrance or exit), the bullet trajectory, and the shooting distance. In general, physical measurements of the bullet hole's geometry combined with visual investigation and sampling, and separate analysis of the trace materials form the basis for such investigations. Its non-destructive nature combined with the ability to scan large surfaces makes macro-XRF scanning a potentially valuable addition to this present pallet of approaches.

With Locard's exchange principle in mind, it is expected that a bullet traveling through a piece of textile leaves one or more traces. From this starting point, a complete T-shirt with various bullet holes was examined with a macro-XRF scanner. As the analysis of clothing and textile materials has proven, certain elements that are commonly present in bullets can be used as proper indicators for the presence of a bullet hole. Copper for example is known to be found around bullet holes as it is used for plating bullets and is one of the alloying elements in brass that in turn is often used as jacketing material on bullets. Figure 7.3 shows the elemental distribution of copper on the T-shirt that was hit by various types of ammunition. Irrespective of the shooting distance, each of the bullet holes 2–8 shows elevated levels of copper being present at the location of the bullet hole. Furthermore, one can observe complete distributions of copper around the bullet hole for short (e.g. 2.5 and 10 cm) shooting distances.

Although copper is commonly used, it can be absent if for example lead bullets are being used. A big advantage of macro-XRF scanning, however, is the simultaneous acquisition of various elements in one scan. It is therefore possible to select another element such as lead for the initial location of the bullet holes. In casework, it is feasible to base the initial selection of elements on case information (e.g. bullets found on the crime scene) or by preselecting a set of candidate elements based on general ammunition knowledge. Subsequently, the location of the bullet holes is the starting point for further analysis as the XRF-scan may constitute a detailed analysis of the elemental composition around the bullet holes to identify the use of different types of ammunition. Such differences indicate that multiple weapons could be involved in the shooting incident.

In the experiment that was performed with the T-shirt, three different types of ammunition that can be encountered in casework were used. Shots 1–5 were

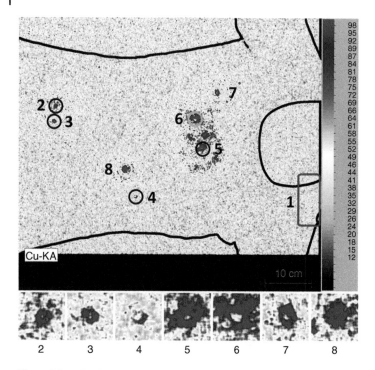

Figure 7.3 XRF Cu image of the T-shirt showing the location of the bullet holes. The different types of ammunition are indicated by different colors, i.e. CBC Magtech (black), Speer Lawman (green), and Makarov (yellow). The inserts show image details near each bullet hole (pixel scan time = 60 ms) [1].

fired with a Glock model 17 firearm using CBC Magtech ammunition that is characterized by a classic sinoxid primer containing lead, barium, and antimony. Shot 6 was fired with the same weapon directly after these first five shots.

To mimic a memory effect where residual GSR is released with a consequent shot, Speer Lawman Clean Fire ammunition was used. This is heavy metal free ammunition with a strontium-based primer composition [9]. The last two shots were fired with a Makarov (PM) type pistol with matching Makarov ammunition that is characterized by a mercury fulminate composition. After finishing the complete scan one can easily obtain elemental information from each shot by selecting the area directly around the bullet hole and identifying the peaks in the corresponding energy-dispersive X-ray spectroscopy (EDS) spectra. Examples of such spectra are shown in Figure 7.4 for several of the close range (i.e. 10 cm) shots in the T-shirts. From these spectra it becomes clear that Cu, Pb, Cl, K, Sb, Ba, Sr, and Hg can be used for screening of the various ammunitions as is illustrated in Figure 7.5.

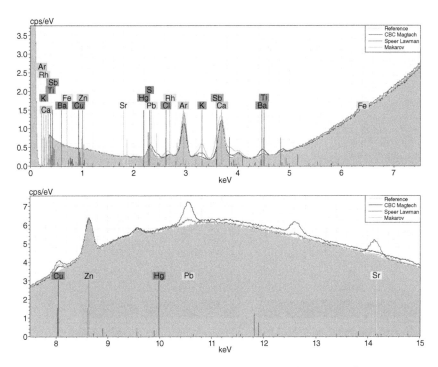

Figure 7.4 Overlay of the XRF spectra for the three ammunition types. Data were obtained from the bullet wipe area from shots within a 10 cm range (Instrument: M6, pixel scan time = 60 ms.)

Despite suboptimal conditions such as overlapping trace patterns and an ambient atmosphere suppressing the detection of low energy X-ray photons, all shots except nos. 3 and 4 can easily by identified. This approach seems therefore limited to short shooting distances as these shots were fired from 50 to 150 cm, respectively. At these distances, the converging gas cloud that emerges from the barrel is diluted, resulting in less trace material being deposited around the bullet hole. The bullet wipe (deposited by the bullet upon penetration of the textile material) is then the only trace material that is detected by the macro-XRF scan. However, under these conditions the data is of insufficient detail to identify the corresponding ammunition. This reveals one of the challenges related to the interpretation of the elemental images and corresponding spectra: how to deal with the combination of unknown parameters such as shooting distance, circumstances, and ammunition type that may affect such elemental images. One potential approach into solving these questions is the combination with other forensic evidence. The information retrieved from a high-resolution scan as shown in Figure 7.6, for example can provide interesting insights if the examination of material retrieved from the

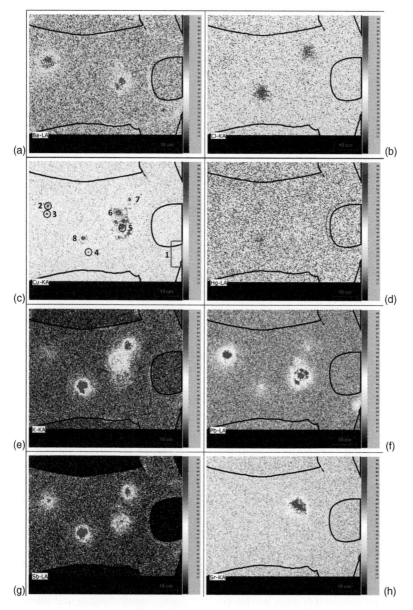

Figure 7.5 Element-specific images to illustrate how bullet holes can be matched to various types of ammunition with MA-XRF (see circles in the Cu image with CBC Magtech (black), Speer Lawman (green), and Makarov (yellow), pixel scan time = 60 ms, a = Ba, b = Cl, c = Cu, d = Hg, e = K, f = Pb, g = Sb, and h = Sr) [1].

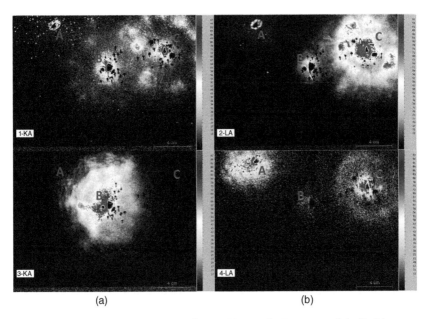

(a)   (b)

Figure 7.6 Elemental hi-res images of a specific area in the center of the T-shirt containing three shots; (a) Makarov ammunition, Russian Makarov pistol, firing range of 10 cm and 45° angle to the T-shirt, (b) Speer Lawman ammunition, Glock 17 pistol, firing range of 2.5 cm and 90° angle to the T-shirt, and (c) CBC Magtech ammunition, Glock 17 pistol fired prior to shot B using the same range and angle (pixel size = 150 μm, pixel scan time = 150 ms, 1 = Cu, 2 = Pb, 3 = Sr, and 4 = Sb).

crime scene indicates that multiple types of ammunition were fired with one gun. Based on this information one can now conclude that shot B was fired after shot C because traces of lead are visible around bullet hole B whereas traces of strontium are absent around hole C. This is a strong indication for the previously described memory effect. With the classical approach of stubbing each bullet hole followed by SEM/EDS analysis such information could easily be lost as both patterns overlap, resulting in collections of similar particles resulting in reduced information on the distribution of elements.

High-resolution elemental images also provide insight in the shooting angle. All elemental distributions around shots B and C in Figure 7.6 are evenly distributed around the bullet hole indicating that the shots were fired perpendicular to the T-shirt. The elemental image of Sb related to shot A is however strongly oriented toward the left side of the image. This is the result of the weapon being tilted toward this side of the T-shirt. As a result, the cloud escaping the weapon needs to travel a shorter distance left from the bullet hole, has less time to diverge, and thus results in a higher concentrated trace pattern. Results like this may provide

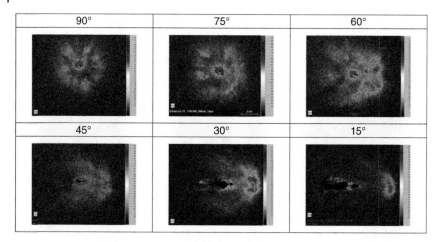

Figure 7.7 The effect of the shooting angle on the distribution pattern of gunshot residues near a bullet hole illustrated by element specific micro-XRF images. Shots were fired from close range (5 cm) at different angles of incidence with respect to the sheetrock. All angles indicate a tilt of the firearm to the right. (Instrument: Bruker M4, pixel size = 200 μm, pixel scan time = 10 ms.)

information regarding the position of the shooter in relation to the victim or target and can therefore contribute to the reconstruction of a shooting incident. Similar results were obtained for short distance shots (5 cm) on sheetrock as shown in Figure 7.7. The lead pattern around the bullet hole changes from a circular shape at a 90° angle of incidence toward a single high concentration region few centimeters apart from the actual hole at a 15° angle of incidence.

One of the main applications of XRF-scanning in GSR analysis is the investigation of potential bullet holes and the corresponding shooting distance estimation [10–13]. These studies, however, focused on scanning one specific bullet hole. In previous work [1], we have shown that data acquired during a single scan provides sufficient detailed information to perform multiple shooting distance determinations on a complete piece of evidence.

The elemental images shown in Figure 7.8 are 15 × 15 cm areas that were extracted directly from the single dataset that was obtained for the T-shirt. The images show that for each of the four elements, short (i.e. 2.5 and 10 cm) shooting distances can easily be recognized by large circular areas with elevated levels of Cu, Pb, Ba, and Sb. These highly concentrated regions are the result of the diverging cloud of materials that leaves the barrel when firing a shot. These patterns, however, fade with increasing shooting distances (i.e. 50 and 150 cm) where only the bullet wipe remains visible. The offset at which the bullet wipe remains the only trace that can be visualized will depend on several other factors in addition to the actual shooting distance.

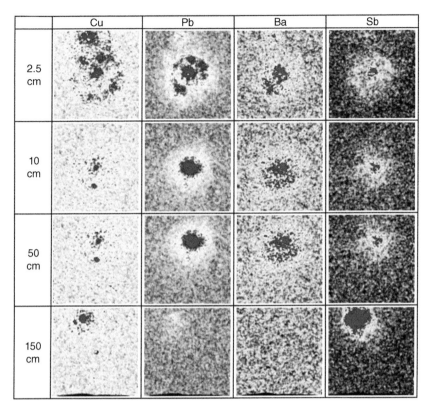

Figure 7.8 Element-specific MA-XRF images (Cu, Pb, Ba, and Sb) as a function of shooting distance. The images were created from a 15 × 15 cm selection of the full MA-XRF dataset encompassing the CBC Magtech bullet holes. Some images contain multiple GSR patterns. The bullet hole relating to the indicated distance is plotted in the center of each image (Instrument: M6, pixel scan time = 60 ms.)

Other important factors include the sensitivity of the instrument (which partially depends on the measurement parameters), the propellant and primer of the ammunition used and the substrate on which the traces deposit. Although short shooting distances could still be easily identified, the transition range to long distance will be more challenging and requires further investigation.

7.4 Using Elemental Markers to Detect and Image Biological Traces

At first sight, XRF imaging does not seem to be a suitable technique for the detection of forensic biological traces. Human biological fluids typically are

Table 7.1 Reported average elemental levels in human
biological fluids (in bold: potential body fluid marker element).

Sample	Element	Typical level (μg/ml)
Blood	**Fe**	500
	K	2000
	Cl	3000
Semen	**Zn**	150
	K	1000
	Cl	1500
Saliva	K	1000
Sweat	K	200
	Cl	1500
Urine	K	2000
	Cl	3500

For references see Supplemental Information of the Nature Scientific
Reports paper [1].

characterized through their complex biochemical composition (e.g. proteins, enzymes, and DNA) and not through their elemental composition. Furthermore, element concentrations are usually quite low and XRF is not known for its sensitivity. So successfully applying MA-XRF in biological trace detection and characterization in forensic case work could be hampered by lack of both selectivity and sensitivity. On the other hand, sensitivity of modern XRF equipment has been greatly increased and human biological samples do contain interesting marker elements as indicated in Table 7.1. To establish whether MA-XRF could successfully be used for human biological trace detection, elemental sensitivity needs to be assessed in terms of surface concentrations for realistic stains as encountered on evidence items such as clothing. If a single drop (roughly 50 μl) of blood, semen, saliva, sweat or urine is considered (either from the perpetrator or a victim) this corresponds to 5–200 μg of the elements listed in Table 7.1. Considering a final surface area of the corresponding dried stain of 1 cm^2 and detection limits that should be at least a factor of 10 lower, this translates to average surface concentrations in the range of 0.5–20 μg/cm^2 (assuming a homogenous distribution) at which the potential marker elements still need to be confidently detected with MA-XRF.

In Figure 7.9, the calibration curve is shown that was obtained with the M4 instrument by applying a fixed volume of a Zn standard solution on a fixed cotton area. Zinc is an interesting marker element for the detection of semen stains as

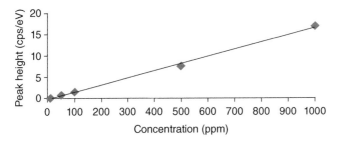

Figure 7.9 Calibration curve for Zn on μ-XRF instrument by applying a fixed volume of 670 ± 28 μl of standard solutions of given concentrations on a white cotton surface of 5.3 cm².

this element is not abundant in other body fluids and in the environment. However, the typical Zn level in human sperm is relatively low (150 ppm). Excellent linearity in the XRF response is observed up to 10.000 ppm and based on the measured noise levels a detection limit of 2.2 μg/ml is estimated which corresponds to a surface concentration of 0.3 μg/cm². Somewhat surprisingly it can be concluded that even marker elements present at relatively low levels in human body fluids can be confidently detected and thus imaged under realistic conditions.

In forensic biology, even the smallest traces can yield full STR (Short Tandem Repeat) DNA profiles that can provide very strong evidence of the involvement of the donor when the profile is matching his/her genetic reference material. Current DNA profiling techniques are so sensitive that full profiles can be retrieved from the DNA extracted from a single human cell (roughly 6 pg of DNA). Therefore, experts sample areas that based on frequently occurring modes of operandi often contain perpetrator material even if no latent traces are detected. From this perspective it is interesting to estimate the minimal trace size that could still be detected with XRF on the basis of the marker element for a given body fluid. As the trace gets smaller than the X-ray bundle interrogating the surface, the elemental signal will be "diluted" leading to increased detection limits. Mimicking this by selecting a reduced area of pixels in the calibration samples resulted in a similar elemental signal but at an increased noise level due to reduced number of counts (noise statistics). At an imaginary stain containing only 25 measurement pixels in the image, the surface concentration detection limit increased sixfold but Zn could still be detected as illustrated in Figure 7.10. It is estimated that the minimal volume of male semen that would still result in a detectable Zn signal in the XRF image is in the range of 10 μl. In a similar fashion, the minimal blood volume which can be detected by monitoring iron (Fe) is estimated to be in the low μl range.

The results described above suggest that MA-XRF can be of value for the detection of biological traces on surfaces that hamper the use of regular techniques

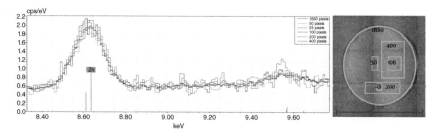

Figure 7.10 M4 XRF signal for Zn as function of stain size as modeled by the number of pixels considered (a volume of $670 \pm 28\,\mu l$ of the 100 ppm calibration solution was applied on white cotton on an area of 5.3 cm^2, pixel scan time = 10 ms).

based on indicative testing (color reactions), fluorescence, and forensic light sources. Such challenging surfaces for instance include darkly colored items which tend to absorb all incoming light or clothing items that generate a strong fluorescence background signal due to the fabric or detergent used. In such situations, the characteristic elemental contrast offered by XRF imaging can be exploited. As long as the substrates do not generate strong background signals for the elements of interest (Fe, Zn, K, and Cl) stains can be visualized through their marker elements. In the Nature Scientific Reports study [1], several proof-of-principle experiments were conducted to test the applicability of MA-XRF for realistic forensic case work conditions. First, typical blood patterns (fingermarks set in blood, passive drops from a bloody nose, expirated blood, and an impact pattern) were applied on a black cotton T-shirt. Due to the dark color, UV/vis-based techniques and visual inspection using forensic light sources do not work very well because of the limited reflectance. However, as long as the cotton does not contain high back-ground levels of Fe, K, or Cl, these elements can be used to detect the blood stains with XRF imaging as is illustrated in Figure 7.11.

Combining the three elemental signals in a single image further improves the contrast and sensitivity clearly revealing the shape and position of the four applied blood stains. This allows the BPA expert to assess and interpret the patterns whereas the forensic biology expert now knows where to sample to obtain the associated DNA profiles to identify the donor. However, with a total scan time of 17 hours, this elemental contrast requires a substantial time investment. But as the scan process is fully automated, MA-XRF images can be obtained overnight and only require operators to monitor the scan progress.

A second example presented in the Nature Scientific Reports article [1] was related to sexual assault cases where garments from female victims could contain semen traces from the perpetrator. The female underwear in this case contained strongly fluorescent fabric that hampered the use of a UV lamp to detect semen through its fluorescent properties. In addition to the semen stain (typically from

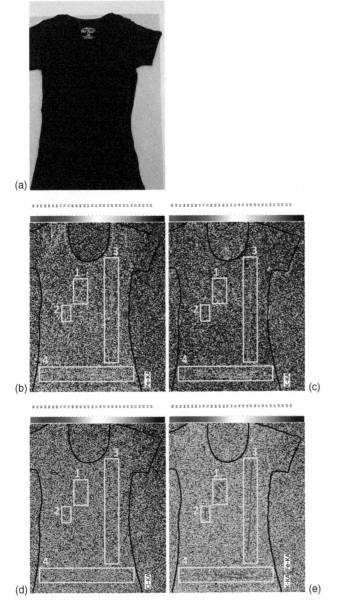

Figure 7.11 MA XRF elemental scans of Fe (b), K (c), and Cl (d) and a combined element map (e) of a black T-shirt (a) containing several blood patterns (i.e. (1) transfer pattern, (2) bloody nose, (3) expirated blood, and (4) impact pattern) applied under controlled conditions (pixel scan time = 35 ms, scan time = 17 hours) [1].

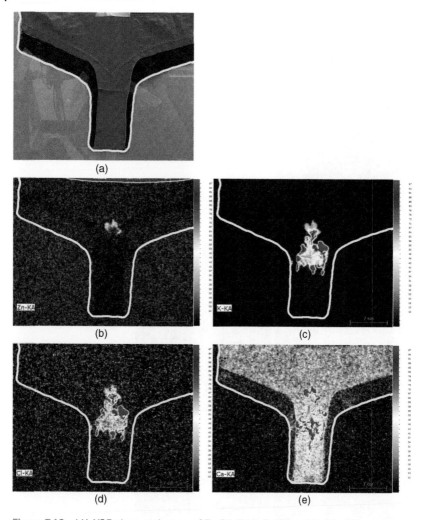

Figure 7.12 MA XRF elemental scans of Zn (b), K (c), Cl (d), and Ca (e) of fluorescent female underwear (a) containing a semen and urine stain applied under controlled conditions (pixel scan time = 35 ms) [1].

the perpetrator) also a urine stain (typically from the victim) was applied to explore the options of differentiating body fluids on the basis of the elemental composition. As illustrated in Figure 7.12, the Zn signal can be used to exclusively image the semen stain. This assists in finding the optimal spot to sample for perpetrator cell material while minimizing the contamination from victim stains.

Interestingly, the Ca image (Figure 7.12e) shows a background signal that follows the contour of the underwear. This garment was washed several times prior to use to mimic realistic conditions (victims most likely will wear clothing that is washed regularly). Depending on the hardness of the water it is well known that calcium hydroxide can precipitate in the washing machine and thus also on the washed clothing. Therefore, one can expect clothing items to contain significant Ca background levels making this element less useful for biological stain detection. However, the stains are still visible because the Ca surface concentration in the stain exceeds the Ca background. But clearly a high background will decrease sensitivity and at similar levels the stain contour will no longer be discernible. Marker elements such as Fe for blood and Zn for semen and K for body fluids in general possess a limited selectivity. Many natural and man-made materials exist that contain significant levels of these elements. In previous work we have studied elemental background levels of various fabric types. Relevant Zn levels were only found for less general fabric types such as imitation leather and checkered shear. But significant iron XRF signals were also obtained from more frequently occurring garments containing black cotton, denim or wool. Although this might hamper body fluid detection, rendering the XRF approach less effective, it will not lead to false positive outcomes as the background levels will typically be homogenous. However, such a false positive result can be instigated by a stain that is not body fluid related but contains a similar elemental profile. Such stains could for instance originate from food or cosmetics or could be the result of work-related activities (e.g. car repair, painting, and construction). Of the most frequently occurring stains, 70% contain potassium, over 50% have measurable amounts of iron, roughly 40% test positive for calcium and chloride, and 14% also contain zinc [1]. However, often the overall elemental profile as measured with XRF is sufficiently different to distinguish body fluids from other stains on clothing as is illustrated in Figure 7.13. Still the elemental profile for a given human biological matrix is certainly not unique; hence a false positive result can occur. In this respect it should be noted that biological fluids from animals, especially mammals, will most likely generate a very similar XRF response to the human equivalents. However, the impact of such a false positive result when a stain is mistakenly identified as a human body fluid on the basis of a similar "elemental signature" is limited. The subsequent DNA analysis will not yield a STR profile in which case the adverse consequence is a waste of expert time and laboratory capacity.

After the discovery of biological traces on physical evidence, it is essential to identify the donors involved. This makes it critical that the technique applied for the detection of these (latent) traces does not negatively affect the subsequent forensic DNA analysis. This is for instance an important aspect to consider for new chemical and physical methods to visualize fingermarks (although for a high-quality fingermark the mark itself can also be used to identify its donor). As it is well known that X-ray irradiation can damage genetic material (high doses of high energy photons can cause cancer in humans), the effect of MA-XRF

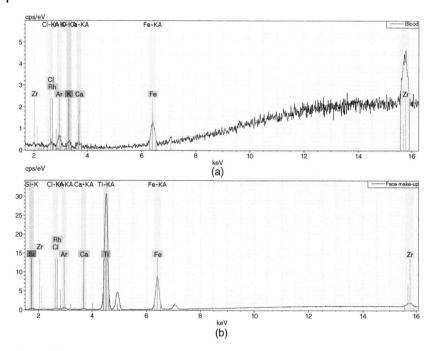

Figure 7.13 XRF elemental scans of blood (a) and foundation make-up (b) as measured on the M4 instrument [1]. For both stains Fe is present but in case of the cosmetic product this is accompanied by a strong signal for Ti.

screening on the subsequent STR DNA analysis was studied. The results shown in Figure 7.14 show that even after 10 subsequent MA-XRF scans the amount of DNA as quantified by the STR DNA analysis remains very comparable to the quantity of genetic material originally present in human blood stains. High quality profiles were obtained after MA-XRF imaging meeting all the quality criteria for forensic comparison and addition in DNA databases. This convincingly demonstrates that MA-XRF can be considered a noninvasive method for biological trace detection and imaging.

On a positive note, it is worth mentioning that sensitivity in XRF is not affected by the oxidation state or chemical binding of the element. This means that XRF is ideally suited to detect aged stains (e.g. in cold cases or when human remains or physical evidence is found long after the crime has been committed). For a blood stain that was nearly five months old, we observed a near identical XRF elemental profile compared to the profile of a fresh, recent blood stain as shown in Figure 7.13a. For severely aged and degraded stains, MA-XRF could still provide an indication of the biological nature of the material even if the genetic material has degraded to a point that forensic DNA analysis is no longer possible. Even

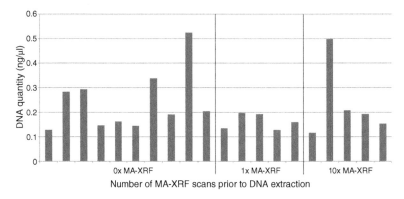

Figure 7.14 Amount of DNA quantified in extracts of sampled blood stains after 0, 1, or 10 consecutive MA-XRF scans and subsequent forensic STR DNA analysis [1].

though a potential donor can thus not be identified, the nature of the stain can be important for authentication, forensic reconstruction, and to establish whether a crime was committed.

7.5 Visualizing Cosmetic and Personal Care Product Stains

In the previous paragraph the limited selectivity of elemental XRF profiles for the detection and visualization of biological stains has been discussed. Figure 7.13 illustrates that personal care products can contain substantial levels of the marker elements for biological fluids. However, cosmetics and personal care products can also leave traces of great forensic interest [14]. Especially in the absence of DNA evidence, finding traces of female make-up on the clothing or hands of a suspect that matches chemically with victim reference material can be very valuable. Of course, the evidential value of such a matching composition is much less than that of a DNA profile match as cosmetics are produced on a massive scale and according to precisely controlled formulations. However, such evidence is clearly incriminating as in everyday life the probability of finding cosmetic traces on a man's clothing is expected to be relatively low especially when considering the specific types and brands that match the chemical composition. The forensic interpretation becomes even more complicated when the suspect and victim knew each other, were in a relation, and were living in the same household. However, these criminalistic aspects do not change the fact that XRF imaging can also be exploited to detect and image forensic traces of cosmetic and personal care products on challenging substrates.

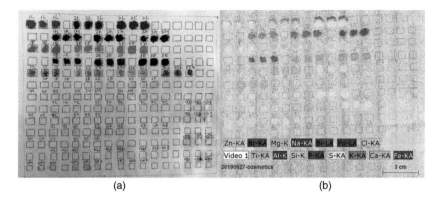

(a)	(b)

Figure 7.15 Photo of personal care sample grid on white cotton (a) and associated multi-elemental map of main elements detected on the M4 micro-XRF (b). Of each product type (on different rows) three to five brands were analyzed in triplicate (L = lipstick, M = mascara, F = foundation, E = Eye liner, S = Eye shadow, P = sunscreen, H = hand cream, C = facial cream, G = hair gel, HL = hair spray, GM = lubricant, and CD = condom lubricant).

To explore the potential of XRF for the characterization of personal care traces, a total of three specimens from different brands/manufacturers were purchased in Dutch pharmacies for various cosmetic product types. These product classes included facial cream, hand cream, sun screen, foundation, eye liner, lipstick, mascara, eye shadow, hair spray, hair gel, and (condom) lubricants and were chosen on the basis of expected occurrence in criminal investigations and forensic relevance. The XRF spectra were recorded in an efficient way utilizing the scanning capabilities of the bench top M4 instrument by applying a small amount of each class/product in a sample grid on a white cotton cloth as is illustrated in Figure 7.15. This figure also includes a multielement map showing the main elements found for the various product types. Details of the elements found in the entire sample set are given in Table 7.2.

The results show that most products contain sufficient elemental levels to be detected with XRF imaging. Furthermore, many products contain multiple elements suitable for XRF analysis indicating that the chemical contrast can be tuned depending on the substrate background levels. This provides methodological flexibility which is beneficial considering the variable, uncontrolled and often unknown conditions of the crime scene and the physical evidence. Another interesting observation is that within a specific product class different brands can have different formulations leading to significant differences in the elemental profile. This indicates that brand and product classification within a given class of materials will be feasible to some extent. This enables class comparison in addition to product identification which strengthens the evidential value of a

Table 7.2 Elements detected in personal care and cosmetic products with micro-XRF.

Type	Brand	Detected elements (in bold: abundant signal)
Lipstick	Etos Color Care – Vampires Kiss	**Ti**, **Fe**, **K**, Si, S, P, Al, Ca, Fe
	Velvet Matte Lipstick 09 – Essence	**Fe**, **Ti**, Br, S, K, Si, Al, Ca, Ba
	Rimmel London – 170 Alarm	**Fe**, **Ti**, S, K, Si, Ca, Al, Sr, Ba
Mascara	Volume Hero – Essence – Cosnova	**Fe**, K, Si, Mn
	Fear Me Fabulous – Etos	**Fe**, K, Si, Mn
	Extra Super Lash – Rimmel London	**Fe**, Mn
Foundation	Long lasting Foundation Stick – Etos	**Ti**, **Fe**, **Bi**, K, Si, Al
	Essence Cover Stick – Cosnova	**Zn**, **Ti**, **Ca**, **Fe**, Si, Na
	Hide the Blemish – Rimmel	**Ti**, **Fe**, Si, S, K, Ca, Mg, Al, Ba
Eye Liner	Essence Superfine – Cosnova	S
	Infallible Stylo Eyeliner – L'oreal	**Fe**, Si, K, Ti, Mn, Al
	Eyeliner pen – Etos	**Br**, S
Eye Shadow	Color Sensational – Maybelline New York	**Fe**, **Ti**, Si, K, Al, Ca
	Zinc about you – Essence – Cosnova	**Fe**, Al, Si, K
	Midnight Blues Eyeshadow Duo – Etos	**Ti**, **Fe**, Si, K, Al **Ti**, **Fe**, Mg, Al, Si, S, K, Bi
Sunscreen	Face Cream – Etos – SPF 50	**Ti**, K, S, Si
	Ambre Solair – Garnier – SPF30	**Ti**, K, Si, P, S, Fe, Mg, Al
	Nivea Sun Protect and Care – SPF30	S, K
Hand Cream	Vaseline Intensive Care Advanced Repair – Unilvr	Si
	Sanex Moisturizing Hand Creme – CP	Si
	Niveau Repair – Beiersdorf	**S**
Facial Cream	Olaz Double Action – Procter and Gamble	**Ti**, Si, K
	Hydra Night Cream – Etos	K
	Diadermine – Schwarzkopf Henkel	**Ti**, Si, S, K, P, Fe
	Dove Derma Spa Intensive – Unilever	**Si**, **Ti**
	Nivea Crème – Bayersdorf	S

(*continued*)

Table 7.2 (Continued)

Type	Brand	Detected elements (in bold: abundant signal)
Hair Gel	Andrelon Men – Matt Clay	**Si**, **Fe**, K, Al
	L'Oreal Paris – Studio Line – Shining Wax (5)	Si
	Schwarzkopf Junior – Style and Control Gel	S
	Studio Line – Invisi Fix Super – Strong (8) – L'Oreal	S
	Nivea Men – Styling Gel Aqua – Mega Strong (6)	S
Hair Spray	Wella – Extra Strong (3)	–
	Andrelon Pink – Pretty Perfect – Extra strong (3)	–
	L'Oreal Paris Elnett	–
Lubricant	Etos Tingle Glijmiddel	K
	Durex Play – Sensitive	–
	Sensilube – Reckitt Benckiser	–
Condom	Durex Classic Natural	**Si**, Ca
	Etos Extra Glide	**Si**, Ca
	Wingman	**Si**

cosmetic stain matching with victim reference material. The possibility to differentiate brands of the same cosmetics product type is illustrated in Figure 7.16 showing the XRF spectra for the three foundation products in the sample set. Visually, these foundation products seem quite similar which makes sense as these products are intended to mimic a certain skin color.

Exception to these positive findings is the elemental profiles of lubricants as such and as applied on condoms. As expected, a strong Si response is obtained for the condom lubricants as siloxane polymers are typically the main lubricating agents used. Interestingly, the stand-alone lubricant products do not exhibit a Si signal and may contain other lubricating compounds. In addition, only relatively low levels of calcium, iron, and zinc are found in most of the formulations. Unfortunately, the "light element" Si provides X-rays of relatively low energy that do not reach the detector under ambient conditions. Whereas the M4 instrument provides the option of XRF imaging at reduced pressure this option does not exist for MA-XRF imaging making it very difficult to map Si over larger areas using the M6.

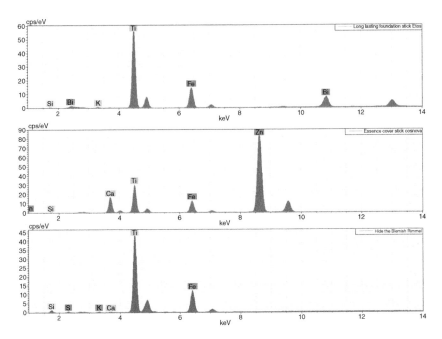

Figure 7.16 XRF spectra of the three foundation products (a = Long lasting Foundation Stick from Etos [Pharmacy Own Brand], b = Essence Cover Stick by Cosnova, and c = Hide the Blemish from Rimmel).

This currently limits the application of XRF imaging of condom lubricants which is unfortunate as this class of materials is often of importance in sexual assault cases especially when a condom has been used by the perpetrator to prevent the transfer of his biological material.

Next, as a proof of principle experiment, make-up was applied to a mannequin in a regular fashion with respect to position and quantity as can be seen in Figure 7.17. Subsequently, the mannequin was smothered with a pillow similar to a modus operandi that is sometimes encountered in criminal investigations. The pillow was pressed firmly onto the mannequin face by a male volunteer who applied substantial force to mimic an attempt to suffocate a victim. Due to the force applied cosmetic traces were transferred from the mannequin to the cotton casing of the pillow. The pillowcase was subsequently imaged with MA-XRF to establish to what extent such traces could successfully be detected on the basis of their elemental composition. The multi-elemental map shown in Figure 7.18 clearly shows the potential of MA-XRF to find, image and characterize transferred cosmetic traces.

In this example, it should be noted that the cosmetic traces can also be detected by visual inspection. However, in case of more challenging substrates such as a

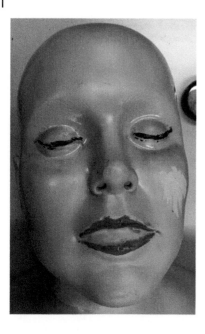

Figure 7.17 Cosmetics applied to the mannequin prior to the smothering experiment including lipstick, mascara, eye shadow, eye liner, foundation, hair gel, facial cream, and sunscreen (all products listed as third item per class in Table 7.2).

dark colored pillowcase or a color matching the cosmetic trace elemental imaging could be beneficial. Due to sensitivity limitations, trace elements as detected with the M4 scans are not easily picked up in the MA-XRF scans. So, detection of cosmetic traces under realistic forensic conditions needs to be based on those elements for which abundant signals are found in the XRF spectra of the products. Such elements include Ti, Fe, Si, K, and to a lesser extent also S. These elements are present in many cosmetic products and the elemental images shown in Figure 7.18 clearly show the presence of transferred traces of lipstick, foundation and sunscreen. Hair gel, facial cream, mascara, eye shadow and eye liner were more difficult to trace via elemental imaging. Partly because limited presence of XRF sensitive elements in these products but also the transfer properties of these products could play a role. The eye liner used in this experiment exhibits a clear Br signal in the product scan but no Br was found in the MA-XRF images. The most plausible explanation of this finding is that during the smothering only a minimal amount of eyeliner transferred from the mannequin to the pillow case. It should be noted, however, that this experiment cannot accurately replicate the actual transfer of cosmetics and creams from human skin to other objects as the transfer will be affected by the adhesion of the material to the original substrate which in this case is a hard plastic material with very different properties. But under the reasonable assumption that in case of violent attacks cosmetic and personal care material from the victim will be transferred, Figure 7.18 clearly demonstrates that

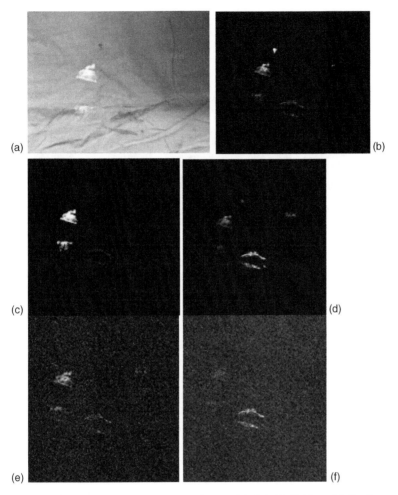

Figure 7.18 MA XRF elemental scans of Fe (b), Ti (c), K (d) Si (e), and S (f) of a cotton pillow case (a) containing cosmetic traces.

MA-XRF could be used to detect and image cosmetic traces which subsequently can be sampled and analyzed in more detail in the forensic laboratory.

7.6 Noninvasive Imaging of Hidden and Concealed Forensic Traces

An interesting feature of XRF is that while the technique is virtually noninvasive, the high-energy photons involved often penetrate the surface and thus yield

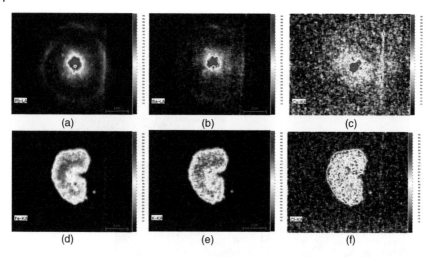

Figure 7.19 MA XRF elemental images of GSR (Pb (a), Ba (b), and Cu (c)) on a piece of black cotton covered with a blood stain (Fe (d), K (e), and Cl (f)) (CBC MAGtech 9 mm PARA ammunition fired from 10 cm distance with a Glock 17 firearm) [1].

information on the elemental composition of hidden layers. This feature of X-rays is of course used extensively in a medical setting and forms the basis of the CT scan technique or regular X-ray photos as used by dentists and doctors on a massive scale. In a forensic setting considering physical evidence, this feature can be exploited in cases where attempts have been undertaken to conceal evidence after a crime has been committed or where trace patterns of different origin show significant overlap. In this chapter, convincing results have been shown regarding the elemental visualization of GSR patterns and human biological stains. However, when a victim gets shot, the GSRs deposited on his/her garments often get covered with blood streaming out of the wounds caused by the bullet entering the body. This "contamination" can limit GSR analysis with conventional techniques (i.e. color reactions and scanning electron microscopy). However, with MA-XRF the elemental selectivity of the different materials can be used to separately image overlapping patterns. An example of this feature is given in Figure 7.19 showing the elemental images of a blood stain covering a GSR pattern. Assuming that the GSR cloud will not be altered by the blood exposure, the pattern can be revealed by mapping those elements that are specific for the GSR particles and the bullet wipe (such a lead, barium, and copper). However, also the blood stain can be visualized by selecting Fe, K, and Cl elemental lines. This approach is of course limited when elements are present at significant levels in both patterns or when the cover layer shields the X-ray irradiation (e.g. at high lead levels).

7.7 Future Outlook

The interpretation of forensic evidence at activity level is gaining more importance and as a result the location, orientation, and distribution of forensic traces on pieces of evidence is becoming more relevant. As most sampling strategies and methods are optimized to minimize pre-processing steps for the actual analysis, detailed spatial information is often permanently lost in this first critical phase of the forensic examination.

The work presented in this chapter has shown the potential for using XRF-scanning to detect and record the location, orientation, and distribution of biological traces as well as GSR. Furthermore, it is even possible to do some preliminary analysis of the various types of traces within these disciplines. Despite these results, the technique only relies on the presences of specific elements and therefore does not use the full potential of the traces in terms of their chemical and physical characteristics. Organic components for example are largely ignored as well as the physical contrast that potentially distinguishes the trace from its background or from other traces. In addition to XRF, there are numerous techniques such as infrared photography [15], hyperspectral imaging [16], and even MS-scanning [17] that are being used to create images from forensic traces.

In our opinion, the sensible and logical next step toward future forensic work would therefore be to combine multiple complementary techniques into one scanning method that allows for noninvasive scanning of a complete piece of evidence in order to record and store the location, orientation, and distribution of a large set of forensics traces. It is in art and archeology that such an approach has already proven to be very valuable in revealing previously unknown traces on antique statues and decorated surfaces [18, 19].

In forensics such scanning methods, first of all would allow for precise and information-driven sampling of the traces that are present, without compromising on other evidence. In the early stages of the forensic investigation, it is then possible to select promising areas that provide relevant and directly available information to assist in the next steps of the investigation. In addition, knowledge on the presence and spatial distribution will help acquiring optimal samples to increase the rate of success in further detailed analysis within each forensic discipline.

Second, the use of a multi-technique scanning method will enable the recording and storing of the exact state of pieces of evidence with all its traces before compromising this pristine state with sampling or further handling of the object. Later in the process of the forensic investigation, this information – combined with the analytical results from the various different forensic disciplines – can be used to evaluate case specific scenarios. By doing so, it is possible to make full use of the evidence that is present on an item of forensic interest.

It is with these benefits in mind that we strongly believe in the successful combination of multiple techniques within one single scanning method in order to get the most out of future trace evidence examinations.

List of Abbreviations

BPA	bloodstain pattern analysis
DNA	deoxyribonucleic acid
CT	computed tomography
DESI	desorption electrospray ionization
GSR	gunshot residue
HSI	hyperspectral imaging
IR	infrared
LA-ICP-MS	laser ablation-inductively coupled plasma-mass spectrometry
MALDI	matrix-assisted laser desorption ionization
NFI	Netherlands Forensic Institute
NIR	near infrared
NSR	nature scientific reports
PIXE	proton-induced X-ray emission
RADAR	radio detection and ranging
SEM/EDS	scanning electron microscopy/energy-dispersive X-ray spectroscopy
STR	short tandem repeat
TOF-SIMS	time-of-flight secondary ion mass spectrometry
UV–vis	ultraviolet–visible range
XRF	X-ray fluorescence

Acknowledgments

Information on stain types and frequency was kindly provided by Erik Krijnen and Arnold de Graaf from Unilever R&D Laundry (Vlaardingen, The Netherlands) who also donated the stain swatches that were used in the Nature Scientific Reports (NSR) study to test the selectivity of the XRF elemental profiles of human biological stains. Most of the practical work that formed the basis of the NSR paper was performed by Kirsten Langstraat as part of her MSc Forensic Science research project at the University of Amsterdam. This project was supported by Gerda Edelman from the Crime Scene Innovation unit of the Netherlands Forensic Institute (NFI). Titia Sijen from the Biological Traces Division of the NFI supervised the DNA profiling work and contributed to the discussion on the potential adverse effects of XRF radiation on subsequent DNA profiling. The DNA profiling and quantitation experiments were conducted by Linda van de Merwe at the laboratories of the Biological Traces Division of the NFI. Rene Cupedo from the Chemical and Physical Traces division assisted and supervised the shooting experiments at NFI's indoor shooting range to produce the various gunshot residue trace patterns on the black cotton T-shirts.

References

1 Langstraat, K., Knijnenberg, A., Edelman, G. et al. (2017). Large area imaging of forensic evidence with MA-XRF. *Nat. Sci. Rep.* 7: 15056. https://doi.org/10.1038/s41598-017-15468-5.

2 Dik, J., Janssens, K., Van der Snickt, G. et al. (2008). Visualization of a lost painting by Vincent van Gogh using synchrotron radiation based X-ray fluorescence elemental mapping. *Anal. Chem.* 80: 6436–6442. https://doi.org/10.1021/ac800965g.

3 Alfeld, M., Janssens, K., Dik, J. et al. (2011). Optimization of mobile scanning macro-XRF systems for the in situ investigation of historical paintings. *J. Anal. At. Spectrom* 26: 899–909. https://doi.org/10.1039/c0ja00257g.

4 Alfeld, M., Vaz Pedroso, J., van Eikema Hommes, M. et al. (2013). A mobile instrument for in situ scanning macro-XRF investigation of historical paintings. *J. Anal. At. Spectrom* 28: 760–767. https://doi.org/10.1039/c3ja30341a.

5 Solé, V.A., Papillon, E., Cotte, M. et al. (2007). A multiplatform code for the analysis of energy-dispersive X-ray fluorescence spectra. *Spectrochim. Acta B: Atom. Spectrosc.* 62: 63–68. https://doi.org/10.1016/j.sab.2006.12.002.

6 Alfeld, M. and Janssens, K. (2015). Strategies for processing mega-pixel X-ray fluorescence hyperspectral data: a case study on a version of Caravaggio's painting supper at Emmaus. *J. Anal. At. Spectrom* 30: 777–789. https://doi.org/10.1039/c4ja00387j.

7 Cook, R., Evett, I.W., Jackson, G. et al. (1998). A hierarchy of propositions: deciding which level to address in casework. *Sci. Justice* 38: 231–239.

8 ENFSI, Best Practice Manual for Chemographic Methods in Gunshot Residue Analysis, ENFSI-BPM-FGR-01 version 01 November 2015, http://enfsi.eu/wp-content/uploads/2016/09/3._chemographic_methods_in_gunshot_residue_analysis_0.pdf.

9 Knijnenberg, A., Stamouli, A., and Janssen, M. (2014). First experiences with 2D-mXRF analysis of gunshot residue on garment, tissue and cartridge cases. *SPIE Scanning Microsc.* 31: https://doi.org/10.1117/12.2066992.

10 Berendes, A., Neimke, D., Schumacher, R., and Barth, M. (2006). A versatile technique for the investigation of gunshot residue patterns

on fabrics and other surfaces: m-XRF. *J. Forensic Sci.* 51: 1085–1090. https://doi.org/10.1111/j.1556-4029.2006.00225.x.

11 Latzel, S., Neimke, D., Schumacher, R. et al. (2012). Shooting distance determination by m-XRF- examples on spectra interpretation and range estimation. *Forensic Sci. Int.* 223: 273–278. https://doi.org/10.1016/j.forsciint.2012.10.001.

12 Brazeau, J. and Wong, R.K. (1997). Analysis of gunshot residues on human tissues and clothing by X-ray microfluorescence. *J. Forensic Sci.* 42: 424–428. https://doi.org/10.1520/JFS14142J.

13 Flynn, J., Stoilovic, M., Lennard, C. et al. (1998). Evaluation of X-ray microfluorescence spectrometry for the elemental analysis of firearm discharge residues. *Forensic Sci. Int.* 97: 21–36.

14 Chophi, R., Sharma, S., Sharma, S., and Singh, R. (2019). Trends in the forensic analysis of cosmetic evidence. *Forensic Chem.* 14: 100165. https://doi.org/10.1016/j.forc.2019.100165.

15 Schotman, T.G., Westen, A.A., van der Weerd, J., and de Bruin, K.G. (2015). Understanding the visibility of blood on dark surfaces: a practical evaluation of visible light, NIR, and SWIR imaging. *Forensic Sci. Int.* 257: 214–219.

16 Edelman, G.J., Gaston, E., van Leeuwen, T.G. et al. (2012). Hyperspectral imaging for non-contact analysis of forensic traces. *Forensic Sci. Int.* 223: 28–39. https://doi.org/10.1016/j.forsciint.2012.09.012.

17 Ifa, D.R., Manicke, N.E., Dill, A.L., and Cooks, R.G. (2008). Latent fingerprint chemical imaging by mass spectrometry. *Science* 321: 805. https://doi.org/10.1126/science.1157199.

18 Alfeld, M., Pedetti, S., Martinez, P., and Walter, P. (2018). Joint data treatment for Vis–NIR reflectance imaging spectroscopy and XRF imaging acquired in the Theban Necropolis in Egypt by data fusion and t-SNE. *C. R. Phys.* 19: 625–635. https://doi.org/10.1016/j.crhy.2018.08.004.

19 Alfeld, M., Mulliez, M., Devogelaere, J. et al. (2018). MA-XRF and hyperspectral reflectance imaging for visualizing traces of antique polychromy on the frieze of the Siphnian Treasury. *Microchem. J.* 141: 395–403. https://doi.org/10.1016/j.microc.2018.05.050.

8

Characterization of Human Head Hairs via Proteomics

Joseph Donfack[1], Maria Lawas[2], Jocelyn V. Abonamah[2], and Brian A. Eckenrode[1]

[1] Research and Support Unit, Federal Bureau of Investigation Laboratory Division, 2501 Investigation Parkway, Quantico, VA 22135, USA
[2] Visiting Scientist Program, Research and Support Unit, Federal Bureau of Investigation Laboratory Division, 2501 Investigation Parkway, Quantico, VA 22135, USA

8.1 Introduction

For over three decades, DNA profiling of nuclear DNA (nuDNA) has been considered the gold standard for human identification in forensic science. Over the years, the technique evolved and took advantage of different kinds of repeat units found within regions of DNA, known as loci. Currently, the variable lengths of forensically selected short tandem repeat (STR) regions within DNA serve as markers for forensic examiners to distinguish individuals [1]. Given the number of STRs at a given locus varies between individuals, and the fact that an individual inherits a set of STRs from each parent, comparison of markers on loci that are inherited independently can lead to high statistical discriminatory power (>1 : 10^{16}, depending on the typing kit used) [2–4]. Furthermore, STR profile databases such as those included in the Combined DNA Index System (CODIS) established by the Federal Bureau of Investigation (FBI) allow for identification of possible sample contributors, even those who were not initially identified as suspects but were involved in past, unsolved cases. When combined with other mathematical approaches such as probabilistic genotyping, DNA profiling continues to be an indispensable method for evidence analysis.

However, analysis of DNA evidence is not without its limitations. For ideal interpretation, DNA must be present in adequate concentration and must be easily distinguished from contaminant DNA. In a forensic setting, DNA evidence is neither abundant nor in pristine condition [5]. Sensitivity of DNA profiling has drastically improved, allowing examiners to obtain partial or full profiles not only from blood samples but also from samples that contain trace amounts of human nuDNA, such

Leading Edge Techniques in Forensic Trace Evidence Analysis: More New Trace Analysis Methods,
First Edition. Edited by Robert D. Blackledge.
© 2023 John Wiley & Sons, Inc. Published 2023 by John Wiley & Sons, Inc.

as touch samples [6]. Yet the caveat to overcoming the limitations in DNA analysis is the detection of multiple DNA profiles from contaminant DNA, which complicates identification of minor contributors in DNA mixtures. Furthermore, harsh environmental factors can degrade DNA, resulting in segments too short for polymerase chain reaction (PCR) amplification [7]. Thus, in cases where high quality and adequate quantities, often 1 ng or more in total yield of nuDNA, cannot be recovered and DNA evidence is uninformative, other evidence types can be used to make an association between a suspect and a victim or a crime scene. One such evidence type is hair, which will constitute the focus of this chapter.

Typical forensic hair analysis workflow often begins with high-magnification microscopy to compare questioned and known hairs for human origin and class characteristics. As such, microscopical hair comparisons performed by trained forensic examiners can be regarded as a very simple, fast, and cost-effective technique for excluding questioned hairs from further examination [8]. However, this type of analysis cannot lead to discriminatory identification [9, 10] and has recently come under scrutiny, in part because it does not rely on frequency data of the microscopic hair variations observed in the general population [11]. The lack of statistics does not invalidate hair comparison via high magnification microscopy as a useful forensic analysis tool, but it would be beneficial to complement the technique with other biochemical and molecular biology approaches. As a result, microscopical hair comparison is routinely used in conjunction with nuDNA and mitochondrial DNA (mtDNA) analysis. The former is currently achieved by capillary electrophoresis (CE) and the later by Sanger sequencing of the polymerase amplified DNA loci of interest. However, in the future both might be done via massively parallel DNA sequencing (MPS) which will be discussed later.

Human hair growth is asynchronous, and each hair region goes through the three growth phases of anagen (actively differentiating and dividing), catagen (cessation of proliferation), and telogen (relatively quiescent). It is sometimes possible to obtain a nuDNA profile from a single hair, most often when root sheath cells adhere to the shaft. However, hairs obtained at crime scenes are most often in the telogen phase, when the hair can be easily shed from the hair follicle [12–14]. Telogen phase hairs are also fully keratinized, and the keratinization process results in minimal and likely degraded nuDNA in the hair shaft, if any [9, 14, 15]. Since nuDNA typing of telogen hairs has a relatively low success rate, telogen hairs instead are often analyzed for mtDNA, typically targeting the hypervariable regions of the control region (CR) [14, 16]. In a human cell, mtDNA exists in a higher copy number compared to nuDNA, and therefore is more likely to be retained in the keratinized region of the hair shaft. However, mtDNA is, in general, inherited maternally (i.e. mitochondrial genome passed down from a mother to all her children), so the analysis may not always be definitive. This is due to the fact an individual may not be uniquely identified via mtDNA typing [17]. The analysis of mtDNA is also susceptible to the same limitations of nuDNA analysis in that with increased sensitivity, there arise complexities in distinguishing authentic heteroplasmy (the presence of more than one type of mitochondrial genome within a cell or individual) from contamination in mixed mtDNA samples [18]. Microscopical hair comparisons and/or mtDNA sequencing cannot constitute a basis for personal identification.

However, when nuDNA is not available, the hair shaft has untapped potential beyond mtDNA or nuDNA typing. Recall that two of the major challenges to DNA evidence analysis deal with abundance and contamination. This is also true of nuDNA found in hair roots. Instead of limiting the analysis of hair evidence to microscopic characteristics and DNA (nuDNA and mtDNA) profiles, the hair shaft can be analyzed via its proteome to avoid DNA limitations. This is due to the fact that a single hair is a self-contained piece of evidence that can be easily separated from other hairs and washed of contamination [19]. Moreover, a single hair shaft is made almost exclusively of abundant proteins, which are more resistant to degradation than DNA [20, 21]. Already, 300 unique proteins have been identified in hair, which give the hair shaft great potential in providing biological markers [22]. Recent studies have also shown that many of these proteins can still be identified in severely damaged hair (post-blast) [23].

Early animal research demonstrated that inbred mouse strains were distinguishable by evaluating the proteogenomics of their hair shaft(s) [24]. This finding paved the way for the possibility that humans could also be differentiated by the genetic information contained and inferred from their hair proteome. This hypothesis was proven in subsequent studies involving identical twins and biological relationships [25–27]. Recently, scientists from Lawrence Livermore National Laboratory's Forensic Science Center and other collaborators [28] further expanded the forensic potential of hair shaft proteins by using the shotgun protein sequencing technique to infer non-synonymous single-nucleotide polymorphisms (nsSNPs), variation in the DNA that does change the amino acid sequence, preserved in hair shaft proteins as single amino acid polymorphisms (SAPs). In theory, an individual has a profile of single-nucleotide polymorphisms (SNPs) that are unique to that individual's genome, analogous to the STR profiles used in DNA typing [29, 30]. As such, a panel of meticulously selected SNPs can also be used for identification, or at least to establish whether two individuals are related when their SNP profiles are compared. Using a panel of 50 SNPs, Yousefi et al. reached a $\sim$7 in 10^{20} probability of identity when full siblings were not included and 1 in 10^{10} when accounting for the presence of full siblings [31].

However, unlike STR analysis, which is based on repeat lengths, SAP profiling involves the sequencing of amino acids present within peptides/proteins detected via mass spectrometry (MS)-based proteomic techniques. Particularly, protein sequence information is required at single amino acid resolution with high sensitivity and reproducibility [28, 32]. Since peptides are encoded by DNA sequences, peptide sequencing can reveal genetically variable peptides (GVPs) which can be used to infer the presence of nsSNPs occurring in an individual's nuclear genome.

Thus, by creating a panel of nsSNPs that result in SAPs, the SAP profile of questioned hairs from a crime scene can be compared to the SAP profile of known hairs to support microscopical comparison analysis. SAP profiles can be used to calculate the power of discrimination and biogeographic background because the population SNPs genotype frequencies are well known. Parker et al. [28] in their earlier work reported a maximum power of discrimination of 1 in 12 500. In their most recent work, they report even a higher discriminatory power of 1 in 10^8 which is a reflection on improvements to the technique [33].

Given all the above, the comparative proteomics technique being developed by Parker and collaborators has strong potential to individualize hairs and provide quantifiable scientific support for hair comparison. Therefore, it should be evaluated further to determine its efficacy and reliability for forensic casework implementation.

8.2 Human Hair

8.2.1 Structure and Role

Hair is a hallmark characteristic of mammals that plays important roles in environment-sensing, thermoregulation, production of pheromones, sweat and sebum distribution, and protection from chemical and physical damage [34]. The hair follicle (Figure 8.1), which serves as an autonomous stem cell repository, is responsible for the regeneration and pigmentation of the nonliving, terminally differentiated keratinized cells known as the hair shaft.

The hair shaft itself is often comprised of three main concentric layers – the cuticle, cortex, and sometimes the medulla, differences in which give humans the variety of hair types encountered on a daily basis (Figure 8.2) [35–37]. The protective cuticle consists of overlapping, transparent scale cells that anchor the hair shaft into the inner root sheath of the hair follicle as the hair shaft continues to grow [38, 39]. In addition to its function as a barrier, the cuticle carries dirt and desquamated cells away from the scalp as the hair shaft lengthens. The cell membrane complex, which contains very little protein and is composed of polysaccharides and lipids, functions to cement the cuticle layers together. Underneath the cuticle layer is the cortex, primarily composed of spindle-shaped keratinocytes that are filled with keratin proteins. The hair shaft's extreme flexibility and tensile strength is a direct result of keratin proteins derived from the cortex. Depending on their proximity to adjacent layers, the keratinocytes which populate the cortex may be of an orthocortical, paracortical, or mesocortical cell type. While all

Figure 8.1 Schematic representation of the structure and anatomy of human hair. Source: Reproduced with permission from Activlong, https://activilong.com/en/content/95-structure-composition-of-the-hair

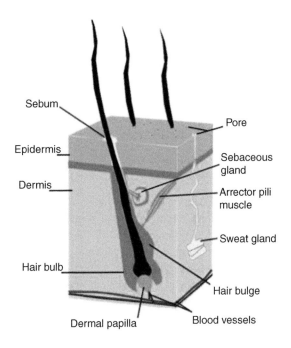

Figure 8.2 Basic structure of hair. Source: Adapted from Deedrick and Koch [35], Figure 1.

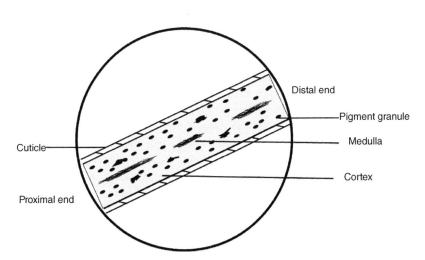

subtypes of cortical cells run parallel to the central axis of the hair shaft [40], the distribution of para- and orthocortices determine the hair fiber's curvature and strength, regardless of an individual's ethnic origin [41]. Additionally, these keratinocytes acquire melanin during hair generation, contributing to the color of the hair shaft. Lastly, unique to thick and coarse hairs, the medulla is the hair marrow which contains citrulline-filled eosinophilic granules that leave an air space in the mature hair shaft. The medulla's biological function is unclear. All layers of the hair shaft are held together by complexes that form between the cell membranes of various hair shaft cell types.

8.2.2 Growth Cycle

Generation of the hair shaft involves a complex and highly coordinated process involving the differentiation and proliferation of stem cells in the hair follicle throughout an individual's entire life [42–44]. While the upper region of the hair follicle has a constant physical presence, the lowermost bulbar region continuously undergoes cycles of hair growth or regeneration.

In the first growth phase, anagen, the hair is actively proliferating for an average of two to eight years in a given follicle [2, 45]. Hair shaft pigmentation is strictly limited to the anagen phase as melanocytes deep within the follicle inject melanin into keratinocytes, both of which are enriched in the bulge along with touch-sensitive Merkel cells [46–50]. In a process called keratinization, subpopulations of stem cells terminally differentiate into keratinocyte lineages that will form the three main layers, extruding organelles in the process [51]. As these pigmented keratinocytes continue to form the hair shaft, double-strand breaks occur in DNA [52]. This process, called cornification, occurs for a duration of approximately six hours and converts living keratinocytes into nonliving corneocytes, leaving the lengthening hair shaft virtually void of intact nuDNA under normal circumstances [53–55]. Recent studies demonstrated that remnants of abundant and highly degraded nuDNA could be extracted from hair shafts [56, 57]. Only in exceptional cases does intact nuDNA remain in the hair shaft [58].

The dystrophic growth phase, catagen, follows anagen. The hair can be in catagen for two to three weeks while the follicle experiences apoptotic bulbar involution [59, 60] and the vitreous membrane thickens around the forming club root (proximal end) [61, 62]. The vitreous membrane is also described as a glassy membrane that is often considered a hallmark of hairs in catagen [63]. Proliferation of the hair shaft and pigmentation of the hair bulb cease as the hair enters a period of regression, resulting in the degeneration of the lower two-thirds of the hair follicle [59, 60, 62].

In the third phase, relatively quiescent growth phase, telogen, the hair root forms a club that is completely depigmented and anchored to the follicle by junction proteins during two subphases of telogen that are refractory, then competent, to the induction of the next anagen phase hair cells [64–69]. Although there is a lack of proliferative activity, signaling activity continues in telogen for approximately three months until follicular cells induce anagen in a follicular space that is distinct from where the telogen club rests [38, 70]. It is in this manner that telogen hairs can coexist adjacent to newly forming anagen hairs in the same follicle, until they are shed passively or exogenically through an anagen-coupled signaling event [70–73]. Thus, telogen hairs are only shed through exogen in early anagen under normal circumstances [73–75], otherwise, they are easily shed with small amounts of mechanical force. Unlike most mammals [64, 69, 71, 76, 77], human hair growth cycles occur in an unsynchronized, mosaic fashion with energy-expensive, prolonged anagen phases [64, 78, 79]. At any given time, an individual without alopecia areata, a condition that causes random hair loss in small patches, will have a significant percentage of hairs in the anagen phase (85–90%), along with approximately 1% in the catagen phase and from 10 to 15% in the telogen phase [80].

The last phase, exogen phase, not discussed often in the literature, is characterized by active hair shedding from the scalp [51, 81]. The exogen phase is part of the resting phase and approximately 50–150 hairs are normally shed daily. The main difference between this last hair growth phase and the three that preceded is that, while hair is still firmly anchored in the anagen, catagen, and telogen phases, hairs in the exogen are not [74, 81]. Hair follicle cycling can be summarized in the figure below (Figure 8.3) [51].

8.2.3 Chemical Composition

In terms of mass, the human hair shaft is mainly composed of proteins (65–90%). The remaining constituents are water, lipids, sugars, pigment (melanins), nucleic acids, and trace elements [82, 83]. The major types of proteins found in human hairs are helical keratin proteins: alpha-keratins (KRTs) and keratin associated proteins (KAPs), encoded by multigene families [84, 85]. Previous research shows that there are about 30 families of keratin proteins coded by 54 functional genes located on chromosomes 12q13.13 and 17q21.2 [86]. Unlike other filamentous proteins, keratins must form heteropolymeric pairs between keratins that have an acidic (type I) or basic (type II) amino acid residue, allowing for complementary dimerization of type I and type II fibers in a 1 : 1 ratio, which collectively run as antiparallel dimers within a tetramere (Figure 8.4) [87–90]. Keratin tetrameres come together as protofilaments to form intermediate filaments up to $90\,\text{Å}$ in diameter, which are thought to consist of an anchoring head domain, interfacing coiled-coil rod domains, and a tail domain that aligns the two dimers [42, 91]. Together with a matrix of KAPs, intermediate filaments aggregate to form macrofibrils (up to $4000\,\text{Å}$) to make the basic unit of cortical cells in hair [92, 93].

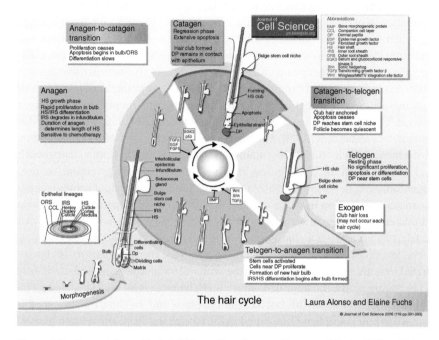

Figure 8.3 An overview of hair follicle cycling. Source: Alonso and Fuchs [51], p. 391. Reproduced with permission from Dr. Elaine Fuchs.

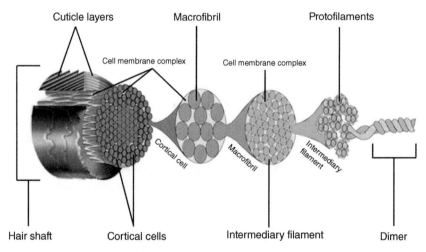

Figure 8.4 Schematic representation of the ultrastructure of a hair shaft. Featuring a highly organized structure, the hair follicle is composed of multiple epithelial layers, creating concentric circles of differentiated cell types. Source: Syed [87]. Reproduced with permission from Dr. Ali N. Syed et al.

The individual peptide chains in hair are held together by various types of covalent cross-links and non-covalent interactions, salt links, cystine bonds, and hydrogen bonds. The disulfide cross-link plays an important role in stabilizing the hair fiber, leading to its relatively high wet strength, moderate swelling, and insolubility. Trace elements commonly detected in human hair include magnesium, sodium, potassium, strontium, calcium, zinc, mercury, phosphorous, iron, manganese, cadmium, selenium, lead, arsenic, and silicon. Most of the trace elements found in human hairs are believed to have extraneous sources.

8.3 Human Hair as Forensic Evidence: The Investigative Value of Hair

An individual can shed anywhere from 100 to 200 scalp hairs a day. "Locard's exchange principle" states that when there is contact between two surfaces, an exchange of materials occurs. This implies that shed scalp hairs can be easily transferred to surfaces or other people [94]. As such, hair can be classified as a type of trace material and is often collected as physical evidence during criminal investigations, with 95% of collected hairs being telogen hairs [95]. Since hairs are prevalent, originate directly from an individual, and are highly resistant to degradation and mechanical wear, they can serve as valuable sources of evidence [96].

Microscopy is commonly used for the forensic examination of recovered human and nonhuman hairs, with emphasis on several features. However, it is often difficult to describe and interpret these microscopic characteristics in a standard, universally accepted language or methodology. Furthermore, since there are inherent variations of these non-discrete traits even within a single hair fiber, definitive frequencies for observable microscopic variations are difficult to establish [37]. As a result, hair analysis based on comparisons of morphological characteristics can be subjective, requiring the involvement of expert hair analysts. Consequently, microscopic hair examinations can be viewed as less objective when compared to DNA analysis, which leaves microscopic hair examinations vulnerable to technical challenges in court [96].

Determining an association of specific individuals (i.e. suspect(s) and victim(s)) to each other or a crime scene(s) via the analysis and comparison of known and questioned samples has long been the primary goal of forensic hair examinations. However, the morphological characteristics of hair features alone (e.g. roots) can provide useful information and answer questions that may be of interest or relevance to a case [97–99]. As previously mentioned, the characteristics of hair roots vary according to hair growth phase, and therefore, can help elucidate whether a hair was naturally shed or forcibly removed [37, 98]. For example, a forcibly removed anagen hair will have a pigmented root surrounded by elongated

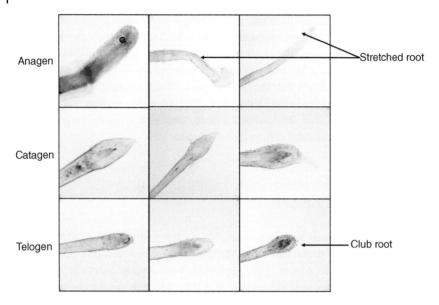

Figure 8.5 Examples of anagen, catagen, and telogen phase hair roots. Examples of anagen phase hair roots, exhibiting some amount of soft tissue surrounding the apical portion of the stretched root (top row). Examples of catagen phase hair roots exhibiting intermediate features of anagen and telogen phase hair roots (middle row). Examples of telogen phase hair roots exhibiting minimal soft tissue around a club root (bottom). Images were taken at 20× magnification under brightfield illumination. Source: Lawas et al. [32, 100], figure 1, p. 4. Reproduced with permission from Elsevier.

soft tissue, whereas a shed telogen hair possesses a club-shaped, depigmented root surrounded by little to no soft tissue (Figure 8.5) [100].

8.3.1 Physical Hair Analysis Workflow

In general, hairs must be collected, analyzed, compared, and then evaluated, to be associated with an individual [101]. Trace evidence is first examined via high magnification microscopy to discern whether collected fibers are indeed hairs [102, 103]. Once identified as a hair, the examiner makes a taxonomic classification and distinguishes between human and animal hair. If human, the body region from which the hair derives is determined (scalp, axillary, facial, limb, or pubic), otherwise the hair is deemed transitional. Several microscopic features are assessed to categorize the hair source as having either African, Asian, or European ancestry, or a combination of the major population groups. Taking all features into account, the suitability of a hair for microscopic comparison is thus determined.

In a full microscopic comparative analysis, a high magnification comparison microscope is used in a thorough side-by-side comparison of qualitative features between evidentiary hairs (questioned) and hairs of established origin (known). These features include, but are not limited to: diameter variation, cross sectional morphology, cuticle thickness variation, overall color variation, size and distribution of pigment granules in the cortex, medulla presence or absence, medulla morphology and fragmentation, medullary index, growth phase, soft-tissue presence, hair treatment, and other less common features [104]. At least 25 head or pubic hairs from a known source are used as references for a single questioned hair [103]. In general, one of three main conclusions can be drawn after a successful hair comparison. The physical characteristics of the questioned hairs may be consistent with the known hairs and an association can be made (inclusion), or they may be dissimilar to known hairs and an association cannot be made (exclusion), or the combination of differences and similarities between questioned and known hairs are not sufficient to either eliminate or include the source of known hairs as a possible source for the questioned hairs (inconclusive).

8.3.2 Microscopy (Physical) in Conjunction with DNA (Chemical) Analysis

Despite its meticulously rigorous nature and required level of expertise, comparison of microscopic characteristics alone does not constitute a basis for personal identification. Facial hairs and transitional hairs may appear to have head or pubic origin, and will be referenced according to assumed body origin, leading to possible false exclusions, or false inclusions in cases where hairs of two individuals share similar features [105]. Nevertheless, the microscopic physical hair comparison workflow requires the examiner to determine whether other techniques may be used for that specific comparison. As such, the inclusion of an individual as a possible source of hair based on microscopic characteristics must be complemented by nuDNA or mtDNA analysis. Several studies have been conducted to assess the potential of hair roots to contain nuDNA, operating under the premise that the hair growth phase impacts nuDNA availability or integrity [15, 16, 95, 106]. However, this premise is based on the idea that the greater tissue amounts are always accompanied by greater amounts of nuclei in the hair root, making the amount of tissue present at the root the primary indicator of nuDNA content [16]. This assumption is furthermore clouded by the fact that epithelial DNA was found to be the primary contributor for nuDNA in some telogen hairs [16]. However, neither hair roots bearing soft tissue nor anagen hairs guarantee successful results.

Additionally, in contrast to other forms of DNA on the hair shaft, nuDNA can be found in high abundance [107], but in degraded forms unsuitable for STR analysis [56]. Developments are underway to assess more quantitative methods to determine the suitability of hair roots for DNA analysis, particularly for recovery of alleles in STR analysis [9, 14, 100, 106].

Despite the potential of shed telogen hairs to provide valuable nuDNA information, success rates in nuDNA analysis of recently collected and aged hairs have proven to be quite low [15, 16, 57, 108]. This is because during cornification, nuDNA is specifically targeted for destruction, leaving only low quantities of varying quality in the hair shaft [55]. Failures to degrade nuDNA in the hair shaft are often an etiological manifestation of faulty enzymes, leading to hair shafts of weakened strength. In contrast, mitochondria and their DNA escape this degradation [105]. Given the fact that nuDNA is found in such scarcity for most evidence hairs, and that methods to predict the amplification success of a hair for nuDNA analysis have yet to be adopted, mtDNA is currently the most practical choice for DNA analysis for several reasons.

Mitochondrial DNA exists in high copy number of approximately 1000–10 000 copies per cell [109], making it an ideal target for extraction in severely degraded, damaged, aged, or limited-quantity hair samples with little to no nuclei. *In vivo*, mtDNA is also more resistant to DNA degradation through protective and immediate destruction of entire mtDNA molecules, even after receiving severe lesions [110] or double-stranded breaks [111], which result in the loss of mtDNA integrity for all cell types. Lower fidelity mitochondrial polymerase, Pol γ, and limited Base Excision Repair (BER) [112] play a role in mtDNA's high mutation rate, up to 10 times as much when compared to nuDNA. Theoretically, reactive oxygen species from oxidative phosphorylation was also believed to contribute to this high mutation rate, but recent studies prove otherwise [113, 114]. Prone to the high rate of mutagenesis are the hypervariable region I (HVI) and hypervariable region II (HVII) regions in the short noncoding sequence of mtDNA designated as the CR. Analysis of these sequences from questioned and known hairs or swabs samples are directly compared for similarity to determine whether or not a suspect can be excluded [115, 116].

The focus of this chapter is on hair analysis and mtDNA typing of hair. In light of the developmental processes for human hair generation, genetic bottlenecks can occur at individual follicles, leading to the presence of multiple mtDNA sequences in a single individual (heteroplasmy). With respect to race, individuals of European and Asian descent exhibit more differential heteroplasmy in their scalp hairs than African individuals [117]. These haplotypes can be shared between individuals of a population – for example, 40–50% and 1.5% of persons of European ancestry share the H and J haplogroups, respectively, with 7% of European-Americans sharing the subgroup haplotype H1 [118]. More importantly, heteroplasmy is known to occur in the CR [119, 120] and can confound results, leading to possible false exclusions. Additionally, in contrast to nuDNA, mtDNA quantity and quality decrease by several fold along the length of the hair shaft, even in recently collected hairs [56, 121, 122]. Taking these limitations into account, there is a clear need for an additional source of genetic information for the purposes of identification.

8.4 Current and Emerging Proteomic Methods for Forensic Human Hair Analysis

8.4.1 Applicability of SAPs and GVPs in Hair Analysis

Keratin proteins of the hair may provide a solution in cases where nuDNA analysis may not be possible or complement mtDNA analysis. Occasionally, a single base mutation changes the codon sequence of a protein such as keratin, thereby coding for a different amino acid [28]. The resulting amino acid, a SAP, yields a GVP, which can be used to infer the presence of a nsSNP occurring in an individual's nuclear genome [28].

In Figure 8.6, the SAP process identification is represented in the reverse order of the central dogma in biology (i.e. DNA duplication, DNA transcription, and translation) because GVPs containing SAPs are first identified at the translation level and nsSNP are subsequently inferred from SAPs.

Depending on the frequency of alleles influencing hair protein composition, the presence of multiple nsSNPs may provide inclusion or exclusion data for an individual within a given population. In their most recent work, Parker and collaborators identified 33 GVPs from hairs of 66 human subjects, most of them derived from keratins and keratin-associated proteins. From those GVPs, they identified 608 nsSNPs residing in 22 genes [28]. Of those nsSNPs 596 (98%) were true positives confirmed via DNA sequencing and 12 (2%) were false positives

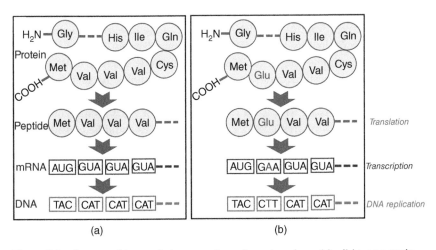

(a)   (b)

Figure 8.6 Concept of human hair comparison via proteomics – (a) wild type protein sequence exhibiting the non-variant peptide sequence [Met-Val-Val-Val-] (b) mutant protein exhibiting the genetically variant peptide sequence [Met-Glu-Val-Val-]. Source: Maria Lawas.

Protein	GVP	SAP	nsSNP rs#	Gene	Chromosomal location		Protein	SAP profile
KRT32	[K].TYLSSSCR.[A]	Q72R	rs3744786	*KRT32*	17q21.2		KRT32	72R
KRT32	[R].ILDDLTLCKADLEAQVEYLKEELMCLK.[K]	S222Y	rs2071561	*KRT32*	17q21.2		KRT32	222Y
KRT32	[R].ARLEGEINMYR.[S]	T395M	rs2071563	*KRT32*	17q21.2		KRT32	395M
KRT33A	[K].QVVSSSEQLQSYQVEIIELRR.[T]	A270V	rs12937519	*KRT33A*	17q21.2		KRT33A	270V
KRT33B	[R].DNAELKNLIR.[E]	E85K	rs12450621	*KRT33B*	17q21.2		KRT33B	85K
KRT35	[R].TNYSPRPICVPCPGGRF.[-]	C441Y	rs12451652	*KRT35*	17q21.2		KRT35	441Y
KRT35	[R].TNCSARPICVPCPGGRF.[-]	P443A	rs2071601	*KRT35*	17q21.2		KRT35	443A
KRT35	[R].VSAMYSSSPCKLPSLSPVAR.[S]	S36P	rs743686	*KRT35*	17q21.2		KRT35	36P
KRT36	[R].MVNALEIELQAQHSMR.[N]	T315M	rs2301354	*KRT36*	17q21.2		KRT36	315M
KRT39	[R].MRDSQECILMETEAR.[Y]	T341M	rs17843021	*KRT39*	17q21.2		KRT39	341M
KRT81	[R].CISACGPRPGR.[C]	S13R	rs79897879	*KRT81*	12q13.13		KRT81	13R
KRT82	[K].GAFLYEPCGVSMPVLSTGVLR.[S]	T458M	rs2658658	*KRT82*	12q13.13		KRT82	458M
KRT83	[K].CQNSKLKAAVAQSEQQGEAALSDAR.[C]	E352K	rs756220670	*KRT83*	12q13.13		KRT83	352K
KRT83	[K].LDNSRDLNMDCMVAEIKAQYDDIATR.[S]	I279M	rs2852464	*KRT83*	12q13.13		KRT83	279M
KRT83	[K].LQFYQNCECCQSNLEPLFAGYIETLRR.[E]	R149C	rs2857663	*KRT83*	12q13.13		KRT83	149C
KRT86	[R].GISCYRGLTGGFGSHSVCGGFQAGSCGR.[S]	R55Q	rs774273613	*KRT86*	12q13.13		KRT86	55Q
(a)							(b)	

Figure 8.7 Example of hair shaft keratin protein variants and associated nsSNPs – (a) GVP sequences depicting in red the amino acid polymorphism. The first GVP depicts a glutamine (Q) to arginine (R) polymorphism at position 72 (Q72R) of human KRT32 protein. rs3744786 indicates the associated nsSNP; (b) SAP profile represented by the string of all 16 SAPs. Note only the variant amino acid is represented for simplicity. Source: Adapted from Lawas et al. [32, 100], figure 2 and table 2, p. 5. Reproduced with permission from Elsevier.

not confirmed via DNA sequencing. By looking at the presence or absence of these specific GVPs, a SAP profile can be developed. Additionally, because these nsSNPs are derived from nuclear DNA (nuDNA), they have the added value of being more discriminatory than mtDNA haplotypes. The SAP profile idea may be conceptually comparable to the STR profile currently used for nuDNA comparisons between questioned and known biological samples. As such, a peptide (a small part of a protein generally between 2 and 50 amino acids in length) could be regarded as an STR locus (a small part of a DNA strand containing nucleotide repeats of interest). SAPs contained in GVPs could be considered as the equivalent of STR alleles, and the collection of SAPs identified from a hair sample would constitute the donor's SAP profile (Figure 8.7) [32]. However, a SAP profile is typically not reported in the same way a conventional STR profile would be reported (Figure 8.8) [123]. For simplicity, only peptides containing the variant amino acid (SAP) of interest is reported in SAP profiles. As such, a typical SAP profile does not include information about the heterozygous or homozygous state of the individual as an STR profile does.

The identification and characterization of peptides containing SAPs can be accomplished using traditional bottom-up mass spectrometric approaches with liquid chromatographic (LC) separation followed by electrospray ionization (ESI).

Locus	GenBank accession	Chromosomal location	Genotype (Alleles)
Amelogenin	M55418.M55419	Xp22.1-22.3Y	X, X
D1S1656	G07820	1q42	18.3, 18.3
D2S1338	G08202	2q35	19, 23
D2S441	AC079112	2p14	10, 14
D3S1358	AC099539	3p 21.31	14, 15
D5S818	G08446	5q23.2	11, 11
D6S1043	AL132766	6q15	12, 18
D7S820	G08616	7q21.11	10, 11
D8S1179	G08710	8q24.13	13, 13
D10S1248	AL391869	10q26.3	13, 15
D12S391	AF076965.1	G08921	18, 20
D13S317	G09017	13q31.1	11, 11
D16S539	G07025	16q24.1	11, 12
D18S51	X91254	18q21.33	15, 19
D19S433	G08036	19q12	14, 15
D21S11	M84567	21q21.1	30, 30
D22S1045	AL022314	22q13.1	11, 14
CSF1PO	X14720	5q33.1	10, 12
FGA	M64982	4q28	23, 24
Penta D	AP001752	21q22.3	12, 12
Penta E	AC027004	15q26.2	12, 13
TH01	D00269	11p15.5	8, 9.3
TPOX	M68651	2p25.3	8, 8
vWA	M25858	12p13.31	17, 18
DYS391	G0913	Yq11.21	–

(a)

Locus	STR profile
Amelogenin	X, X
D1S1656	18.3, 18.3
D2S1338	19, 23
D2S441	10, 14
D3S1358	14, 15
D5S818	11, 11
D6S1043	12, 18
D7S820	10, 11
D8S1179	13, 13
D10S1248	13, 15
D12S391	18, 20
D13S317	11, 11
D16S539	11, 12
D18S51	15, 19
D19S433	14, 15
D21S11	30, 30
D22S1045	11, 14
CSF1PO	10, 12
FGA	23, 24
Penta D	12, 12
Penta E	12, 13
TH01	8, 9.3
TPOX	8, 8
vWA	17, 18
DYS391	–

(b)

Figure 8.8 Example of STR loci and chromosomal locations. (a) STR loci typed for positive control (Control DNA 9947A) DNA control included in the Applied Biosystems AmpFlSTR® Identifiler® PCR Amplification Kit (b) STR profile represented by the string of all STR alleles. Note that both alleles are represented for each loci. Source: Adapted from Zhang et al. [123], table 1, p. 64. Reproduced with permission from Elsevier.

Briefly, the bottom-up methodology involves several steps. The process begins with protein extraction from hair shafts. It is important to note that a single hair strand is often analyzed for DNA or proteomic typing. Typically, less than 1 cm of the root for nuDNA typing and about 2 cm of hair shaft for proteomic or mtDNA typing. Therefore, unlike real world touch evidence that are often composed of shed skin cells from multiple human and nonhuman contributors, hair evidence can be regarded as a single source sample since a fraction of a single hair strand is used for analysis. After extraction, the purified proteins or complex mixture of protein are digested into peptides using trypsin or a cocktail of proteolytic enzymes. Following proteolytic cleavage, the complex mixture of peptides is separated by liquid chromatography and the peptide products are subsequently analyzed by MS. Peptides are identified from the generated mass spectra using database searching [124]. Several chromatographic techniques to characterize proteins and peptides are available, but we will focus on widely used techniques.

8.5 Current and Emerging Methods for Forensic Human Hair Analysis

8.5.1 Nano-Liquid Chromatography (nLC) with Electrospray Ionization and MS/MS

Liquid chromatography using nanoliter flow rates, specifically nLC, with tandem mass spectrometry (MS/MS) or high-resolution mass spectrometry (HRMS) analysis has proven successful in many aspects of protein analysis in forensic applications [125]. The stationary LC phases used in recent research, primarily discussed in this chapter, stem from investigations into the true forensic utility provided by analysis of a single hair shaft [126–129]. The phases used in these studies were vendor proprietary in nature and were used for both the pre-column and analytical column; however, these phases were primarily a variation of reverse-phase compositions. In reverse-phase mode, LC (C-18 phase or slight modifications thereof) concentrates and separates analytes, such as peptides, according to their hydrophobicity. Over the years, LC systems have evolved to improve the concentration of analytes delivered to the MS and thereby improve sensitivity, which is related to the rate of change in the magnitude of the signal produced by the analyte in the MS analyzer. This is very important in hair analysis as the keratin (KRT) and keratin-associated proteins (KAPs) dominate the sample [28, 127]. However, KAPs may still be under-represented due to their relatively small size and limited number of trypsin cleavage sites [22]. It is important to characterize and maximize peptide information derived from these high abundant proteins, however, it is also critically important to characterize low abundance proteins. The large and complex mixture of peptides in a hair shaft necessitates a technique with high resolving power (both in retention time and mass), low detection limits for low abundance peptides (lowest detectable peptide concentration), high sensitivity (high signal magnitude per unit concentration), adequate precision (minimal technical replication variations) given the bio-chemical nature of human physiology, and high selectivity (precursor as well as product ion mass resolution) for quantitative assessments. The major advantage of using nLC with tandem MS or HRMS is the increased selectivity provided by MS/MS, which translates into lower detection limits when compared to traditional high-performance liquid chromatography (HPLC), CE, and capillary electrochromatography (CEC) front ends. MS instrumentation can detect tens of thousands of peptides generated in one separation from a small amount of sample [130]. The nLC–MS interface and nLC–MS/MS analytical methodology has improved peptide limits of detection to the picomolar [131] and even femtomolar level [132]. Single hair proteomics analyses require detection limits in the femtomole range and therefore nLC–MS/MS has become the method of choice.

In general, when there is a need to detect low abundance peptides (and their associated proteins), there are two requirements that have been successfully achieved with nLC. The first requirement involves the improvement of the separation of peptides (retention time resolution) and the second requirement involves decreasing the peak width of the eluting analytes (to increase the signal-to-noise ratio) prior to the ionization process. Improvements to the post-column emitters have been successful in enhancing the overall performance of the nLC front end to the mass spectrometer. It has been suggested that nLC mobile phase flow paths are more homogenous due to the more uniform cross-sectional packing structures of small inner diameter (I.D.) columns. The decrease in packing structure variation leads to a more uniform retention factor, which reduces column band broadening [133].

nLC systems can facilitate the analysis of hair protein digests by providing longer analytical time windows, improved ionization, and improved analyte transfer to the MS. In the late 1990s, Shevchenko et al. [134] injected ~1 μl of sample to achieve an analytical time of one to two hours, which allowed enough time to acquire good quality MS spectra, to select many ions, and to successively generate MS/MS spectra. Nano-LC is a concentrating technique, while MS is a concentration-dependent instrument when using constant flow rates. Narrowing the time that the concentration of peptides is delivered to the MS improves (lowers) detection limits and improves sensitivity. A peptide solution that reaches the tip of the nanoelectrospray emitter is electrosprayed into the MS at stable nanoliter per minute flow rates. Typically, with 1 μl of sample, analysis times will range from about 30 minutes to 1.5 hours. Separation via liquid chromatography, ahead of the spray system, offers the advantage (over direct injection techniques) of concentrating the analytes into shorter and distinct analytical time windows (typically 1–30 seconds). It is during these windows in time that all MS measurements for an analyte must be performed.

Reducing the size an LC column I.D. by a factor of 10 potentially to concentrate analytes by approximately 100-fold. The (concentration of equally abundant components in the LC mobile phase is proportional to the square root of the

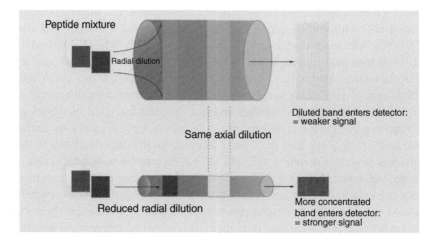

Figure 8.9 nLC concentration effect. Source: Wilson et al. [136], figure 1, p. 1801. Reproduced with permission from Future Science Ltd.

column I.D. [135]. Reduction of the radial dilution within chromatographic bands is responsible for the enhanced sensitivity observed in nLC (see Figure 8.9) [136]. A secondary advantage of nLC is decreased solvent consumption, which can decrease analysis costs and eliminate the need to change solvent reservoirs as frequently as is typically required when using conventional LC and large bore columns.

As peptides elute from the nLC, they are either preformed ions or they require a charge for entry into the mass spectrometer. Placing a charge on the analytes is typically accomplished using the technique of ESI. ESI provides a direct interface between the MS and the outlet of the nLC which is at atmospheric pressure. ESI was invented by John Fenn and Masamichi Yamashita in 1984 [137] whereby ionization is achieved by applying an electric field between the tip of a small emitter and the entrance to a mass spectrometer. ESI is a soft ionization technique that allows molecules in the liquid phase to be transferred directly into ions in the gas phase. Soft ionization techniques are advantageous in protein and peptide analysis because they produce little to no fragmentation and thus preserve ionized macromolecules. The electric field forces the charged liquid at the end of the emitter to form a cone, called the Taylor cone, that minimizes the charge/surface ratio. Analytes and LC solvent leaving the LC form droplets that rapidly traverse a repetitive process of solvent evaporation and charge density changes. The droplets become increasingly smaller and more charged as solvent evaporates. When the electrostatic repulsion of like charges becomes more powerful than the surface tension of the droplet, it explodes, creating many smaller, more stable droplets.

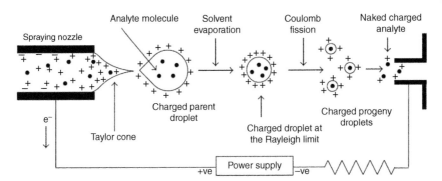

Figure 8.10 The electrospray process. Source: Banerjee and Mazumdar [138], figure 6, p. 7. Reproduced with permission from the authors.

This process continues until eventually the solvent is eliminated and only charged analyte gas phase ions remain (see Figure 8.10) [138].

At the lower flow rates used with nLC, ESI generates smaller droplets, which leads to improved transfer of ions into the MS [139]. Smaller droplet formation also eliminates the need for sheath gas, which aids in solvent evaporation from the larger droplets formed in typical LC–MS. The pH in the droplets decreases which leads to the formation of charged and multiply charged analytes. ESI produces a rich set of multiply charged ions that then provide adequate MS/MS spectra with a sufficient signal to noise ratio (S/N). For peptides, it is advantageous to obtain +2 ions as +1 ions typically do not produce information rich MS/MS spectra. Also, the multiply charged peptides can be effectively fragmented by collision-induced dissociation (CID) in the high energy collision dissociation (HCD) trapping region of a tandem MS such as a hybrid MS.

Within the last decade, ESI has been found to be advantageous for the analysis of low-level protein digests. Working with small sections or segments of human hair can yield digests that are in the low μg levels. Using nLC methods for hair proteomics allows sample amounts to be as small as 1 mm [126] or 1 in. [128]. With nanoliter flow rates provided by nLC, the flow requirement (50–500 nl/min) can be achieved while maintaining a stable flow rate by reducing the internal and external diameter of the sprayers. For example, in the studies mentioned above, involving segmented hair analyses conducted for proteomic determinations, a nanospray needle/emitter was used that had been tapered down to approximately 10 μm in diameter. This approach is necessary for the low-level (typically only a few μg of protein) samples that we experience in forensic hair analysis. Forensically relevant hair shaft samples sizes (e.g. <2 cm in length) can be analyzed by completely dissolving the matrix; however, it is important to avoid hydrolysis of peptide bonds that hold proteins together when employing this technique.

It is important to select a highly reliable nLC system that can handle the required pressures with reliable and stable flow rates both with and without a gradient to reduce the limitations and issues described above. A nLC from any of several vendors (Easy nLC – Thermo Scientific; Nanoflow LC System for MS – Agilent; Prominence nano – Shimadzu; Nano Acquity UPLC System – Waters) can be used due to their updated and advanced features, such as:

- nanoliter flow rate stability,
- pressure capability and stability,
- automated high-pressure injections,
- solvent handling and low dead volume mixing in microchambers under high pressure,
- high pressure valves and flow switching dynamics, and
- easy to use front-end control, maintenance, and diagnostics.

Reduction of dead volume (regions within a pumping and elution system that are larger than the analytical flow path or unnecessary gaps within the flow path) is an important capability and performance factor because when using low flow rates of only a few hundred nanoliters per minute, any dead volume will act as a small mixing chamber and cause reductions in sensitivity and separation efficiency [135].

8.5.2 Proteomics Analysis with Tandem/Hybrid Mass Spectrometry

There is an increasing number of MS-based analytical methods for protein and peptide analysis, including top-down, middle-down, and bottom-up, with bottom-up being most widely used, especially for hair proteomics. In the bottom-up protein analysis strategy, characterization of the structure can be performed by MS/MS or MS^n (multiple stages of precursor ion selection and subsequent generation of product ions) of the intact protein or by chemical and enzymatic reactions aided by MS analysis. The top-down approach provides the molecular mass data of intact proteins, which can be used to identify post-translation protein modifications or to disclose the presence of any isoforms [140]. However, the top-down approach cannot be used at the peptide scale due to the lack of effective intact protein fractionation strategies. Bottom-up analysis involves the digestion of protein mixtures, fractions, or two-dimensional electrophoresis (2DE) purified proteins by chemical or enzymatic proteolysis (e.g. trypsin digest) and analyzed by MS directly or after a separation method is applied. The basic analysis scheme is shown in Figure 8.11 where the resulting MS/MS profiles can be used for quantitation.

The MS/MS profiles of the digests are then analyzed in bioinformatics databases by matching for protein identifications. This method is called "shotgun

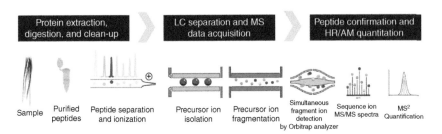

| Protein extraction, digestion, and clean-up | | LC separation and MS data acquisition | | Peptide confirmation and HR/AM quantitation | |

Sample | Purified peptides | Peptide separation and ionization | Precursor ion isolation | Precursor ion fragmentation | Simultaneous fragment ion detection by Orbitrap analyzer | Sequence ion MS/MS spectra | MS² Quantification

Figure 8.11 Basic analysis scheme for parallel reaction monitoring. Source: Adapted with permission from Thermo Scientific.

proteomics" [130, 140]. An expanded overview of a bottom-up analysis scheme from the front-end sample processing through nLC and then tandem MS to the back-end data analysis and bioinformatics is shown below in Figure 8.12 [141].

In the hair analysis scheme, typically 1 µl or less of the protein digest is pressure loaded onto the nLC system. To accomplish this in a practical sense requires the smaller I.D. columns and lower flow rates now available commercially as accessories to nLC systems. For example, the column and precolumn configuration used in this study is shown below in Figure 8.13.

In most cases, as shown in Figure 8.13, a precolumn was used to pre-concentrate the peptide samples and effectively wash the trapped peptides prior to their release onto the analytical column for separation. Use of a trapping or precolumn can help avoid problems of column overloading, peak broadening, and poor separation that may arise from direct on-column injection of a larger, more dilute sample. Injecting a larger amount of sample onto a precolumn with a low elution power mobile phase can preconcentrate, clean-up, and focus the sample in one step [135].

In hair analysis, a gradient of increasing hydrophobicity is typically used with acetonitrile as the solvent and formic acid (0.1%) added to improve retention. The peptides typically elute in the order of their hydrophobicity. They are charged and multiply charged in the ESI interface and then accelerated into the MS vacuum system and ion optics. The eluting analytes successively trigger the MS to select one of the analytes (hopefully well separated from other charge competing peptides) and to execute and obtain the MS/MS spectra. CID followed by product-ion scanning can provide systematic fragment information of amino acid sequences [142]. This is done by generating oligopeptide "ladders" inside the MS that are used to determine partial amino acid sequences of those peptides. Trypsin digestion is often used to create sequence-specific cleavage reactions that produce peptides that are unique enough to identify their source proteins from the mass spectra produced by CID [142]. Fragmentation spectra can give additional information for structural peptide identification. In addition, single-ion monitoring (SIM), multiple/selected-reaction monitoring (MRM/SRM), or parallel-reaction

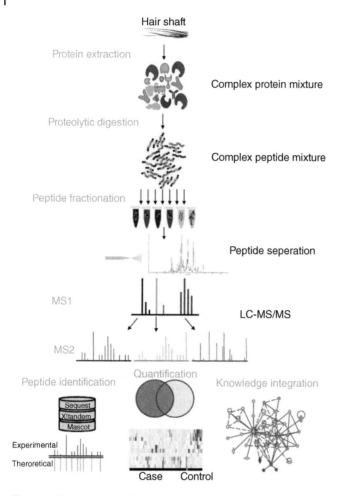

Figure 8.12 Overview of bottom-up proteomics with human hair sample processing and proteome discovery scheme. Source: Adapted from Angel et al. [141], figure 1, p. 3194. Reproduced with permission from The Royal Society of Chemistry.

monitoring (PRM) scan modes increase selectivity and sensitivity by avoiding isobaric contaminants and nearly eliminating background noise [140, 143]. The MS/MS spectra generated are then used to search protein databases.

ESI is often used in conjunction with a hybrid quadrupole-Orbitrap MS. There are several advantages to using nLC and ESI as an interface to an HRMS such as the hybrid Orbitrap system described here and shown in Figure 8.14.

With the new advanced hybrid quadrupole-Orbitrap mass spectrometers now commercially available combined with an improved concentration effect enabled

Figure 8.13 nLC precolumn and analytical column with an ESI interface to a hybrid quadrupole-Orbitrap MS. Source: Brian Eckenrode.

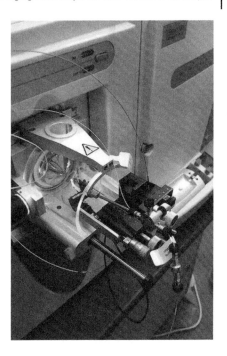

by nLC, and with improved control of the different MS/MS experiments, both data-dependent analysis (DDA) and data-independent analysis (DIA) approaches can be conducted. Each approach utilizes automated acquisition, however with DDA the user specifies how MS/MS data are to be acquired, whereas with the DIA approach the user captures as much information as possible and works on the back-end, typically via software, to sort the results. Each approach has its advantages and disadvantages and will be only briefly described here.

When conducting DDA, the typical choices and options an analyst has within an instrument's data system are:

(1) automated selection of precursor ions based on a threshold or a S/N ratio,
(2) the nth-most intense ion selectable for fragmentation as predefined by the user,
(3) a static exclusion list of ions (pre-definable),

Figure 8.14 Hybrid quadrupole-Orbitrap Exploris mass spectrometer used in proteomics. Source: Adapted with permission from Thermo Scientific.

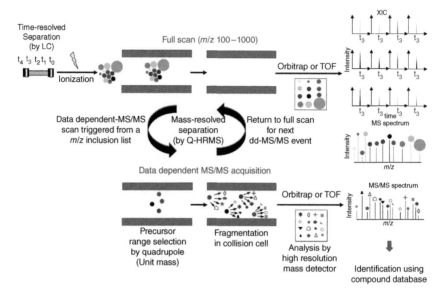

Figure 8.15 Automated instrument control software enables DDA for efficient and effective MS/MS analysis in a hybrid quadrupole-Orbitrap/TOF mass spectrometer. Source: Wong et al. [144], figure 2, p. 9575. Reproduced with permission from the American Chemical Society.

(4) the ions for which x number of MS/MS spectra have been generated can be added to a dynamic exclusion list until their intensity in MS mode falls below a threshold value or until a selectable time has elapsed,

(5) an option to add isotopic ions into the exclusion list in a dynamic fashion,

(6) an option to exclude specific charge states, and

(7) an ability to use an automated selection of fragmentation energy, typically approximately 30 eV for peptides.

The static exclusion list (3) allows for the exclusion of known contaminants, for example, tryptic autolytic peptides, clusters, and any other known peptides. The dynamic exclusion of ions (4) allows the selection of the next most intense ion for the next round of MS/MS spectra. The ability to dynamically add isotopic ions into the exclusion list (5) is particularly important for higher resolution instruments. This feature ensures that the isotope peaks from an ion do not trigger the acquisition of an MS/MS spectrum. With the ability to exclude specific charge states, the analyst can eliminate singly charged (+1) ions, which can generate less informative MS/MS spectra. Exclusion of these less informative MS/MS spectra decreases analysis time and file size. A diagram of the ion flow path through a hybrid quadrupole mass spectrometer for DDA is shown below in Figure 8.15 [144].

Full MS scan (*m/z* 100–1000)

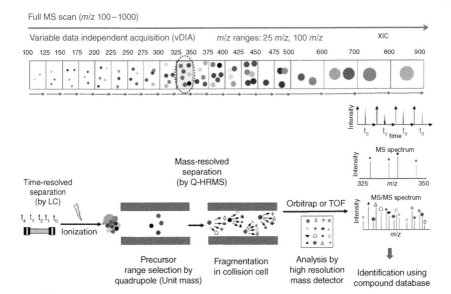

Figure 8.16 Analysis scheme for DIA using either an Orbitrap or a time-of-flight MS. Source: Wong et al. [144], figure 3, p. 9576. Reproduced with permission from the American Chemical Society.

Even though DDA has been used successfully by many researchers in proteomics, it is important to discuss briefly the DIA approach for completeness as this is gaining in popularity with the advent of large data storage capabilities and more sophisticated software. DIA allows for the detection and identification of lower abundant peptides that may be otherwise not recorded with conventional DDA methods. DIA approaches are not new but they are receiving greater attention since the advent of techniques such as MS^E (Waters) and sequential window acquisition of all theoretical fragment ion spectra (SWATH, SCIEX) [145]. In SWATH-type DIA, precursor selection windows are defined in the quadrupole (MS1) and all ion m/z within each window are fragmented in the collision cell (HCD) and collected in a subsequent composite spectrum generated from the MS2. The ion path and scheme are shown in Figure 8.16 below [144].

In SWATH-type DIA techniques, all precursor ions are fragmented in a series of quadrupole isolation windows. The complete precursor/product ion data of all the peptide ions and their product ions are recorded through the whole chromatogram. The isolation windows can be sequential or staggered. Sequential windows can be either fixed (e.g. 20 Da) or variable (i.e. the window width is not uniform) and are selected depending on the selectivity required and the cycle time, which must be as short as possible when combined with nLC due to signal-to-noise constraints. In MS^E analysis, all precursor ions in the entire specified mass range are fragmented to acquire MS^2 spectra. DIA is now widely used in proteomics because of its reproducibility, speed, compound coverage, and quantitation accuracy. However, DIA spectra can be difficult to interpret without stable isotope labeled standards and/or staggered isolation windows. This may be problematic in forensic science or courtroom settings if these limitations are not considered [143].

HRMS instruments are also useful for both CE and nLC as they have a fast scan rate compatible with both methods. Mass spectrometric parameters have been shown to affect protein and peptide identification rates for bottom-up proteomic analysis [129]. It is important to optimize both the separation method and the MS method for peak identification performance.

Overall, there are some limitations and experimental nuances with nLC when used to characterize the complex mixture of hair shaft proteins. For example, reproducibility and repeatability are not as robust as traditional LC methods often targeting a very limited set of peptides. nLC typically generates more unique high-confidence sequences but with less reproducibility [146]. Carryover is sometimes encountered, and the washing step between runs is time- and resource-consuming. The use of an elution gradient can also produce inconsistent electrospray conditions throughout the separation [147]. Hydrophilic peptides can get lost during the washing step of the precolumn, and phosphopeptides can be suppressed by coeluting peptides [147].

8.5.3 Hair Proteome Sequencing Via CE-MS/MS

CE is a complimentary, orthogonal technique to protein analysis via HPLC, and capillary zone electrophoresis (CZE) in particular offers a high separation efficiency, low carryover, small sample volume, short analysis time, high resolution, inexpensive capillaries, compatibility with many volatile buffers, and a setup with a low risk of analyte loss [129, 130, 140, 148]. While forensic hair analysis using CE for protein and peptide identification has not been widely explored, it has been shown to be a viable technique [149] and is currently widely used for protein analysis of other sample types (e.g. blood, urine, and soft tissue).

The average number of spectra per unit of time is slightly higher for CE than nLC, but the total number of spectra obtained is typically much lower [146]. A greater number of spectra per unit time requires an MS analyzer that is capable of fast scanning, such as a time-of-flight (TOF) or Orbitrap instrument. Like nLC, CE also consumes less solvent than traditional LC methods. Costs are further reduced because unmodified fused silica capillaries are used and can be reused several times. Fused silica capillary tubing is significantly less expensive than chromatographic columns used for both LC and nLC [140]. Solvent consumption can be further lessened by employing sheathless CE-ESI interfaces. These interfaces operate at nanoflow rates and do not require sheath liquid (see Figure 8.17 below

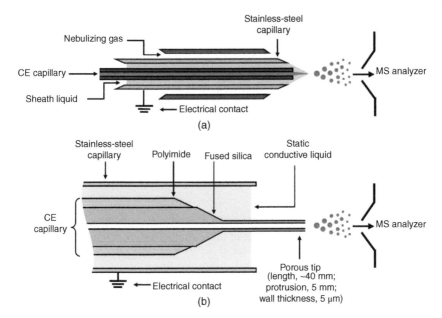

Figure 8.17 Schematic representation of a sheath-liquid (a) and a sheathless (b) CE-MS interface. Source: Domínguez-Vega et al. [150], figure 1, p. 27. Reproduced with permission from Springer Nature.

[150]). Not only do sheathless methods decrease solvent consumption, but they can also reduce ion suppression and increase electrospray sensitivity because the sample is not diluted in the sheath spray [130, 151].

CE does have limitations though as sheathless interfaces can be fragile and require a separation buffer that supports ESI [130]. In addition, many buffers typically used for CE are not volatile enough to carry proper separation and CE-MS is less robust and is less reproducible compared to HPLC-MS. Migration times and peak areas can be inconsistent, which is problematic for robust lab work or MS interfacing [148]. Proteins and peptides may adsorb to silanol (-SiOH) on the CE capillary wall, leading to sample loss, peak broadening and tailing, and poor reproducibility. This can be overcome by using background electrolytes (BGEs) at high or low pH in uncoated capillaries. Typically, a low pH BGE is used to generate positively charged peptides. Static or dynamic coating of the capillary wall can also be used to overcome this problem [152]. Sample volumes are also limited in CE to limit band broadening, and low abundance components may not be detected [130]. Concentration detection limits for CZE-MS is relatively poor due to the small injection volumes [153]. However, this can be overcome with the use of sample pre-concentration techniques such as solid phase microextraction (SPME), liquid–liquid extraction, pH-mediated stacking, field-amplified stacking, and transient isotachophoresis (ITP) [152]. High electroosmotic flow (EOF) produced by silanol groups on the inner surface of the separation capillary results in short separation windows that limit the number of MS and MS/MS spectra that can be obtained. Using state-of-the-art higher scan rate HRMS instruments can assist in overcoming this limitation.

As CE instrumentation improvements in HRMS instrumentation continue, CE-MS may become a valuable alternate method to nLC–MS, particularly in cases where speed of analysis is an issue.

8.6 Challenges to Implementing Protein Sequencing in Forensic Casework

SAP profiling must overcome a wide array of practical challenges to be adopted into routine forensic casework. Some of these challenges are discussed below.

8.6.1 Triage of Evidence and Prioritization for Examination

When it comes to forensic analysis of biomolecules, it is essential to maximize the extraction of the target molecules from biological evidence. This is true even for DNA analysis which is regarded as the gold standard of forensic science. Indeed, successful DNA typing begins with the proper collection and DNA extraction, and the same is true for protein extraction. It is obvious that DNA or protein extraction protocols are destructive to the biological sample. As a result, a forensic examiner will have to choose between DNA extraction or protein extraction from the same hair evidence because samples are often very limited. In addition, validated DNA extraction protocols used in forensic laboratories do not preserve proteins for further analysis.

On the other hand, and contrary to common belief that no nuDNA is present in hair shafts due to cornification (DNA degradation process), recent research studies argue otherwise [56, 57]. For example, Brandhagen et al. demonstrated enough highly fragmented nuDNA can be extracted form hair shafts or telogen hairs [56]. Brandhagen et al. research further demonstrated nuDNA is present in high quantity when compared to mtDNA. Furthermore, DNA techniques are well-established and validated in forensic DNA laboratories and hair proteome sequencing is an emerging technique that needs substantial improvement and protocol development. In other words, because hair evidence is often in limited supply, DNA extraction will likely always be given the priority, thus posing an immense challenge for hair proteome sequencing as a useful tool in forensic casework. To add another layer of complication, forensic laboratories are quickly embracing MPS which generates a massive amount of genetic information. In fact, Brandhagen et al. used a Whole Shotgun Sequencing method (i.e. nontargeted

PCR amplification followed by DNA sequencing of PCR amplicons) to achieve their overarching goal. Brandhagen MD et al. randomly fragmented the DNA extracted from hair shafts and then cloned the resulting small fragments into a vector for sequencing. This strategy facilitated the characterization of nuDNA in telogen hairs and hair shafts that was not possible using other techniques and provided evidence that nuDNA existed in high quantity but low quality. The road to forensic application of hair protein sequencing in casework is further hindered by the commercial availability of only a few panels targeting the amplification and sequencing of the whole mitochondrial genome. Acknowledging that hair protein sequencing and DNA analysis are complementary techniques for forensic investigation, it is essential that the sample preparation methods used for protein sequencing are also compatible with downstream DNA analyses. Thus, for the full promise of hair proteomics to be realized for human hair comparison, a dual protein–DNA extraction protocol from the same evidence must optimized for forensic size samples.

8.6.2 Need for Validated Protocols and Appropriate Quality Assurance/Quality Control Procedures

To date, there is no consensus about which hair protein extraction protocol yields the highest concentration of protein with minimal external chemical amino acid modifications (nonbiological). Indeed, differences in sample preparation and protein extraction workflows are known to influence protein sequencing results, thus hindering the protein sequencing adoption as a tool for hair comparison [22, 25, 27, 28, 129, 154, 158].

When it comes to hairs, a wide array of protein extraction methods have been evaluated including: buffers containing sodium deoxycholate detergent, high/low urea concentrations, and ProteaseMax™ surfactant [28], thio-urea/urea mixtures [126, 127, 155, 156], sodium dodecyl sulfate (SDS)-based buffers [22, 25, 154]; sodium hydroxide and SDS-based buffers [157]; microwave-assisted extraction via mortar and pestle pulverization; ceramic bead beating [22, 25, 28, 117, 156] and direct microwave-assisted extraction [129]. Clearly, there is no consensus about which hair protein extraction method should be adopted for forensic hair protein sequencing.

In addition to issues related to the lack of hair shaft protein extraction protocols standardization of, preliminary peptide profiling results demonstrated the need for substantial improvement to the proteomic technique [28, 100, 126, 127]. For example, a core set of candidate proteins of interest to forensic examiners need to be empirically established. This task can be further hindered by the fact that is a malleable and the use of some cosmetic products can have a negative impact on the detection of GVPs. In addition, these proteins need to be experimentally verified to exhibit a robust detection across a large sample population via MS.

The primary challenge during the discovery phase of a core set of hair proteins of interest is that some proteins will be expressed in high abundance and others in low abundance and the difference between the two classes could be on the order of 10 or more orders of magnitude, further making it difficult to detect less abundant hair proteins [127]. However, candidate proteins must exhibit sufficient allelic variability between individuals (i.e. not population specific) and must be present in sufficient quantity in hair for reliable detection with maximum sequence coverage during examination to avoid the stochastic effects. Further, candidate proteins must also exhibit consistent peptide sequence across an individual. Thus, protocol development of low abundant proteins (i.e. mainly non-keratin proteins) needs to be developed [125]. This is especially important when considering the unavoidable stochastic effects which arise from low amounts of hair shaft samples commonly encountered in forensic settings. In addition, there is a need to identify and validate a panel of both common (>1% frequency in the population) and rare (<1% frequency in the population) SAP/GVPs that are robustly and reproducibly detected by mass spectrometry. Sequencing of peptides derived from low-abundant proteins should increase the number of independent (minimal genetic linkage) SAPs to be included into this panel of SAPs. Another way to improve SAP detection is to boost protein sequencing coverage. This can be done through various means, including improvement of existing protein extraction methods and the use of proteinase cocktails during protein digestion that should help modulate the average peptide length [32, 127, 129]. More importantly, data acquisition should reveal the presence of additional SAPs if acquisition methods are adjusted to overcome limitations of LC–MS/MS. Additionally, choosing a more appropriate database size can help identify more peptides in mass spectral data. Once suitable peptides have been determined, the custom database may be refined to represent a collection of core genetically variant proteins. While steps have been made to address some of these issues [158], until more work is done to address the complexities of peptide profiling in a forensic context, the technique is not yet mature for forensic casework use.

8.6.3 Not Amenable to Databasing vis-a-vis CODIS

As indicated previously, protein sequencing can potentially be used when standard STR-based analysis is not possible. Protein sequencing will detect GVPs that are associated with nsSNPs thus linking SNP genotyping and SAP typing. Consequently, SAP typing will share the same biological and database issues that hinder SNP genotyping implementation into forensic casework. The biology at any given genetic locus shows that SNPs are biallelic and thus are not as informative as STR loci. As a result, more forensically selected SNPs are needed to reach STR match probabilities for individual identification. In addition, forensically selected STR loci have little to no linkage disequilibrium among them. This is the case for the keratin proteins that constitute 65–90% of the hair shaft mass [82, 83]. In Figure 8.7, for example, the hair keratin proteins identified reside on two main independent loci as opposed to 26 independent loci for STR in Figure 8.8.

Another major challenge to the adoption of autosomal SNPs in forensic casework is the nature of existing forensic databases. Currently, STR-based databases are well established and reliably used for individual identification. Furthermore, the ability to database DNA profiles is highly regulated and there currently is no legal authority to database an individual's SNP data. While SNP typing holds significant promise for forensic human DNA typing, databasing SNP information will require considerable changes to legal authorities, laboratory operations and workflow, database structure, searching algorithms, and many other related topics.

8.6.4 Variable Protein Expression

Another challenge that must be addressed is inter-individual biological variability in protein expression which can result in false exclusion. While the DNA copy number (genotype) in a cell is constant, the number of proteins (phenotype) in a cell could vary depending of the physiological state of the cell [64]. The daily to yearly changes in protein expressions and quantity could pose a problem for protein sequencing and forensic tools for hair comparison, because it would be very difficult to determine if the non-detection of a GVP of interest is the result of biochemical proprieties or the natural (due to biology) abundance of related proteins. In addition to these quantitative changes in protein expressions, other qualitative changes can occur in the biological samples without a clear knowledge of the conditions that produced them. Examples of such changes include: chemical amino acid modifications due to the chemical composition of buffers used for hair shaft protein extraction; posttranslational modifications (biochemical modification that occurs to one or more amino acids on a protein after the protein has been translated by a ribosome); mono-allelic expression (only one of the two copies of a gene is active, while the other is silent); and epigenetic effects (a change in phenotype without a change in genotype). All these confounding factors can hinder the association between the expressed protein and a GVP, ultimately resulting in false leads and further complicating the adoption of protein sequencing as a tool for hair comparison.

Protein sequencing must meet the standard of other forensic disciplines such as fingerprint analysis, DNA profiling, and forensic toxicology via assay internal validation, all of which have demonstrated expected performance in the laboratory from an accumulation of test data within the laboratory. "Prior to using a procedure for forensic applications, a laboratory shall conduct internal validation studies" [159]. Protein sequencing validation as a hair comparison technique should be done in accordance with standard recommendations in the forensic science's arena such as those established by the American Academy of Forensic Sciences Standards Board (ASB), the Scientific Working Group on DNA Analysis Methods (SWGDAM), and FBI's Quality Assurance Standards for Forensic DNA Testing Laboratories. These recommendations address issues associated with sensitivity and stochastic effects, precision and accuracy, mixture analysis, stability, carryover, and species specificity (Figures 8.16 and 8.17) [159, 160].

8.7 Conclusion

Protein sequencing via mass spectrometry is among the most promising methods in the emerging field of chemical-based hair comparison (not individual identification) which leverages genetic information contained in GVPs to facilitate hair comparison. Although the field of proteomics remains largely underutilized by the forensic science community, it has the potential to play a critical role by expanding the tool box of forensic techniques used to evaluate common forensic evidence for which intact nuDNA may no longer be available for STR analysis, such as hair. In addition, protein sequencing may be even more valuable by providing investigative lead in other cases where there a no need for individual identification such forensic serology, microbial forensics, toxins, authentication of human growth hormones, and food species, just to name a few applications.

In this book chapter, an emphasis was laid on hair protein sequencing as a potential forensic hair comparison tool which must stand up to the scrutiny of the courtrooms, and validations to ensure method performance is consistent and robust while identifying limitations under normal operating conditions. Ultimately, protein sequencing should follow the path already set forth by human DNA analysis to facilitate its acceptance in the forensic arena. Until more work is done to address the complexities of peptide profiling in a forensic context, the technique is not yet mature for forensic human identification and conviction.

Acknowledgments

The authors would like to thank Leslie D. Mccurdy and Jason Bannan for their careful review of this chapter. The authors would also like to thank Linda M. Otterstatter for being an excellent consultant about forensic hair casework examination, and Robert D. Blackledge for inspiring the conception of this chapter. This is publication number 21–23 of the Laboratory Division of the FBI. Names of commercial manufacturers are provided for information only and inclusion does not imply endorsement by the FBI or the U.S. Government. The views expressed are those of the authors and do not necessarily reflect the official policy or position of the FBI or the U.S. Government. This research was supported in part by appointments to the Visiting Scientist Program at the FBI Laboratory Division, administered by the Oak Ridge Institute for Science and Education, through an interagency agreement between the US Department of Energy and the FBI.

References

1 Butler, J.M. (2006). Genetics and genomics of core short tandem repeat loci used in human identity testing. *J. Forensic Sci.* 51 (2): 253–265.

2 Krause, K. and Foitzik, K. (2006). Biology of the hair follicle: the basics. *Semin. Cutan. Med. Surg.* 25 (1): 2–10.

3 Reilly, P. (2001). Legal and public policy issues in DNA forensics. *Nat. Rev. Genet.* 2 (4): 313–317.

4 Tautz, D. (1989). Hypervariability of simple sequences as a general source for polymorphic DNA markers. *Nucleic Acids Res.* 17 (16): 6463–6471.

5 Butler, J.M. (2014). *Advanced Topics in Forensic DNA Typing: Interpretation.* Academic Press.

6 Martin, B., Blackie, R., Taylor, D., and Linacre, A. (2018). DNA profiles generated from a range of touched sample types. *Forensic Sci. Int. Genet.* 36: 13–19.

7 Whitaker, J.P., Clayton, T.M., Urquhart, A.J. et al. (1995). Short tandem repeat typing of bodies from a mass disaster: high success rate and characteristic amplification patterns in highly degraded samples. *Biotechniques* 18 (4): 670–677.

8 Kolowski, J.C., Petraco, N., Wallace, M.M. et al. (2004). A comparison study of hair examination methodologies. *J. Forensic Sci.* 49 (6): JFS2003430-3.

9 Bourguignon, L., Hoste, B., Boonen, T. et al. (2008). A fluorescent microscopy-screening test for efficient STR-typing of telogen hair roots. *Forensic Sci. Int. Genet.* 3 (1): 27–31.

10 Oien, C. (2009). Forensic hair comparison: background information for interpretation. *Forensic Sci. Commun.* 11 (2): 1–25.

11 National Research Council (2009). *Strengthening Forensic Science in the United States: A Path Forward.* National Academies Press.

12 Cotsarelis, G., Sun, T.-T., and Lavker, R.M. (1990). Label-retaining cells reside in the bulge area of pilosebaceous unit: implications for follicular stem cells, hair cycle, and skin carcinogenesis. *Cell* 61 (7): 1329–1337.

13 Lavker, R.M., Sun, T.T., Oshima, H. et al. (2003). Hair follicle stem cells. *J. Investig. Dermatol. Symp. Proc.* 8 (1): 28–38.

14 Lepez, T., Vandewoestyne, M., Van Hoofstat, D., and Deforce, D. (2014). Fast nuclear staining of head hair roots as a screening method for successful STR analysis in forensics. *Forensic Sci. Int. Genet.* 13: 191–194.

15 Opel, K.L., Fleishaker, E.L., Nicklas, J.A. et al. (2008). Evaluation and quantification of nuclear DNA from human telogen hairs. *J. Forensic Sci.* 53 (4): 853–857.

16 Hellmann, A., Rohleder, U., Schmitter, H., and Wittig, M. (2001). STR typing of human telogen hairs–a new approach. *Int. J. Leg. Med.* 114 (4–5): 269–273.

17 McNevin, D., Wilson-Wilde, L., Robertson, J. et al. (2005). Short tandem repeat (STR) genotyping of keratinised hair: part 1. Review of current status and knowledge gaps. *Forensic Sci. Int.* 153 (2–3): 237–246.

18 Just, R.S., Irwin, J.A., and Parson, W. (2015). Mitochondrial DNA heteroplasmy in the emerging field of massively parallel sequencing. *Forensic Sci. Int. Genet.* 18: 131–139.

19 Gilbert, M.T., Menez, L., Janaway, R.C. et al. (2006). Resistance of degraded hair shafts to contaminant DNA. *Forensic Sci. Int.* 156 (2–3): 208–212.

20 Adav, S.S., Subbaiaih, R.S., and Kerk, S.K. (2018). Studies on the proteome of human hair - identification of histones and deamidated keratins. *Sci. Rep.* 8 (1): 1599.

21 Wadsworth, C. and Buckley, M. (2014). Proteome degradation in fossils: investigating the longevity of protein survival in ancient bone. *Rapid Commun. Mass Spectrom.* 28 (6): 605–615.

22 Lee, Y.J., Rice, R.H., and Lee, Y.M. (2006). Proteome analysis of human hair shaft: from protein identification to posttranslational modification. *Mol. Cell. Proteomics* 5 (5): 789–800.

23 Chu, F., Mason, K.E., Anex, D.S., and Jones, A.D. (2020). Proteomic characterization of damaged single hairs recovered after an explosion for protein-based human identification. *J. Proteome Res.* 19 (8): 3088–3099.

24 Rice, R.H., Rocke, D.M., Tsai, H.-S. et al. (2009). Distinguishing mouse strains by proteomic analysis of pelage hair. *J. Invest. Dermatol.* 129 (9): 2120–2125.

25 Laatsch, C.N., Durbin-Johnson, B.P., Rocke, D.M. et al. (2014). Human hair shaft proteomic profiling: individual differences, site specificity and cuticle analysis. *PeerJ* 2: e506.

26 Wu, P.W., Mason, K.E., Durbin-Johnson, B.P. et al. (2017). Proteomic analysis of hair shafts from monozygotic twins: expression profiles and genetically variant peptides. *Proteomics* 17 (13–14): 1600462.

27 Karim, N., Plott, T.J., Durbin-Johnson, B.P. et al. (2021). Elucidation of familial relationships using hair shaft proteomics. *Forensic Sci. Int. Genet.* 54: 102564.

28 Parker, G.J., Leppert, T., Anex, D.S. et al. (2016). Demonstration of protein-based human identification using the hair shaft proteome. *PLoS One* 11 (9): e0160653.

29 Kidd, K.K., Kidd, J.R., Speed, W.C. et al. (2012). Expanding data and resources for forensic use of SNPs in individual identification. *Forensic Sci. Int. Genet.* 6 (5): 646–652.

30 Kidd, K.K., Pakstis, A.J., Speed, W.C. et al. (2006). Developing a SNP panel for forensic identification of individuals. *Forensic Sci. Int.* 164 (1): 20–32.

31 Yousefi, S., Abbassi-Daloii, T., Kraaijenbrink, T. et al. (2018). A SNP panel for identification of DNA and RNA specimens. *BMC Genomics* 19 (1): 90.

32 Lawas, M., Jones, K.F., Mason, K.E. et al. (2020). Assessing single-source reproducibility of human head hair peptide profiling from different regions of the scalp. *Forensic Sci. Int. Genet.* 50: 102396.

33 Plott, T., Karim, N., Durbin-Johnson, B. et al. (2020). Age-related changes in hair shaft protein profiling and genetically variant peptides. *Forensic Sci. Int. Genet.* 47: 102309.

34 Erdogan, B. (2017). Chapter 2. Anatomy and physiology of hair. In: *Hair and Scalp Disorders* (ed. Z. Kutlubay and S. Serdaroglu). IntechOpen https://doi.org/10.5772/63002.

35 Deedrick, D. and Koch, S. (2004). Microscopy of hairs part 1: practical guide and manual for human hairs. *Forensic Sci. Commun.* 6 (1): 1–50.

36 Harding, H. and Rogers, G. (1999). Physiology and growth of human hair. In: *Forensic Examination of Hair*, 1e, 1–78. Imprint CRC Press.

37 Ogle, R.R. Jr., and Fox, M.J. (2017). *Atlas of Human Hair: Microscopic Characteristics*. CRC Press.

38 Messenger, A., De Berker, D., and Sinclair, R. (2010). *Disorders of hair*. In: *Rook's Textbook of Dermatology*, 8e, vol. 1 (ed. B. Burns, Cox and Griffiths), 1–100. Blackwell Publishing.

39 Wolfram, L.J. (2003). Human hair: a unique physicochemical composite. *J. Am. Acad. Dermatol.* 48 (6 Suppl): S106–S114.

40 Yang, F.C., Zhang, Y., and Rheinstädter, M.C. (2014). The structure of people's hair. *PeerJ* 2: e619.

41 Kajiura, Y., Watanabe, S., Itou, T. et al. (2006). Structural analysis of human hair single fibres by scanning microbeam SAXS. *J. Struct. Biol.* 155 (3): 438–444.

42 Arwert, E.N., Hoste, E., and Watt, F.M. (2012). Epithelial stem cells, wound healing and cancer. *Nat. Rev. Cancer* 12 (3): 170–180.

43 Oshima, H., Rochat, A., Kedzia, C. et al. (2001). Morphogenesis and renewal of hair follicles from adult multipotent stem cells. *Cell* 104 (2): 233–245.

44 Tumbar, T., Guasch, G., Greco, V. et al. (2004). Defining the epithelial stem cell niche in skin. *Science* 303 (5656): 359–363.

45 Buffoli, B., Rinaldi, F., Labanca, M. et al. (2014). The human hair: from anatomy to physiology. *Int. J. Dermatol.* 53 (3): 331–341.

46 Narisawa, Y., Hashimoto, K., Nakamura, Y., and Kohda, H. (1993). A high concentration of Merkel cells in the bulge prior to the attachment of the arrector pili muscle and the formation of the perifollicular nerve plexus in human fetal skin. *Arch. Dermatol. Res.* 285 (5): 261–268.

47 Nishimura, E.K., Granter, S.R., and Fisher, D.E. (2005). Mechanisms of hair graying: incomplete melanocyte stem cell maintenance in the niche. *Science* 307 (5710): 720–724.

48 Slominski, A. and Paus, R. (1993). Melanogenesis is coupled to murine anagen: toward new concepts for the role of melanocytes and the regulation of melanogenesis in hair growth. *J. Invest. Dermatol.* 101 (1 Suppl): 90s–97s.

49 Slominski, A., Pisarchik, A., Zbytek, B. et al. (2003). Functional activity of serotoninergic and melatoninergic systems expressed in the skin. *J. Cell. Physiol.* 196 (1): 144–153.

50 Slominski, A., Wortsman, J., Plonka, P.M. et al. (2005). Hair follicle pigmentation. *J. Invest. Dermatol.* 124 (1): 13–21.

51 Alonso, L. and Fuchs, E. (2006). The hair cycle. *J. Cell Sci.* 119 (Pt 3): 391–393.

52 Eckhart, L., Lippens, S., Tschachler, E., and Declercq, W. (2013). Cell death by cornification. *Biochim. Biophys. Acta* 1833 (12): 3471–3480.

53 Fischer, H., Eckhart, L., Mildner, M. et al. (2007). DNase1L2 degrades nuclear DNA during corneocyte formation. *J. Invest. Dermatol.* 127 (1): 24–30.

54 Fischer, H., Szabo, S., Scherz, J. et al. (2011). Essential role of the keratinocyte-specific endonuclease DNase1L2 in the removal of nuclear DNA from hair and nails. *J. Invest. Dermatol.* 131 (6): 1208–1215.

55 Linch, C.A. (2009). Degeneration of nuclei and mitochondria in human hairs. *J. Forensic Sci.* 54 (2): 346–349.

56 Brandhagen, M.D., Loreille, O., and Irwin, J.A. (2018). Fragmented nuclear DNA is the predominant genetic material in human hair shafts. *Genes (Basel)* 9 (12): 1–20.

57 Grisedale, K.S., Murphy, G.M., Brown, H. et al. (2018). Successful nuclear DNA profiling of rootless hair shafts: a novel approach. *Int. J. Leg. Med.* 132 (1): 107–115.

58 Bengtsson, C.F., Olsen, M.E., Brandt, L.O. et al. (2012). DNA from keratinous tissue. Part I: hair and nail. *Ann. Anat.* 194 (1): 17–25.

59 Kligman, A.M. (1959). The human hair cycle. *J. Invest. Dermatol.* 33 (6): 307–316.

60 Stenn, K.S. and Paus, R. (2001). Controls of hair follicle cycling. *Physiol. Rev.* 81 (1): 449–494.

61 Myung, P. and Ito, M. (2012). Dissecting the bulge in hair regeneration. *J. Clin. Invest.* 122 (2): 448–454.

62 Schneider, M.R., Schmidt-Ullrich, R., and Paus, R. (2009). The hair follicle as a dynamic miniorgan. *Curr. Biol.* 19 (3): R132–R142.

63 Langbein, L., Rogers, M.A., Praetzel, S. et al. (2003). K6irs1, K6irs2, K6irs3, and K6irs4 represent the inner-root-sheath-specific type II epithelial keratins of the human hair follicle. *J. Invest. Dermatol.* 120 (4): 512–522.

64 Geyfman, M. et al. (2015). Resting no more: re-defining telogen, the maintenance stage of the hair growth cycle. *Biol. Rev. Camb. Philos. Soc.* 90 (4): 1179–1196.

65 Greco, V., Plikus, M.V., Treffeisen, E. et al. (2009). A two-step mechanism for stem cell activation during hair regeneration. *Cell Stem Cell* 4 (2): 155–169.

66 Hsu, Y.C., Pasolli, H.A., and Fuchs, E. (2011). Dynamics between stem cells, niche, and progeny in the hair follicle. *Cell* 144 (1): 92–105.

67 Kimura-Ueki, M., Oda, Y., Oki, J. et al. (2012). Hair cycle resting phase is regulated by cyclic epithelial FGF18 signaling. *J. Invest. Dermatol.* 132 (5): 1338–1345.

68 Oshimori, N. and Fuchs, E. (2012). Paracrine TGF-β signaling counterbalances BMP-mediated repression in hair follicle stem cell activation. *Cell Stem Cell* 10 (1): 63–75.

69 Plikus, M.V. (2012). New activators and inhibitors in the hair cycle clock: targeting stem cells' state of competence. *J. Invest. Dermatol.* 132 (5): 1321–1324.

70 Paus, R., Müller-Röver, S., and Botchkarev, V.A. (1999). Chronobiology of the hair follicle: hunting the "hair cycle clock". *J. Investig. Dermatol. Symp. Proc.* 4 (3): 338–345.

71 Chase, H.B. and Montagna, W. (1951). Relation of hair proliferation to damage induced in the mouse skin. *Proc. Soc. Exp. Biol. Med.* 76 (1): 35–37.

72 Herman, B. (1954). Growth of hair. *Physiol. Rev.* 34: 113–126.

73 Milner, Y., Kashgarian, M., Sudnik, J. et al. (2002). Exogen, shedding phase of the hair growth cycle: characterization of a mouse model. *J. Invest. Dermatol.* 119 (3): 639–644.

74 Higgins, C.A., Richardson, G.D., Westgate, G.E., and Jahoda, C.A. (2009). Exogen involves gradual release of the hair club fibre in the vibrissa follicle model. *Exp. Dermatol.* 18 (9): 793–795.

75 Sato-Miyaoka, M. et al. (2012). Regulation of hair shedding by the type 3 IP3 receptor. *J. Invest. Dermatol.* 132 (9): 2137–2147.

76 Plikus, M.V., Mayer, J.A., de La Cruz, D. et al. (2008). Cyclic dermal BMP signalling regulates stem cell activation during hair regeneration. *Nature* 451 (7176): 340–344.

77 Plikus, M.V., Widelitz, R.B., Maxson, R., and Chuong, C.-M. (2009). Analyses of regenerative wave patterns in adult hair follicle populations reveal macro-environmental regulation of stem cell activity. *Int. J. Dev. Biol.* 53 (5–6): 857.

78 Halloy, J., Bernard, B.A., Loussouarn, G., and Goldbeter, A. (2000). Modeling the dynamics of human hair cycles by a follicular automaton. *Proc. Natl. Acad. Sci. U. S. A.* 97 (15): 8328–8333.

79 Lin, K.K., Chudova, D., Hatfield, G.W. et al. (2004). Identification of hair cycle-associated genes from time-course gene expression profile data by using replicate variance. *Proc. Natl. Acad. Sci. U. S. A.* 101 (45): 15955–15960.

80 Berker, D., Higgins, C., Jahoda, C., and Christiano, A.M. (2004). Biology of hair and nails. *Dermatologia* 29: 1075–1093.

81 Van Neste, D., Leroy, T., and Conil, S. (2007). Exogen hair characterization in human scalp. *Skin Res. Technol.* 13 (4): 436–443.

82 Robbins, C. (2009). The cell membrane complex: three related but different cellular cohesion components of mammalian hair fibers. *J. Cosmet. Sci.* 60 (4): 437–465.

83 Robbins, C.R. (1994). *Chmical and Physical Behavior of Human Hair*, 3e. Springen Science + Business Media, LLC.

84 Rogers, M.A., Winter, H., Langbein, L. et al. (2000). Characterization of a 300 kbp region of human DNA containing the type II hair keratin gene domain. *J. Invest. Dermatol.* 114 (3): 464–472.

85 Rogers, M.A., Winter, H., Wolf, C. et al. (1998). Characterization of a 190-kilobase pair domain of human type I hair keratin genes. *J. Biol. Chem.* 273 (41): 26683–26691.

86 Rogers, M.A., Langbein, L., Praetzel-Wunder, S. et al. (2006). Human hair keratin-associated proteins (KAPs). *Int. Rev. Cytol.* 251: 209–263.

87 Syed, N.A., *The structure of hair - Part 2: The Cortex*. http://www.dralinsyed .com/blog, 2015.

88 Crewther, W., Dowling, L., Steinert, P., and Parry, D. (1983). Structure of intermediate filaments. *Int. J. Biol. Macromol.* 5 (5): 267–274.

89 Fraser, R., MacRae, T., Sparrow, L., and Parry, D. (1988). Disulphide bonding in α-keratin. *Int. J. Biol. Macromol.* 10: 106–112.

90 Moll, R., Divo, M., and Langbein, L. (2008). The human keratins: biology and pathology. *Histochem. Cell Biol.* 129 (6): 705–733.

91 Bray, D.J., Walsh, T.R., Noro, M.G., and Notman, R. (2015). Complete structure of an epithelial keratin dimer: implications for intermediate filament assembly. *PLoS One* 10 (7): e0132706.

92 Randebrook, R. (1964). Neue erkenntnisse über den morphologischen aufbau des menschlichen haares. *J. Soc. Cosmet. Chem.* 15: 691–706.

93 Robbins, C.R. and Robbins, C.R. (2012). *Chemical and Physical Behavior of Human Hair*, vol. 4. Springer.

94 Locard, E. (1930). The analysis of dust traces. *Am. J. Police Sci.* 1: 276.

95 Edson, J., Brooks, E.M., McLaren, C. et al. (2013). A quantitative assessment of a reliable screening technique for the STR analysis of telogen hair roots. *Forensic Sci. Int. Genet.* 7 (1): 180–188.

96 Taupin, J. (2004). Forensic hair morphology comparison—a dying art or junk science? *Sci. Justice* 44 (2): 95–100.

97 Koch, S.L., Michaud, A.L., and Mikell, C.E. (2013). Taphonomy of hair—a study of postmortem root banding. *J. Forensic Sci.* 58: S52–S59.

98 Petraco, N., Fraas, C., Callery, F., and De Forest, P. (1988). The morphology and evidential significance of human hair roots. *J. Forensic Sci.* 33 (1): 68–76.

99 Tafaro, J.T. (2000). The use of microscopic postmortem changes in anagen hair roots to associate questioned hairs with known hairs and reconstruct events in two murder cases. *J. Forensic Sci.* 45 (2): 495–499.

100 Lawas, M., Otterstatter, L.M., Forger, L.V. et al. (2020). A quantitative method for selecting a hair for nulear DNA analysis. *Forensic Sci. Int. Genet.* 48: 102354.

101 Hicks, J.W. (1977). *Microscopy of Hairs: A Practical Guide and Manual.* Washington, DC: Federal Bureau of Investigation, U.S. Government Printing Office.

102 Bisbing, R. (2002). *Forensic hair comparisons.* In: *Forensic Science Handbook,* 1. Westmont, IL: McCrone Associates, Inc.

103 Bisbing, R.E. (2020). *The forensic identification and association of human hair.* In: *Forensic Science Handbook,* 184–221. CRC Press.

104 Rowe, W. (1997). *Biodegradation of hairs and fibers.* In: *Forensic Taphonomy: The Postmortem Fate of Human Remains* (ed. D. Hagelund and M. Sorg). Boca Raton: CRC Press.

105 Linch, C.A. and Prahlow, J.A. (2001). Postmortem microscopic changes observed at the human head hair proximal end. *J. Forensic Sci.* 46 (1): 15–20.

106 Brooks, E., Cullen, M., Sztydna, T., and Walsh, S. (2010). Nuclear staining of telogen hair roots contributes to successful forensic nDNA analysis. *Aust. J. Forensic Sci.* 42: 115–122.

107 Parson, W., Huber, G., Moreno, L. et al. (2015). Massively parallel sequencing of complete mitochondrial genomes from hair shaft samples. *Forensic Sci. Int. Genet.* 15: 8–15.

108 Muller, K., Klein, R., Miltner, E., and Wiegand, P. (2007). Improved STR typing of telogen hair root and hair shaft DNA. *Electrophoresis* 28 (16): 2835–2842.

109 Copeland, W.C. (2008). Inherited mitochondrial diseases of DNA replication. *Annu. Rev. Med.* 59: 131–146.

110 Shokolenko, I., Venediktova, N., Bochkareva, A. et al. (2009). Oxidative stress induces degradation of mitochondrial DNA. *Nucleic Acids Res.* 37 (8): 2539–2548.

111 Moretton, A., Morel, F., Macao, B. et al. (2017). Selective mitochondrial DNA degradation following double-strand breaks. *PLoS One* 12 (4): e0176795.

112 Gredilla, R. (2010). DNA damage and base excision repair in mitochondria and their role in aging. *J. Aging Res.* 2011: 257093.

113 Harman, D. (1972). The biologic clock: the mitochondria? *J. Am. Geriatr. Soc.* 20 (4): 145–147.

114 Miquel, J., Economos, A.C., Fleming, J., and Johnson, J.E. Jr., (1980). Mitochondrial role in cell aging. *Exp. Gerontol.* 15 (6): 575–591.

115 Carracedo, A., Bar, W., Lincoln, P. et al. (2000). DNA commission of the international society for forensic genetics: guidelines for mitochondrial DNA typing. *Forensic Sci. Int.* 110 (2): 79–85.

116 Parsons, T.J., Muniec, D.S., Sullivan, K. et al. (1997). A high observed substitution rate in the human mitochondrial DNA control region. *Nat. Genet.* 15 (4): 363–368.

117 Roberts, K.A. and Calloway, C. (2011). Characterization of mitochondrial DNA sequence heteroplasmy in blood tissue and hair as a function of hair morphology. *J. Forensic Sci.* 56 (1): 46–60.

118 Nilsson, M., Andreasson-Jansson, H., Ingman, M., and Allen, M. (2008). Evaluation of mitochondrial DNA coding region assays for increased discrimination in forensic analysis. *Forensic Sci. Int. Genet.* 2 (1): 1–8.

119 Gill, P., Ivanov, P.L., Kimpton, C. et al. (1994). Identification of the remains of the Romanov family by DNA analysis. *Nat. Genet.* 6 (2): 130–135.

120 Tully, L.A., Parsons, T.J., Steighner, R.J. et al. (2000). A sensitive denaturing gradient-Gel electrophoresis assay reveals a high frequency of heteroplasmy in hypervariable region 1 of the human mtDNA control region. *Am. J. Hum. Genet.* 67 (2): 432–443.

121 Andreasson, H., Nilsson, M., Budowle, B. et al. (2006). Nuclear and mitochondrial DNA quantification of various forensic materials. *Forensic Sci. Int.* 164 (1): 56–64.

122 Desmyter, S., Bodner, M., Huber, G. et al. (2016). Hairy matters: MtDNA quantity and sequence variation along and among human head hairs. *Forensic Sci. Int. Genet.* 25: 1–9.

123 Zhang, S., Bian, Y., Tian, H. et al. (2015). Development and validation of a new STR 25-plex typing system. *Forensic Sci. Int. Genet.* 17: 61–69.

124 Kertesz-Farkas, A., Myers, M.P., Reiz, B., and Pongor, S. (2012). Database searching in mass spectrometry based proteomics. *Curr. Bioinf.* 7: 221–230.

125 Merkley, E.D., Wunschel, D.S., Wahl, K.L., and Jarman, K.H. (2019). Applications and challenges of forensic proteomics. *Forensic Sci. Int.* 297: 350–363.

126 Carlson, T.L., Moini, M., Eckenrode, B.A. et al. (2018). Protein extraction from human anagen head hairs 1-millimeter or less in total length. *Biotechniques* 64 (4): 170–176.

127 Jones, K.F., Carlson, T.L., Eckenrode, B.A., and Donfack, J. (2019). Assessing protein sequencing in human single hair shafts of decreasing lengths. *Forensic Sci. Int. Genet.* 44: 102145.

128 Mason, K.E., Paul, P.H., Chu, F. et al. (2019). Development of a protein-based human identification capability from a single hair. *J. Forensic Sci.* 64 (4): 1152–1159.

129 Zhang, Z., Burke, M., Wallace, W. et al. (2020). Sensitive method for the confident identification of genetically variant peptides in human hair keratin. *J. Forensic Sci.* 65 (2): 406–420.

130 Sun, L., Zhu, G., Yan, X., and Dovichi, N.J. (2013). High sensitivity capillary zone electrophoresis-electrospray ionization-tandem mass spectrometry for the rapid analysis of complex proteomes. *Curr. Opin. Chem. Biol.* 17 (5): 795–800.

131 Haskins, W.E., Wang, Z., Watson, C.J. et al. (2001). Capillary LC– MS2 at the attomole level for monitoring and discovering endogenous peptides in microdialysis samples collected in vivo. *Anal. Chem.* 73 (21): 5005–5014.

132 Emmett, M.R. and Caprioli, R.M. (1994). Micro-electrospray mass spectrometry: ultra-high-sensitivity analysis of peptides and proteins. *J. Am. Soc. Mass Spectrom.* 5 (7): 605–613.

133 Myers, P. and Bartle, K. (2002). *Miniaturization in LC–MS. Recent Applications in LC-MS.* LCGC Europe.

134 Shevchenko, A., Chernushevich, I., Ens, W. et al. (1997). Rapid 'de novo'peptide sequencing by a combination of nanoelectrospray, isotopic labeling and a quadrupole/time-of-flight mass spectrometer. *Rapid Commun. Mass Spectrom.* 11 (9): 1015–1024.

135 Mitulović, G., Smoluch, M., Chervet, J.-P. et al. (2003). An improved method for tracking and reducing the void volume in nano HPLC–MS with micro trapping columns. *Anal. Bioanal. Chem.* 376 (7): 946–951.

136 Wilson, S.R., Vehus, T., Berg, H.S., and Lundanes, E. (2015). Nano-LC in proteomics: recent advances and approaches. *Bioanalysis* 7 (14): 1799–1815.

137 Yamashita, M. and Fenn, J.B. (1984). Negative ion production with the electrospray ion source. *J. Phys. Chem.* 88 (20): 4671–4675.

138 Banerjee, S. and Mazumdar, S. (2012). Electrospray ionization mass spectrometry: a technique to access the information beyond the molecular weight of the analyte. *Int. J. Anal. Chem.* 2012: 282574.

139 Schmidt, A., Karas, M., and Dülcks, T. (2003). Effect of different solution flow rates on analyte ion signals in nano-ESI MS, or: when does ESI turn into nano-ESI? *J. Am. Soc. Mass Spectrom.* 14 (5): 492–500.

140 Desiderio, C., Iavarone, F., Rossetti, D.V. et al. (2010). Capillary electrophoresis-mass spectrometry for the analysis of amino acids. *J. Sep. Sci.* 33 (16): 2385–2393.

141 Angel, T.E., Aryal, U.K., Hengel, S.M. et al. (2012). Mass spectrometry-based proteomics: existing capabilities and future directions. *Chem. Soc. Rev.* 41 (10): 3912–3928.

142 Ishihama, Y. (2005). Proteomic LC–MS systems using nanoscale liquid chromatography with tandem mass spectrometry. *J. Chromatogr. A* 1067 (1–2): 73–83.

143 Goecker, Z.C., Legg, K.M., Salemi, M.R. et al. (2021). Alternative LC-MS/MS Platforms and Data Acquisition Strategies for Proteomic Genotyping of Human Hair Shafts. *J. Proteome Res.* 20 (10): 4655–4666.

144 Wong, J.W., Wang, J., Chow, W. et al. (2018). Perspectives on liquid chromatography-high-resolution mass spectrometry for pesticide screening in foods. *J. Agric. Food Chem.* 66 (37): 9573–9581.

145 Wang, R., Yin, Y., and Zhu, Z.J. (2019). Advancing untargeted metabolomics using data-independent acquisition mass spectrometry technology. *Anal. Bioanal. Chem.* 411 (19): 4349–4357.

146 Klein, J., Papadopoulos, T., Mischak, H., and Mullen, W. (2014). Comparison of CE-MS/MS and LC-MS/MS sequencing demonstrates significant complementarity in natural peptide identification in human urine. *Electrophoresis* 35 (7): 1060–1064.

147 Faserl, K., Kremser, L., Müller, M. et al. (2015). Quantitative proteomics using ultralow flow capillary electrophoresis–mass spectrometry. *Anal. Chem.* 87 (9): 4633–4640.

148 Ahmed, F.E. (2009). The role of capillary electrophoresis–mass spectrometry to proteome analysis and biomarker discovery. *J. Chromatogr. B* 877 (22): 1963–1981.

149 Jelínková, D., Deyl, Z., Mikšík, I., and Tagliaro, F. (1995). Capillary electrophoresis of hair proteins modified by alcohol intake in laboratory rats. *J. Chromatogr. A* 709 (1): 111–119.

150 Domínguez-Vega, E., Haselberg, R., and Somsen, G.W. (2016). *Capillary zone electrophoresis-mass spectrometry of intact proteins*. In: *Capillary Electrophoresis of Proteins and Peptides: Methods and Protocols*, Methods in Molecular Biology, vol. 1466 (ed. N.T. Tran and M. Taverna), 25–41.

151 Wang, Y., Fonslow, B.R., Wong, C.C. et al. (2012). Improving the comprehensiveness and sensitivity of sheathless capillary electrophoresis–tandem mass spectrometry for proteomic analysis. *Anal. Chem.* 84 (20): 8505–8513.

152 Mikšík, I. (2019). Coupling of CE-MS for protein and peptide analysis. *J. Sep. Sci.* 42 (1): 385–397.

153 Zhang, Z., Qu, Y., and Dovichi, N.J. (2018). Capillary zone electrophoresis-mass spectrometry for bottom-up proteomics. *TrAC Trends Anal. Chem.* 108: 23–37.

154 Rice, R.H. (2011). Proteomic analysis of hair shaft and nail plate. *J. Cosmet. Sci.* 62 (2): 229–236.

155 Han, M.O., Chun, J.A., Lee, W.H. et al. (2007). A simple improved method for protein extraction from human head hairs. *J. Cosmet. Sci.* 58 (5): 527–534.

156 Shin, S., Lee, A., Lee, S. et al. (2010). Microwave-assisted extraction of human hair proteins. *Anal. Biochem.* 407 (2): 281–283.

157 Wong, S.Y., Lee, C.C., Ashrafzadeh, A. et al. (2016). A high-yield two-hour protocol for extraction of human hair shaft proteins. *PLoS One* 11 (10): e0164993.

158 Goecker, Z.C., Salemi, Z.C., Karim, N. et al. (2020). Optimal processing for proteomic genotyping of single human hairs. *Forensic Sci. Int. Genet.* 47: 102314.

159 SWGDAM. (2012). *Validation Guidelines for DNA Analysis Methods.* Available at: http://swgdam.org/SWGDAM_Validation_Guidelines_APPROVED_Dec_ 2012.pdf (accessed on 04 May 2015).

160 AAFS Standards Board. (2019). *Standard Practices for Method Validation in Forensic Toxicology.* Available at: http://www.asbstandardsboard.org/wp-content/uploads/2019/11/036_Std_e1.pdf (accessed 19 Jul 2022).

9

Photo-induced Force Microscopy

Padraic O'Reilly, Graceson Aufderheide, and Sung Park

Molecular Vista, Inc., San Jose, CA, USA

9.1 Introduction

Photo-induced force microscopy (PiFM) is a hyperspectral imaging method that has nanoscale spatial resolution (<10 nm). It combines atomic force microscopy (AFM) and infrared (IR) spectroscopy/spectroscopic imaging in one instrument to obtain both topography and chemical information concurrently. This chapter will present the principles and basic components of PiFM and show how it can characterize different types of trace evidence.

9.2 Working Principle and Instrumentation

IR spectroscopy and spectroscopic imaging are utilized frequently in forensic science since they provide robust and label free characterization of samples with a high chemical specificity and practical sensitivity [1]. However, extending these techniques to trace amounts of samples that are small in lateral size (microns or less) and thin (less than 100 nm) is difficult due to the diffraction limit and penetration depth (averaging the response of the sample deep into the sample) of the infrared light. Overcoming these limitations to be able to characterize chemical signature of ultrathin coatings on nano-particles, micro-fibers, hair samples, and other trace evidence would prove to be useful for forensic science.

Recently several techniques that combine AFM and optical spectroscopy have been demonstrated to overcome the diffraction limit. They include tip-enhanced Raman spectroscopy (TERS) [2–6], scanning scattering near-field optical microscopy (sSNOM) [7–10], and photothermal-induced resonance microscopy (PTIR) [11–15]. To varying degrees, these techniques enable optical

Leading Edge Techniques in Forensic Trace Evidence Analysis: More New Trace Analysis Methods, First Edition. Edited by Robert D. Blackledge.

spectroscopy with the nanoscale spatial resolution inherent to AFM, thereby providing nanochemical analysis of a specimen. Here, we discuss PiFM, a recently developed technique for nanoscale optical spectroscopy. PiFM measures the attractive forces acting between an AFM tip and IR illuminated sample to detect wavelength-dependent polarization within the sample to generate absorption spectra [16–21]. This approach enables sub-10 nm spatial resolution with spectra that show excellent correlation with bulk Fourier-transform infrared (FTIR) spectra. Unlike other techniques, PiFM achieves this high resolution with virtually no constraints on sample or substrate properties. Table 9.1 compares the various hybrid AFM/IR spectroscopy techniques; those interested in TERS can refer to these review articles [22, 23].

Figure 9.1 shows a schematic drawing for PiFM. In PiFM, the AFM is operated in dynamic mode (noncontact AFM) where the response of one of the eigenmodes of the AFM cantilever is used to generate the topographic data by keeping a constant Van der Waals force between the tip and the sample while scanning; this is equivalent to keeping the tip-sample gap spacing constant. The amplitude of the feedback oscillation is kept at $\sim$1 nm by using the second mechanical mode, which has a force constant of $\sim$1200 N/m compared to a $\sim$30 N/m for the first mechanical mode. This allows the tip-sample spacing to be maintained at $\sim$5 nm, which allows the system to efficiently detect the tip-enhanced photo-induced attractive force by using one of the other eigenmodes of the cantilever. To produce the tip-enhanced photo-induced force (PiF), the laser is p-polarized with the electric field oriented along the tip axis and modulated or pulsed at f_m, where f_m is equal to the difference between the first (f_0) and second (f_1) resonance frequencies. The modulated signals at f_1 (changing tip-enhanced field due to tip-sample gap spacing) and f_m (the laser intensity) will mix and produce the PiF at two sidebands at $f_1 \pm f_m$; given the choice of f_m, one of the sidebands equals f_0, where the PiF signal is amplified by the quality factor Q of the resonance and measured by a lock-in amplifier.

In PiFM, at least two attractive forces are at play – a dipole–dipole force (which measures the real part of the polarizability) and the modulated Van der Waals force due to the changing tip-sample gap spacing arising from thermal expansion caused by IR absorption (which measures the imaginary part of the polarizability) [24]. The steep force gradient of the attractive forces allows PiFM to generate exceptional spatial resolution, on the order of the enhanced field profile, which is typically smaller than the tip's radius of curvature. With gold coated AFM tips, PiFM produces chemical maps with spatial resolution of $\sim$5 nm. By sweeping the laser wavelength while recording the PiF signal, one can generate a PiF spectrum. For homogeneous materials with well-defined IR absorption modes, the imaginary component dominates, and the measured PiF spectra agree well with the bulk FTIR spectra (see Figure 9.1). On samples with nanoscale heterogeneity, PiF spectra will display subset of peaks seen in bulk FTIR spectra since PiFM probes

Table 9.1 Comparison of near-field infrared optical techniques.

Specification	PiFM	PTIR	s-SNOM
Laser type	QCL or OPO/DFG for imaging and nano IR spectra	QCL or OPO/DFG for imaging and nano IR spectra	QCL for imaging; DFG and SC for nano FTIR
AFM operation	Non-contact/light TM	Contact/hard TM	Large amplitude TM
Near-field detection	Attractive force	Repulsive force	Scattered photons
Background suppression	PiFM: heterodyne detection	TM PTIR: heterodyne detection	Pseudo-heterodyne
Data quality (organics)	Excellent	Excellent	Good
Data quality (inorganics)	Excellent	Poor	Good
Data quality (E-field)	Very good	Poor	Excellent
Dielectric constant imaging	Yes	No	Yes
Wavelength specific imaging	QCL: Yes OPO/DFG: Yes	QCL: Yes N-band OPO/DFG: Yes	QCL: Yes OPO/DFG: No
Spatial resolution	<10 nm	~100 nm (contact); <10 nm (TM)	~20 nm
Spectrum type	Sweep	Sweep	Nano FTIR
Spectrum sensitivity (thickness)	Monolayer	>100 nm (contact)	Monolayer
Spectrum quality	Excellent	Excellent on organics	Very good
Spectrum acquisition time	QCL: <1 second OPO/DFG: <1 s/1000 cm^{-1}	QCL: <1 second N-band OPO/DFG: ~30 s/800 cm^{-1}	DFG: ~2 s/400 cm^{-1} (nano FTIR)

N-band, narrowband; OPO/DFG, optical parametric oscillator/difference frequency generation; QCL, quantum cascade laser; SC, supercontinuum; TM, tapping mode.

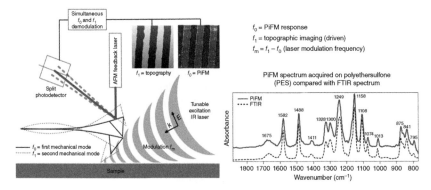

Figure 9.1 PiFM concept. The laser intensity is modulated at a frequency, f_m, and mixes with the carrier frequency (driving frequency for noncontact AFM), f_1, that creates a sideband to coincide with the eigenmode, f_0. Demodulation at f_0 outputs the photo-induced force (PiF) related to chemical imaging. Amplitude of f_0 output versus wavenumber generates a PiF-IR spectrum that agrees well with bulk FTIR on homogeneous samples. Tunable excitation IR laser is typically a quantum cascade laser (QCL), which has a narrow linewidth of $\sim$1 cm^{-1} and is tunable over 760 to $\sim$1900 cm^{-1} range.

about $10\,\mathrm{nm} \times 10\,\mathrm{nm} \times 20\,\mathrm{nm}$ (depth) of sample volume; the subset of peaks will describe the local chemical makeup at the nanoscale (see Figure 9.2). On 2D and other types of materials where the real component can become large due to effects such as surface phonon/Plasmon polaritons (SPP), PiFM can map out the field patterns associated with polaritons efficiently via the measured dipole–dipole force. The relative contribution of different attractive forces for different types of samples is discussed in this Ref. [24]. Since the topography and PiF signals arise from the same point, there is perfect correspondence between IR absorption and topography, highlighting the correlation between structure and chemical information.

As an example of PiFM application, in Figure 9.3 we analyze a polymer blend of poly(lactic acid) (PLA)/acrylic rubber (ACM), which combines a biodegradable thermoplastic and a natural elastomer to create a usable and more eco-friendly material [23]. When a sample's constituents are unknown, AFM topography (Figure 9.3a) may be acquired first. Then a few PiF spectra can be acquired at unique features in topography, which will generate spectra that are like those seen in Figure 9.3f. Seeing that there are strong peaks at $\sim$1730 cm^{-1}, the laser can be tuned to $\sim$1730 cm^{-1} (in this case, 1750 cm^{-1}) to acquire the AFM topography and PiFM image at 1750 cm^{-1} (images 9.3a and b) concurrently. The process can be repeated at 1070 cm^{-1} to highlight a third unexpected component (Figure 9.3c) and at 1720 cm^{-1} to highlight ACM (Figure 9.3d). Different PiFM images can be artificially colored and then combined to form a chemical map as shown in Figure 9.3e where three PiFM images at 1750, 1720, and 1070 cm^{-1} are colored

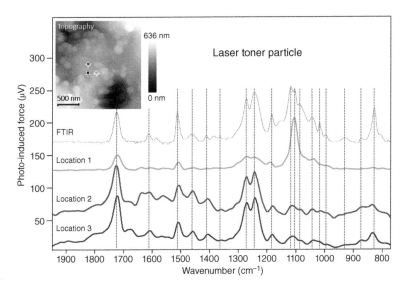

Figure 9.2 Comparison of PiF-IR spectra (obtained on a single toner particle) with a bulk FTIR spectrum (red curve, acquired on many toner particles that are crushed). On samples that contain nanoscale inhomogeneity, $\sim 10^9$ PiF-IR spectra will need to be averaged to show a spectrum that closely matches the FTIR spectrum; another way to state the same fact is that PiF-IR will display a subset of peaks observed in the FTIR spectrum on such samples.

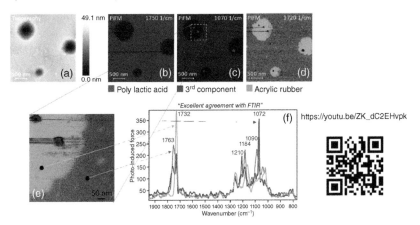

Figure 9.3 (a) Topography and PiFM images at (b) $1750\,\text{cm}^{-1}$, (c) $1070\,\text{cm}^{-1}$, and (d) $1720\,\text{cm}^{-1}$ for PLA, third component, ACM, respectively. A zoomed in (e) composite image consisting of three PiFM images and (f) three spectra from locations shown in (e).

blue, green, and red to represent PLA, acrylic rubber, and a third component, respectively. Note that three distinct spectra are acquired from three locations shown in Figure 9.3e and displayed in Figure 9.3f; PiFM spectra can be reliably acquired from regions as small as ~10 nm in size. Use the QR code (or the URL) in Figure 9.3 to see a video of how this sequence of data is acquired.

9.3 Trace Evidence Examples

As an example of analyzing trace evidence, PiFM is used to analyze different protective coatings that are applied to single fiber threads to (i) establish the chemical signature associated with two different coatings and (ii) identify the coatings on other threads.

9.3.1 Fibers

Figure 9.4 shows six PiFM spectra acquired on a single cotton knit fiber (A2) that is glued onto a glass substrate by a thin layer of adhesive; the bottom inset figure shows the optical image seen in the instrument with the red square denoting the general area from where the PiFM measurements are taken. The upper inset figure shows the AFM topography along with six locations (300 nm apart from each other) where the PiFM spectra were acquired at. Since the fiber is uncoated, it should be quite homogeneous even in the nanoscale, and the similar spectra

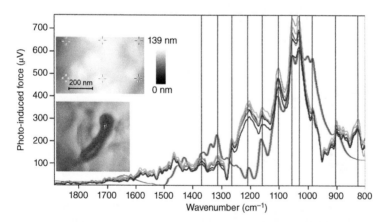

Figure 9.4 Six PiF-IR spectra from six locations on a single bare cotton knit fiber that is held down by epoxy along with the FTIR spectrum for cotton (red curve, from the web). The bottom inset shows the optical view of the fiber and from where the spectra were acquired. The upper inset shows the topography of the fiber and the six locations from where the PiF-IR spectra were acquired.

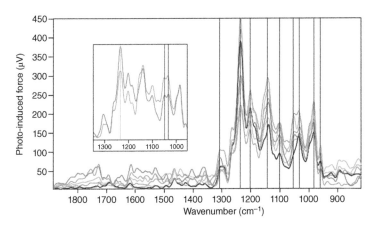

Figure 9.5 Six PiF-IR spectra from six locations on a single cotton fiber coated with 3M protection layer. The inset compares just two spectra for clarity.

reflect such homogeneity. The red FTIR spectrum for cotton (gleaned from the web) shares many of the peaks observed in the spectra. All the other fibers analyzed and discussed below were prepared in a similar manner and six spectra were acquired with 300 nm spacing between each other. Figure 9.5 shows six spectra acquired on a single cotton knit fiber that is coated with a 3M protective coating (A1). Unlike the uncoated cotton knit fiber (Figure 9.4), the spectra are more varying especially in the relative strength of peaks due to the uneven coating thickness. In the inset, two spectra are shown for more clarity and to highlight the effect of uneven coating thickness. Looking at Figure 9.6, which compares the average spectra of the uncoated and coated fibers with the FTIR spectrum, we can see that the coated fiber has the strongest peak at 1237 cm^{-1} whereas the bare cotton has the strongest peaks at 1030 and 1056 cm^{-1}. Looking at the two spectra in the inset of Figure 9.5, we can see that the orange and blue spectra have opposing strengths at these wavenumbers, that is, the orange (blue) spectrum has higher (lower) 1237 cm^{-1} peak strength and lower (higher) 1030 and 1056 cm^{-1} peak strengths. That is due to the fact that PiFM is a surface sensitive technique whose signal strength is quickly attenuated with any intervening layer however thin. Therefore, the location from where the orange spectrum was acquired is covered by a thin layer of 3M protective coating whereas the location from where the blue spectrum was acquired is most likely exposed (in such a case, the IR signature of the protective coating seen in the blue spectrum is likely from the neighboring regions since PiFM collects signals from about 10 nm region). It also could be that there is a thinner layer of 3M protective coating at that location. In Figure 9.6, the green dotted lines identify the PiF-IR peaks associated with the coating while the black dotted line identifies peaks associated with the cotton.

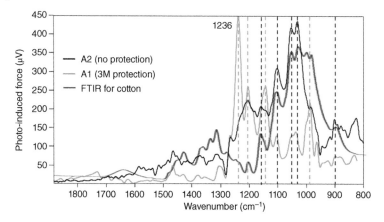

Figure 9.6 Comparison of the averaged PiF-IR spectra for bare cotton (labeled as A2) and 3M coating protected cotton (labeled as A1) against a FTIR spectrum for cotton.

Figures 9.7–9.9 show the results on a bare cotton woven fiber, a fiber with a 3M protective coating, and a fiber with a non-3M protective coating, respectively. Like other samples, six spectra were taken from each fiber strand type, with 300 nm spacing between the spectra. Figure 9.7 compares a spectrum from a bare cotton woven fiber (B2) with two spectra from a fiber that is coated with the 3M protection coating (B1). Interestingly, all three spectra display strong signature of silicone oil (FTIR of silicone oil is shown in green) along with some representative signature from cotton and 3M coating; for cotton, the peak at 1030 cm^{-1} is obscured by

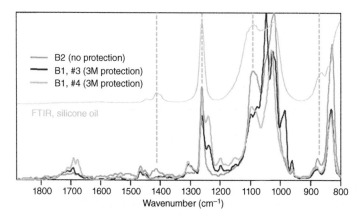

Figure 9.7 Comparison between a spectrum from a bare cotton woven fiber (B2) with two spectra from a fiber that is coated with the 3M protection coating (B1). All three spectra display strong signature of silicone oil (FTIR of silicone oil shown in green) along with some representative signature from cotton and 3M coating.

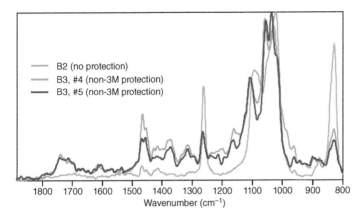

Figure 9.8 Comparison between a spectrum from a bare cotton woven fiber (B2) with two spectra from a fiber that is coated with non-3M protection coating (B3). All three spectra display signature of silicone oil even though it is weaker with B3.

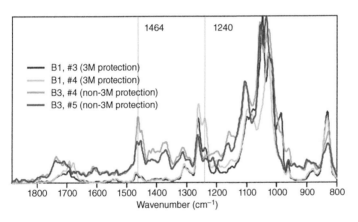

Figure 9.9 Comparison of spectra from single fibers that are coated with 3M or non-3M protection coating.

the $1020 \, cm^{-1}$ peak associated with the silicone oil while for 3M coating, the peak at $1240 \, cm^{-1}$ is seen as a smaller peak adjacent to the $1260 \, cm^{-1}$ peak associated with the silicone oil. In a manner like the uneven coating on A1 (Figure 9.5), the two spectra (purple and yellow) on B1 in Figure 9.7 allow us to decipher the coating condition. First, the silicone peaks at 820 and $1020 \, cm^{-1}$ are similar in strength for both the purple and yellow spectra, suggesting a similar silicone oil coverage at both locations. However, the purple spectrum shows a stronger peak at $1048 \, cm^{-1}$ (associated with cotton) but a weaker peak at $1240 \, cm^{-1}$ (associated with 3M coating, $1237 \, cm^{-1}$ from Figure 9.4) compared to the yellow spectrum, suggesting that the location associated with the purple spectrum has lesser amount of 3M coating. Figure 9.8 compares a spectrum from a bare cotton woven fiber (B2) with two spectra from a fiber that is coated with a non-3M protection coating (B3). Compared to B2 and B1, the spectra associated with B3 has lesser contribution from silicone oil as indicated by much weaker peaks at 820 and 1260 cm^{-1} and noticeable cotton peaks at 1030 and 1060 cm^{-1}. Comparing the spectra between B1 and B3, the most noticeable change occurs between ~1350 and 1470 cm^{-1} with a prominent peak appearing at 1464 cm^{-1}. Figure 9.9 compares the spectra associated with B1 (with 3M protection) and B3 (with non-3M protection), showing that the prominent peaks associated with 3M coating and non-3M coating are 1240 and 1464 cm^{-1}, respectively.

9.3.2 Nanoscale Chemical Mapping

To demonstrate the nanoscale chemical mapping capability of PiFM, we can image B1 (with 3M protection) at 1048 and 1241 cm^{-1} since we noticed varying amount of coating (see discussion for Figure 9.7 above). Figure 9.10 shows PiFM images at 1048 cm^{-1} (for visualizing cotton) and 1241 cm^{-1} (for visualizing 3M coating) colored green and red, respectively along with the topography and the spectra from two locations, each from a location that shows stronger signal strength. It also includes a combined PiFM image where the green color represents cotton molecules and red color represents the 3M coating molecules. We can see that the coating is not covering the cotton surface uniformly.

Finally, we look at fibers that are 80%/20% polyester/rayon that are coated with 3M and non-3M protective coatings. Figure 9.11 shows a PiF-IR spectrum associated with a bare fiber (C2) along with FTIR spectra for polyester and rayon gleaned from the web. The agreement between the PiF-IR spectrum and the combined FTIR spectra is reasonable. Figure 9.12 compares a spectrum each from C2 (bare) and C1 (coating A); the spectrum on the fiber C1 with coating A (wine) shows similar peak positions to the unprotected fiber (green). However, there is a peak at 1294 cm^{-1} on the protected fiber that is not prominent on the unprotected fiber. There is also an overall decrease in absorption between 1200 and 1000 cm^{-1}; this may be due to a lower concentration of rayon in this part of the fiber since rayon shows a strong absorption in this band. Figure 9.13 compares a spectrum from C2 (bare) and two spectra from C3 (coating B); the spectra for the fiber C3 with coating B (pink and purple) are dominated (to a different degree) by silicone peaks. The large absorption at 1244 cm^{-1} in the pink spectrum may be due to the coating protection process since other peaks associated with the silicone oil is weaker in the pink spectrum compared to the purple spectrum.

Figures 9.14 and 9.15 compare the spectra from the two polyester/rayon fibers to the two types of cotton fibers; Figure 9.14 compares the spectra from A1 (cotton

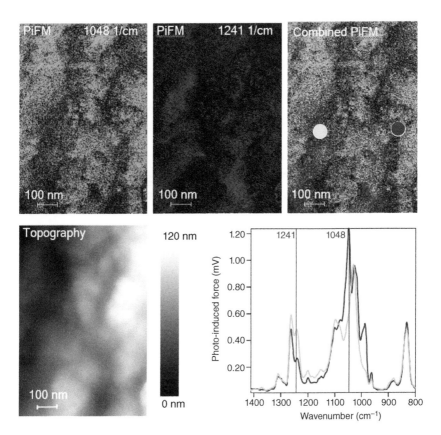

Figure 9.10 Two PiFM images at 1048 and 1241 cm^{-1} show the location of cotton (colored green) and 3M coating (colored red) molecules. The combined PiFM image clearly shows that 3M coating is not uniform; presumptive positions of the two spectra shown are indicated with the matching color. Topography of the same area is also shown.

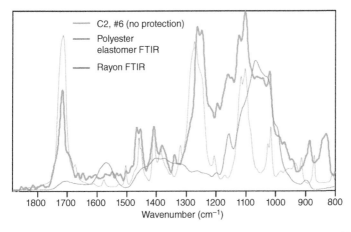

Figure 9.11 Comparison of a PiF-IR spectrum associated with a bare fiber (C2) to FTIR spectra for polyester and rayon gleaned from the web.

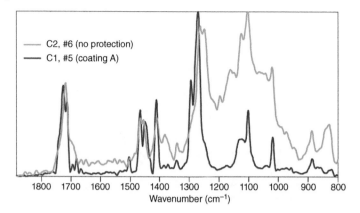

Figure 9.12 Comparison of PiF-IR spectra from C2 (bare) and C1 (coating A) fiber samples.

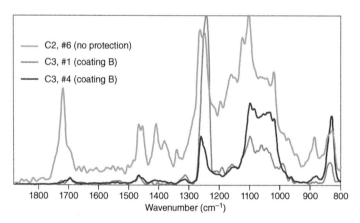

Figure 9.13 Comparison of PiF-IR spectra from C2 (bare) and C3 (coating B) fiber samples.

knit fiber) and B1 (cotton woven fiber) with the 3M protective coating with the spectra from C3 with coating B while Figure 9.15 compares the spectra from B3 (cotton woven fiber) with the non-3M protective coating with the spectrum from C1 with coating A. For C3 (coating B), the most notable peak is at $1240\,cm^{-1}$, which is shared with A1 and B1 and highlighted by the black dashed oval; the sharing of this major peak suggests that C3 is coated with the 3M protection. For C1, it shares a peak around $1460\,cm^{-1}$ with B3, highlighted by black dashed oval in Figure 9.15. For C1 (coating A), it is possible that this is due to polyester in the fiber, which also has a peak here, but it is more likely due to the non-3M protective coating since the dominant doublet near $1100\,cm^{-1}$ for the polyester is greatly reduced

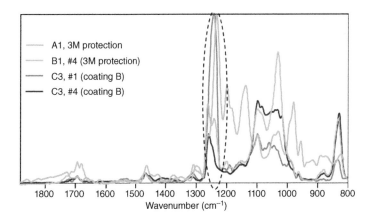

Figure 9.14 Comparison of the spectra from A1 (cotton knit fiber) and B1 (cotton woven fiber) with the 3M protective coating to the spectra from C3 with coating B.

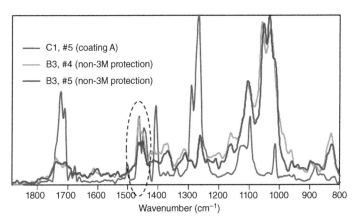

Figure 9.15 Comparison of the spectra from B3 (cotton woven fiber) with the non-3M protective coating to the spectra from C1 with coating A.

by the silicone oil and the protective coating. In summary, IR PiFM can chemically analyze thin protective coatings on and the chemical make-up of individual fiber fragments to provide a valuable capability to forensic community to utilize individual fiber fragments as trace evidence.

In Chapter 3, XPS was shown to be capable of detecting and characterizing the presence on single fibers of very thin surface modifying chemicals. However, for the purpose of characterizing unknown fibers, XPS and PiFM are complementary. While both techniques are surface sensitive and provide useful chemical bonding information, XPS probes over a large area (10 μm to 2 mm), while PiFM probes a much smaller area (0.01 μm to 0.1 mm) with much higher spatial resolution.

9.3.3 Individual Glitter and Shimmer Particles

As the next example of analyzing potential trace evidence, PiFM is used to analyze single glitter and shimmer particles. Glitter particles are larger than shimmer. Glitter contains no mica band and consists of hundreds of very thin (thinner than the diffraction limit) layers of polymer (often polyethylene terephthalate, PET), and if opaque will contain a thin layer of vapor-deposited aluminum to create a mirror effect [25]. Shimmer, also known as special effect pigment, is a component of many cosmetic products, found in lipstick, eye shadow, eye liner, colored mascara, temporary hair highlights, and lotions. Visual properties of shimmer are created frequently from mica, which are coated typically with titanium dioxide (or other metal oxides such as iron oxides and tin oxide) during the calcine process at variable high temperatures. The color is determined by the thickness of the coating on the mica substrate. Because cosmetic traces may be exchanged in crimes during contact between assailant and victim or left behind in automotive accidents where airbags are deployed, shimmer may provide a useful form of trace evidence [26].

Figure 9.16 shows four sample types (two glitters, one shimmer, and one biodegradable shimmer) that were analyzed along with the general location from where multiple PiFM spectra were collected. ATR FTIR spectra on all four sample types (from a large cluster of particles) were collected for comparison with PiFM spectra acquired on individual particles of the four sample types. Figures 9.17–9.20 show multiple PiFM spectra on each type of sample along with the measured ATR FTIR spectrum and the FTIR spectrum (from spectrabase .com) of the corresponding substrate materials (PET, cellulose, and mica). For Tropical Lagoon (Figure 9.17), there is a good agreement between the location of all the absorption peaks between the FTIR and the PiFM spectra so that it would be straightforward to identify the source type with PiFM spectra from a single glitter sample. For Aphrodite (Figure 9.18), PiFM spectra shows a few peaks that

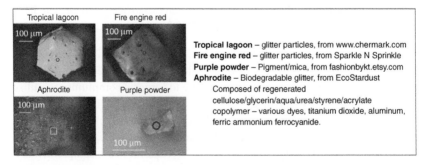

Figure 9.16 Optical views of the glitter and shimmer samples with general area from where the PiF-IR spectra are acquired denoted. The commercial names of the particles and the descriptions are also provided.

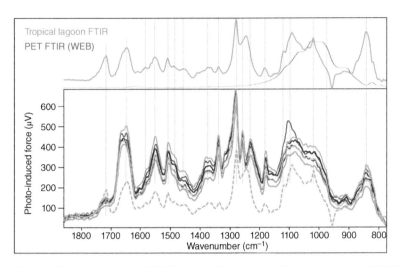

Figure 9.17 Comparison of PiF-IR spectra (bottom panel) to the measured ATR-FTIR spectrum (solid and dotted green curves) of Tropical Lagoon. The FTIR spectrum for PET, the substrate material, is shown in the upper panel (blue curve).

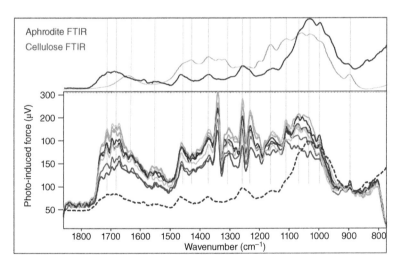

Figure 9.18 Comparison of PiF-IR spectra (bottom panel) to the measured ATR-FTIR spectrum (solid and dotted purple curves) of Aphrodite. The FTIR spectrum for PET, the substrate material, is shown in the upper panel (brown curve).

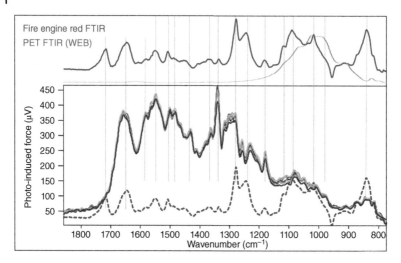

Figure 9.19 Comparison of PiF-IR spectra (bottom panel) to the measured ATR-FTIR spectrum (solid and dotted red curves) of Fire Engine Red. The FTIR spectrum for PET, the substrate material, is shown in the upper panel (blue curve).

are not as obvious on the ATR FTIR spectrum but corresponds well with the peaks from PET, the substrate material; with this sample, ATR-FTIR is not able to detect the weak PET peaks whereas PiFM spectra can. Again, PiFM spectrum from a single particle should allow correct identification of the particle when compared to the combined FTIR spectra of the base material and the shimmer. For the glitter, Fire Engine Red, the measure ATR-FTIR spectrum looks identical to the ATR-FTIR spectrum for the other glitter, Tropical Lagoon. Even though the PiFM spectra show most of the peaks associated with the ATR-FTIR, they are distinct from the PiFM spectra for Tropical Lagoon; while this difference may not be critically important since the two glitters can be distinguished by its different colors, it shows the potential usefulness of PiFM spectra when the samples may show same FTIR spectra and appear visually same also. In the case of Purple Powder (Figure 9.20), the measured ATR-FTIR spectrum resembles the mica FTIR spectrum with very small absorption observed above 1100 cm^{-1}. Unlike with other particles, much more variation among the PiFM spectra is seen, with some sharp peaks in a few spectra at wavenumbers that may have hints of absorption in the FTIR spectrum. Figure 9.21 shows the topography and PiFM images at 1550 cm^{-1} (one of the peaks observed in Figure 9.20) at two magnifications for a Purple Powder particle. From the PiFM images, we can see that the sample is not homogeneously covered by pigment particles, which led to varying PiFM spectra when the tip randomly landed on different sites for spectrum acquisition. Even though the pigment particles are only about 100 nm in size, PiFM can acquire

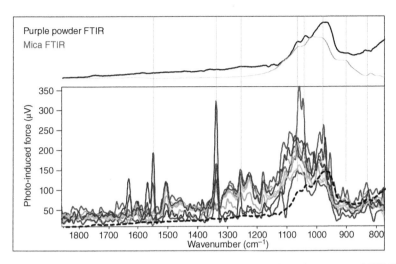

Figure 9.20 Comparison of PiF-IR spectra (bottom panel) to the measured ATR-FTIR spectrum (solid and dotted black curves) of purple powder. The FTIR spectrum for PET, the substrate material, is shown in the upper panel (brown curve). The PiFM spectra show much variation indicating inhomogeneity of material at nanoscale.

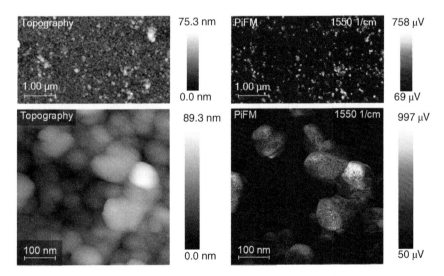

Figure 9.21 Topography and PiFM (at 1550 cm^{-1}) images at the same locations at two different magnifications. Assuming that the active pigment particles will exhibit IR peaks that are different from mica, we see that the active pigment particles, on the order of 100 nm in size, are not homogeneously distributed on the surface; this explains the variations observed in the PiFM spectra in Figure 9.20 since the tip was placed at random locations for spectral acquisition.

clear chemical signature that can be used for analysis of trace evidence. The small size and the low coverage density of the pigment particles explain the mica-like spectrum for FTIR. One last comment about the different strengths of peaks observed for the FTIR and PiFM spectra: the difference most likely arises from the fact that the ATR FTIR spectrum was collected from hundreds of particles whereas the PiFM spectra are acquired from a single particle.

In summary, IR PiFM opens new opportunities in forensic analysis of trace evidence by providing nanoscale chemical analysis from samples that are too small to be analyzed by current techniques.

References

1 Ewinga, A.V. and Kazarian, S.G. (2017). *Analyst* 142: 257.

2 Stöckle, R.M., Suh, Y.D., Deckert, V., and Zenobi, R. (2000). *Chem. Phys. Lett.* 318: 131.

3 Hayazawa, N., Saito, Y., and Kawata, S. (2004). *Appl. Phys. Lett.* 85: 6239.

4 Hartschuh, A., Anderson, N., and Novotny, L. (2003). *J. Microsc.* 210: 234.

5 Berweger, S., Atkin, J.M., Olmon, R.L., and Raschke, M.B. (2010). *J. Phys. Chem. Lett.* 1: 3427.

6 Jiang, N., Foley, E.T., Klingsporn, J.M. et al. (2012). *Nano Lett.* 12: 5061.

7 Ocelic, N., Huber, A., and Hillenbrand, R. (2006). *Appl. Phys. Lett.* 89: 101124.

8 Cvitkovic, A., Ocelic, N., and Hillenbrand, R. (2007). *Nano Lett.* 7: 3177.

9 Fei, Z., Rodin, A.S., Andreev, G.O. et al. (2012). *Nature* 487: 82.

10 Raschke, M.B. and Lienau, C. (2003). *Appl. Phys. Lett.* 83: 5089.

11 Dazzi, A., Prazeres, R., Glotin, F., and Ortega, J.M. (2005). *Opt. Lett.* 30: 2388.

12 Dazzi, A., Glotin, F., and Carminati, R. (2010). *J. Appl. Phys.* 107: 124519.

13 Dazzi, A., Prater, C.B., Hu, Q. et al. (2012). *Appl. Spectrosc.* 66: 1365.

14 Lu, F., Jin, M., and Belkin, M.A. (2014). *Nat. Photonics* 8: 307.

15 Kjoller, K., Felts, J.R., Cook, D. et al. (2010). *Nanotechnology* 21: 185705.

16 Rajapaksa, I., Uenal, K., and Wickramasinghe, H.K. (2010). *Appl. Phys. Lett.* 97: 073121.

17 Nowak, D., Morrison, W., Wickramasinghe, H.K. et al. (2016). *Sci. Adv.* 2: e1501571.

18 Jahng, J., Brocious, J., Fishman, D.A. et al. (2014). *Phys. Rev. B* 90: 155417.

19 Jahng, J., Brocious, J., Fishman, D.A. et al. (2015). *Appl. Phys. Lett.* 106: 083113.

20 Kim, B., Jahng, J., Khan, R.M. et al. (2017). *Phys. Rev. B* 95: 075440.

21 Tumkur, T.U., Yang, X., Cerjan, B. et al. (2016). *Nano Lett.* 16: 7942.

22 Kumar, N., Mignuzzi, S., Su, W., and Roy, D. (2015). *EPJ Tech. Instrum.* 2: 9.

23 Deckert-Gaudig, T., Taguchi, A., Kawata, S., and Deckert, V. (2017). *Chem. Soc. Rev.* 46: 4077.

24 Jahng, J., Potma, E.O., and Lee, E.S. (2019). *Proc. Natl. Acad. Sci. USA* 116: 26359.

25 Blackledge, R.D. and Jones, E. (2007). Chapter 1: All that glitters is gold. In: *FORENSIC ANALYSIS ON THE CUTTING EDGE: New Methods in Trace Evidence Analysis*. Wiley Scientific [And references therein].

26 Griggs, S., Hahn, J., and Bonner, H.K.S. (2011). *Global Forensic Sci. Today* 10: 19.

10

Raman and Surface-Enhanced Raman Scattering (SERS) for Trace Analysis

Cyril Muehlethaler

Department of Chemistry, Biochemistry and Physics, University of Quebec, Trois-Rivières, QC, Canada

10.1 Introduction

Raman spectroscopy is a relatively new technique which has found very specific applications in the field of forensic science. It was initially seen as a complementary method to infrared (IR) spectroscopy but has since acquired an important reputation of its own in laboratories. Recent technological developments have further increased the scope of analyses that it can bring to each forensic field, ranging from analysis of inks, fibers, paints, explosives, fire accelerants, gunshot residues (GSRs), biological traces, illicit drugs, to virtually any other type of physical evidence. Raman spectroscopy is particularly successful for identifying dyes and pigments in nondestructive contactless manner, and has since found a place of choice compared to other instrumental techniques, because of its many advantages:

- Is nondestructive and requires minimal sample preparation
- Is insensitive to the signal of water
- Allows measurements through containers (e.g. glass vial, plastic bags)
- Provides cleaner spectra with narrower and more resolved bands than mid-IR spectroscopy
- Is complementary to most analytical methods (infrared in particular) and does not present any direct alternative in the analysis of the polarizability of molecules

Despite these advantages the technique has suffered from some shortcomings that have minimized its impact in the laboratory. The Raman signal is quite weak, and the instruments require high-performance components to compensate from this lack of sensitivity for trace analysis. On top of that, the main reason hindering the use of Raman spectroscopy more widely is the stray fluorescence of certain molecules, which can mask the relevant signals and prevent their identification.

Leading Edge Techniques in Forensic Trace Evidence Analysis: More New Trace Analysis Methods, First Edition. Edited by Robert D. Blackledge.
© 2023 John Wiley & Sons, Inc. Published 2023 by John Wiley & Sons, Inc.

The last few decades have greatly improved this situation through technological developments but also thanks to the discovery of the surface-enhanced Raman scattering (SERS) effect. SERS is a derived technique that allows to drastically increase the Raman signal intensity for molecules adsorbed on the surface of conductive materials, such as noble metals (i.e. gold, silver, and copper) or semiconductors (e.g. CdSe, MoS_2, and GaN). The proximity with the conductive substrate that acts as an antenna amplifying the electromagnetic field, results in increase of the signal up to millions/billions of times while completely dampening the fluorescence at the same time. The two main drawbacks of Raman spectroscopy have thus become the main advantages of using SERS. SERS discovery and the successful applications to various forensic trace analyses have opened up hundreds of new avenues of research. The main advantage of the technique is its ultra-sensitivity, capable of detecting molecules down to femtograms, making it ideal for trace analysis, development of sensors, or semiquantitative dosages. The technique works beautifully with dyes, explosives, and various other physical evidences that will be presented later on. The following sections aim at introducing the concepts and theory behind Raman spectroscopy and SERS, presenting the instrumentation needed, and discussing forensic applications to various type of traces.

10.2 Theory

10.2.1 Raman Spectroscopy

Raman spectroscopy is based on the inelastic scattering of a monochromatic light provided by a laser. The molecule being analyzed absorbs photons from the laser and reemits them at a different frequency. Two concomitant phenomena happen at the same time. (i) The vast majority of the photons are absorbed and returned with no change in their frequency, an elastic interaction called Rayleigh scattering. (ii) A tiny proportion of the absorbed photons (approximately 1 out of 10^4) will be reemitted with a slightly different frequency ($v_0 \pm v_m$) either up or down of the original frequency (v_0). These inelastic interactions are respectively called Stokes and anti-Stokes and compose the Raman signal (Figure 10.1). For Stokes scattering the molecule gains energy and emits radiation at a wavelength longer (lower frequency) than that of the incident radiation. On the contrary, for anti-Stokes scattering the radiation has a shorter wavelength (higher frequency) due to losing vibrational energy from an initial excited state. This Raman scattering is very weak and its measurement requires very intense excitation sources (lasers) and an almost perfect filtering of the main Rayleigh scattering in order to see and detect the (anti-)Stokes radiation(s). At room temperature, a majority

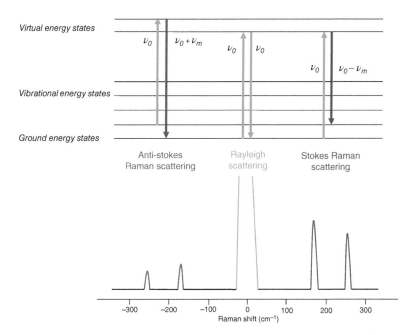

Figure 10.1 Schematic illustration of Rayleigh, Stokes, and anti-Stokes scattering. A Raman spectrum is usually constructed from the shift in wavelength (in units of cm^{-1}) from Rayleigh to the Stokes peaks.

of photons populate the ground level and Stokes scattering is more intense than anti-Stokes scattering. In practice, the Raman spectrum is represented by measuring the Stokes shifts in wavelength from the Rayleigh scattering, with units of wavenumbers (cm^{-1}). The Rayleigh scattering is always centered at $0\,cm^{-1}$, with Stokes Raman scattering bands that form the Raman spectrum usually appearing in the range of 100–4000 cm^{-1} (Figure 10.2).

The intensity of the Raman scattering is directly proportional to the polarizability of the molecule (α), which is the ease with which the electron cloud around the molecule can be distorted under the influence of an electric field (E). Exciting the cloud of electrons with a strong monochromatic light (such as a laser) creates an induced dipole moment in the molecule (P) which causes it to vibrate, with some functional groups more susceptible than others (these functional groups are called Raman-active). The Raman intensity is proportional to the square of the electric dipole moment and either a molecule with better polarizability, or excited by superior laser intensity can provide better Raman scattering through the relation $P = \alpha \cdot E$. This formula also means that a molecule with fewer electrons or densely packed electrons around the nucleus will tend to have a lower polarizability and an inferior Raman signal.

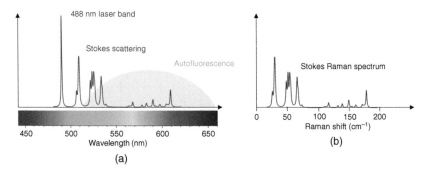

(a)

(b)

Figure 10.2 Schematic illustration of the Raman scattering over wavelength (nm) and Raman shift (cm^{-1}) scales. (a) A molecule is excited with a 488 nm laser and the Stokes scattering peaks are emitted at longer wavelength. A fluorescence from the molecule itself can sometimes appear as well and overlap the Stokes radiation. (b) The Stokes peaks are represented as Raman shifts from the Rayleigh radiation (488 nm = 0 cm^{-1}).

10.2.2 Enhancement Mechanism in SERS

The principle behind SERS is to place the probe molecule in the vicinity of a metal that acts as a resonance antenna, drastically increasing the induced dipole moment of the molecule, and leading to an enormous increase in the Raman scattering intensity. The metal proximity also acts as a fluorescence dampener (Figure 10.3). The fluorescence and the low signal intensity, two of the most challenging drawbacks of Raman spectroscopy that slowed down its use for years, are suddenly improved in a single step, almost magically.

It was first discovered in 1973 by Fleischmann that a molecule placed in contact with a silver electrode exhibited an extraordinary increase in signal intensity [1], and the theory was later confirmed by Van Duyne and Albrecht in 1977 [2, 3]. This discovery was left relatively untouched for many years due to the complexity in repeating the experiments and providing reproducible signals. The 90's marked an important step in the development of the technique. Technological advances supported by new dispersive spectrometers, efficient light filtering, and intense laser powers greatly improved the sensitivity of standard Raman spectroscopy. SERS has since almost surpassed standard Raman spectroscopy, and has seen a huge increase in popularity and avenues of research.

The improvement of the signal intensity underlying the SERS phenomenon in metals is the result of three main mechanisms [4]: (i) the surface plasmon resonance (SPR) (that is the collective oscillation of the electrons on the metal surface), (ii) the molecular resonance originating from the molecule itself, and (iii) the charge transfer taking place between the newly formed complex of the molecule and the metal (Figure 10.4). While these mechanisms are now

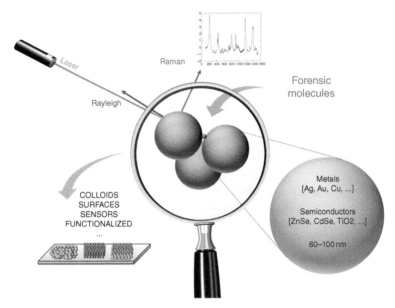

Figure 10.3 Schematic illustration of the surface-enhanced Raman spectroscopy (SERS) mechanism. Molecules adsorbed on noble metals or semiconductors benefit from the increase in resonance conditions to provide enhanced Raman signals up to $10^8 - 10^{10}$. Metal colloids are the most common type of substrates, but other type of functionalized surfaces can act as SERS substrates as well.

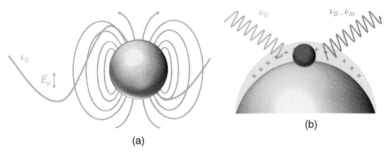

Figure 10.4 (a) Noble metals such as silver or gold act as antennas and locally enhance the electro-magnetic (EM) field around them when excited by an electromagnetic radiation (e.g. laser). (b) Surface plasmon resonance (SPR) is the collective oscillation of surface electrons when excited by EM radiation, hence increasing the polarizability of any molecule adsorbed onto its surface that gives rise to enhanced Raman signals.

accepted and demonstrated, recent years have seen the emergence of new types of semiconductors substrates. For these, two additional resonances can be distinguished: (iv) the exitonic resonance and (v) the Mie scattering [5]. In practice, the enormous enhancement factors (EFs) observed in SERS (up to billions of times) are always due to a combination of two or more of the resonances and their coupling according to the Herzberg–Teller model [4]. Each resonance has a maximum efficiency in a specific range (in nm). Understanding the wavelengths that maximize scattering, where they are located, and being able to excite multiple resonances together at the same time is the key to sustain high EFs. The choice of the excitation wavelength (laser) is therefore essential to target as many (or the most efficient) resonances as possible. An adequate superposition of resonances and excitation wavelength can lead to a signal up to 10^{12} times more intense in certain (ideal) conditions, with the average being in the range of 10^4–10^8 (Figure 10.5). The highest enhancement factor observed is 10^{12} for crystal violet or rhodamine 6G dyes [6, 7]. Exciting crystal violet molecules adsorbed on silver colloids with a 633 nm laser is perfectly located over the SPR, molecular, and charge transfer resonances, making it possible to detect a single molecule in solution [8]. This huge increase opened up new avenues of research and a whole new subdiscipline called single-molecule SERS [6–10].

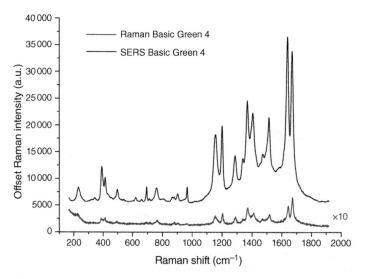

Figure 10.5 Standard Raman (blue) and SERS (black) spectra of the dye Basic Green 4 with silver colloids, illustrating the enormous increase in intensity. The calculated enhancement factor, taking into consideration the absolute intensities but also the number of contributing molecules, is approximately 10^4.

While providing the extremely valuable enhancement we discussed above, SERS also induces a change in the selection rules. The molecule is not by itself anymore but forms a complex with the metal, influencing its polarizability as a whole. This makes the technique complementary to standard Raman spectroscopy, with new peaks appearing and others disappearing.

10.2.3 SERS Substrates

Depending on the application, several types of substrates can sustain the required enhancement and act as a support/binder for the probe molecule. Silver nanoparticle colloids (Ag-NPs) are the most popular and easiest to produce [11] (Figure 10.6). They are particularly well suited for analysis of liquid samples. These nanoparticles offer the possibility to detect many classes of molecules, due to their affinity with the silver surface (cationic), or with the citrates used to stabilize the colloids (anionic). The possibilities offered for functionalizing their surface makes it an almost ideal substrate to start with in the vast world of SERS. An aggregating agent (i.e. KNO_3, NaCl) is usually added to force the formation of agglomerates, thus promoting the creation of hot spots between the molecule and the substrate. Although commercially available, these colloids can also be produced at lower cost in the laboratory [11].

The second most popular type of substrates are nano-surfaces (Figure 10.6). The literature if full of examples, each one fancier that the other [12–15]: nanospheres, nanowires, nanorods, nanopyramids, popcorn-shaped, etc. This nano-structuring is essential to maintain SPR. A smooth surface would not provide SPR with

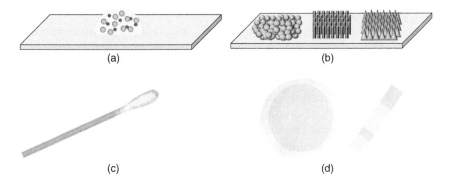

Figure 10.6 Most common type of SERS substrates that are of interest in forensic science: (a) aqueous colloids based on metal or semiconductor nanoparticles, (b) functionalized surfaces, e.g. shaped as nanospheres, nanorods, nanopyramids, (c) silver loaded cotton swabs, and (d) silver-coated filter paper to be used as collection pad or as migration medium using an eluent solution for concentrating the target molecules.

sufficient resonance. Further, the nano-structuration has the double advantage of being able to retain the molecules without the necessity to have a strong chemical bond. Any molecule deposited on the substrate will benefit from the increased electromagnetic field generated by the metal. The metal surface can also be grafted with ligands for increased reactivity with specific types of molecules [15, 16]. The ability to functionalize surfaces to target groups or classes of molecules has greatly contributed to the development of sensors based on SERS technology and paving the way for (semi-) quantitative applications. Recent examples reported in the literature include functionalized papers and filters [17–19], micro-fluidic sensors [19], vapor dosimeters [20], and remote sensing [21].

Semiconductor substrates are quite recent in the SERS landscape. Their enhancement mechanism does not rely much on the SPR, but rather on the chemical enhancement provided by the charge transfer between the substrate and the molecule (and an additional excitonic resonance). At the difference of metal substrates, the semiconductor substrates can be all flat and nonetheless provide significant enhancements [22]. They do require a chemical bond with the molecule however. This specificity in binding and resonance conditions makes their use particularly adapted for sensors and captors. The best demonstration of charge transfer resonances is provided by 2D materials (graphene, molybdenum disulfide, etc.) composed of a monolayer of atoms as thin as 3 Å, that could sustain enhancements factors up to 10^6 [23, 24].

Each and every substrate provides an enhancement mechanism that is unique. Some are more versatile and offer modest to large enhancements with a broad range of molecules. Others form particular complexes with the probe molecule that only resonate under a very particular laser excitation. These properties make them extremely useful for captors or sensors, but do not offer the capacity to detect multiple molecule types.

The rule of thumb regarding substrates is that highly structured and repeatable surfaces, such as nanopillars, nanorods, or nanopyramids provide consistent enhancement all over the surface, with reasonable EF generally up to 10^8. The measures are repeatable and reproducible. On the contrary, colloids are less organized and less structured, and their measurement rely on Brownian motion of the agglomerates in the drop and require longer integration time to provide the enhanced Raman signal. The measurements are less reproducible and show great variation in absolute intensity. Obtaining a highly intense spectrum sometimes require a few consecutive measurements, as suggested by the uncertainty principle. Yet, when it happens, the signal is usually very intense due to dimer hot spots of colloids that can provide record EF up to 10^{12}. The highest enhancement factors, such as single-molecule detection, are obtained with colloids but are difficult to reproduce and rely much on statistical considerations and sufficient replicated measurements.

10.2.4 Probe Molecules

One of the main drawbacks of standard Raman spectroscopy is the fluorescence of some molecules that prevents obtaining successful spectra in some occasions. When present, the fluorescence masks all the relevant peaks and sometimes even saturates the detector. However, under certain conditions when the laser frequency coincides with electronic transition in the molecule, some substances can emit a strong resonance scattering instead of the fluorescence. The resonance is maximal when the laser frequency corresponds to the first or second electronic excited states transitions and provides spectra with an approximately fivefold enhancement in intensity [25].

Dyes and pigments are very effective Raman scatterers and can provide resonance conditions in the visible. Dyes and pigments have greatly contributed to the initial popularity of Raman spectroscopy in forensic science. The complementarity the technique offers with infrared spectroscopy is ideal for full physicochemical characterization. For example, typical paint binders, extenders, and additives can be analyzed with fourier transform infrared (FTIR), while the pigment composition shows up beautifully in Raman spectroscopy.

The resonance conditions being unique for each molecule, Raman instruments should be able to adjust their excitation wavelength to the molecule being analyzed. For this reason instruments with multiple laser bands are common in forensic laboratories. They have the double advantage of being able to move away from any parasitic fluorescence, and to target resonance conditions to improve the Raman signal.

Any molecule can be a Raman scatterer as long as its polarizability is sufficient to induce a vibration. Even molecules without intrinsic polarizability can become Raman active when placed in contact with SERS substrates, which forms a complex with the probe molecule and increases the EM field. It is nonetheless difficult to predict beforehand if a molecule will or will not be a good SERS scatterer. A good probe molecule for SERS is a combination of: (i) having functional groups or reactive species able to provide binding with the substrate, (ii) sufficient polarizability of the complex formed by the molecule + substrate, and (iii) the adequate superposition of at least one resonance (preferably more) in the complex with an excitation laser.

Any type of molecule encountered in the current forensic laboratory might naturally be a good Raman scatterer, or be designed to work with specific substrates through functionalization. The literature is sufficiently abundant at this time to have a preliminary idea of what will or will not work [26–28]. Some molecules will work better under specific conditions (e.g. explosives such as trinitrotoluene [TNT] have great affinity with gold substrates through their nitro groupings and are almost ineffective with silver). Examples include but are not limited to:

- Dyes and pigments from paints, fibers, inks, cosmetics, etc.
- Explosives
- Gunshot residues
- Biological traces (blood, semen, and saliva)
- Fire accelerants
- Fingermark secretions
- Drugs
- Environmental contaminants
- Chemical warfare agents

10.3 Instrumentation

The theory of Raman spectroscopy tells us that the intensity of the measured signal is proportional to the fourth power of the laser frequency. This means that lasers with lower wavelengths are theoretically providing the most intense signals. However, since the Raman signal is so weak with approximately 1 out of 1000 emitted photons being Raman, it implies that the instrument should be able to efficiently filter the elastic Rayleigh scattering in order to be able to measure them. The current Raman instruments typically consist of some excitation lasers in the UV, Visible and near-IR, optics to conduct and filter out radiation, a spectrometer, and a detector (Figure 10.7).

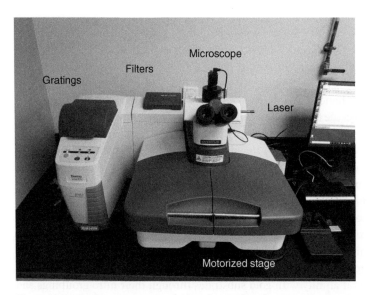

Figure 10.7 Basic components of a benchtop micro-Raman spectrometer.

10.3.1 Spectrometer

Two main types of spectrometers exist: Fourier transform (FT) and dispersive instruments. Fourier transform instruments are less frequent in laboratories but represent a solution of choice if the complete avoidance of fluorescence is desired. The spectrometer is composed of a near-infrared interferometer (similar to FTIR instrument) and a Nd:YAG NIR laser at 1064 nm. FT-Raman spectrometers largely overcome fluorescence, as many molecules encountered in forensic analyses do not fluoresce anymore in the near-infrared wavelength range. FT-Raman instruments have a greater spectral quality with improved resolution compared to dispersive instruments that are relying on monochromators. FT-Raman spectrometers do require a nitrogen-cooled detector however, contributing to more bulky installations and greater needs.

Dispersive instruments make use of monochromators to disperse the collected light and separate the individual wavelengths. A monochromator itself is composed of an entrance slit, a diffraction grating and some focusing mirrors to direct the light along its path. The diffraction grating is the central part separating the polychromatic light into its individual wavelength components. It is composed of a set of equally spaced grooves that refracts the light with path length differences that are dependent on the incident and reflected angles. Only in-phase radiations can enter the exit slit, thereby allowing it to separate the individual wavelengths. The grating can be moving (scanning instruments) or fixed (snapshot over a specific wavelength range). In many laboratories, dispersive instruments are preferred due to their ability to quickly change between multiple lasers ranging from deep-UV (244 nm) up to near infrared (830 nm). Dispersive instruments also have a higher sensitivity which leads to a lower detection limit due to the v^4 scattering efficiency dependency.

10.3.2 Excitation Lasers

During the initial discovery of the Raman effect by Sir C.V. Raman, lasers were not available and only appeared on the market almost 40 years later. The Raman scattering being so weak, it requires strong lasers to provide a sufficiently intense signal to be detected. The shorter the wavelength the higher the Raman scattering due to the v^4 dependency. The laser should also be as monochromatic as possible to provide the necessary spectral resolution and to allow adequate filtering of the Rayleigh scattering. The usual cutoff value of most Raman spectrometers is around 50 cm^{-1} with the help of notch, dielectric band filters, or long-pass filters. Current lasers to be used in Raman spectroscopy cover UV (300–400 nm), visible (400–700 nm), and near-infrared wavelengths (700–1100 nm). This versatility offers many possibilities to vary the excitation wavelengths to avoid fluorescence or to target particular molecular or substrate resonances.

Table 10.1 Most common lasers used in Raman spectroscopy.

Wavelength (nm)	Type	Particularities
488 (blue)	Ar	High risk of fluorescence. Excellent resonance with silver
532 (green)	Nd: YAG (second haromonic)	High risk of fluorescence. Good resonance with silver
632.8 (red)	HeNe	Risk of fluorescence. Excellent resonance with gold
785 (near-IR)	Diode	Good intensity and low fluorescence
830 (near-IR)	Diode	Low fluorescence
1064 (near-IR)	Nd: YAG (first harmonic)	Used in FT-Raman instruments. Absence of fluorescence

The most common laser lines to be used in Raman spectrometer are given in Table 10.1.

10.3.3 Detector

The most common type of detectors are the charged-coupled devices (CCDs). The photosensitive elements of the CCD (pixels) each record a different spectral line, thereby providing instant and accurate readings over the whole spectral range (usually around $100–4000\,cm^{-1}$). Typical CCD pixels have a 16 bit depth, hence being able to detect up to 65'536 photons ($=2^{16}$) before reaching saturation. The intensity of the Raman signal is expressed in units of Raman counts per second (CPS).

10.3.4 Microscope

The coupling of Raman spectrometer with laboratory microscopes has been very successful and an enormous advantage for forensic scientists dealing with small and heterogeneous specimens. Using a microscope permits to obtain focal spots around 1–3 μm in diameter, ideal for particle or contaminant analysis. Microscopic traces such as fibers, paints, or various types of unknown powders are easily measured with 10× up to 100× magnifications.

The Raman sampling area is defined by the laser spot size, which is dependent on the objective numerical aperture (NA), and on the laser wavelength (λ) through the following equation:

$$\text{Laser spot size} = \frac{1.22 \cdot \lambda}{\text{NA}}$$

Table 10.2 Typical laser spot size for Raman measurements using the most common combinations of objectives and excitation wavelengths.

Laser spot size	532 nm	633 nm	785 nm
10× (0.3 NA)	2.16 µm	2.57 µm	3.19 µm
20× (0.5 NA)	1.30 µm	1.54 µm	1.91 µm
50× (0.75 NA)	0.86 µm	1.03 µm	1.27 µm
100× (dry) (0.9 NA)	0.72 µm	0.85 µm	1.06 µm

Some indicative values of laser spot size are given in Table 10.2.

Focusing a very intense laser (mW) on a micrometer-sized focal spot is not without risks, and special care must be taken not to damage the specimen. It is not unusual to observe a burned crater at the measurement location (usually invisible to the naked eye), but that can nevertheless impact the spectral data obtained.

Confocal experiments such as depth profiling or line mapping require the use of a pinhole in the optical path, which blocks out the slightly out-of-focus signal from the otherwise larger interaction volume. Confocal measurements increase resolution and can successfully reject signals from lower or higher layers allowing contactless measurements to be taken through containers (e.g. transparent plastic or glass). Using confocal microscopy, successful measurements can be obtained for liquids in vials, unknown powders in plastic bags, or metallic paints below a varnish layer.

Combining microscopic measurements with a motorized stage also has offered many applications in cartographic reconstruction (mapping) of heterogeneous surfaces. A whole spectrum can be recorded for each pixels of a preselected grid, and later rendered as an artificial map of peak intensity, peak area, or any mathematical transformation of multiple peaks.

10.3.5 Portable Spectrometers

The relative simplicity of the technique – that requires only a laser, a set of filters, a spectrometer and a detector – has led to the development of portable Raman instruments. Two main categories currently share the market: (i) the portable instruments that can be installed in the back of a minivan or in a trunk and (ii) the handheld instruments that are destined to field measurements.

Portable instruments are a bit more bulky than handheld instruments but offer performances that tend to compare well to lab-grade equipments. The lasers are just as intense as those in the laboratory, and the few concessions made to make it portable usually concern the spectral resolution and spatial resolution. While lab-grade instruments can offer spectral resolutions down to $2\,cm^{-1}$, portable instruments usually range in the $10\text{--}12\,cm^{-1}$. They usually do not offer magnification, and the focal spot is consequently much larger ($120\text{--}150\,\mu m$). Other than that, portable instruments are good alternatives to laboratory instruments, and even offer the possibility to have multiple lasers (usually a combination of two, 532 nm/785 nm being the most popular). The instruments can be fitted with many accessories such as microscopic slide holder, cuvettes for liquid analyses, or remote fiber light excitation.

Handheld instruments are much more compact and destined to field use in the form of a point-and-shoot device. The spectral resolution is comparable to portable instruments ($10\text{--}12\,cm^{-1}$), but usually come in single-wavelength. Handheld instruments are particularly well suited for identification of unknown substances during patrols or border security controls.

10.4 Forensic Applications

The following section presents examples of application in certain fields of forensic science where Raman spectroscopy plays a major role. While standard Raman spectroscopy has been around for quite some time, it was only during the last decade that it made its way in courts and for routine cases. While we can now consider Raman as a validated technique (in the sense of Daubert criteria), SERS – even though currently exploding in the numbers of publications – remains at the first levels of validation, and more work is required be able to exploit it fully in routine criminal cases [29].

10.4.1 Questioned Document

Questioned document examination includes the comparison of handwriting and signatures, as well as technical examinations of paper or inks. The technical aspect of document examination often relies on the analysis of inks, whether it comes from pens, printers or, although less frequent nowadays, from typewriters. Thin-layer chromatography (TLC) and LC–MS analyses are the two most reported techniques for ink analysis. Raman spectroscopy has since its discovery proven to be a very valuable technique as a means for identifying and comparing the inks of suspicious entries. Being nondestructive, it allows identification of the pigments and dyes in situ without requiring any ink collection or extraction. Recent reviews on the analysis of questioned documents cover most of the recent literature on that topic [30–32]. The particularities of the SERS studies are discussed below.

The simplest and easiest way of performing a SERS analysis of handwritten ink entries consists of in situ deposition of silver colloids [33, 34]. Pen inks composition being mostly based on dyes and pigments, it makes them the ideal SERS molecules. Basically, any type of dye (i.e. acid, basic, mordent, direct, solvent, and along with those not necessarily used in inks formulation) is expected to have good affinity for the silver colloids and to provide the necessary chemical bond, either to the positively charged silver surface, or to negative citrate groups coating. Although considered semi-destructive – a brown spot of dried colloids approximately a half millimeter in diameter will cover the ink entry after analysis – this procedure proved as successful as the traditional ink extraction and TLC separation, while also permitting the precise identification of the dyes and pigments based on their spectral components [33, 34]. The paper only has a minor interference in SERS while the in situ measurement using standard Raman spectroscopy could be more problematic with strong overlapping bands at 1000–1100 cm^{-1}.

Another sampling strategy that is expected to become an extremely rich avenue of research is the non-destructive extraction of the dyes through silver loaded gels. Although extremely simple – a silver-doped agarose gel is placed in contact with small portions of strokes and its surface extracts and traps the dyes – this strategy works really well and preserves the dyes for several days (Figure 10.8). The gel material does not damage the paper, and the amount of dye collected is imperceptible to the naked eye. Some gel materials (agar–agar, cellulose, etc.) are directly loaded with the silver colloids (the drying and shrinkage provides an additional advantage of creating hot spots), while others are more successfully used in a two-step procedure (extracting the dyes, flipping it, and adding silver colloids on the gel surface). Combining TLC with SERS is another very powerful way to take advantage of the separation and high detection capability of the dyes [35]. Globally, SERS performs better than standard Raman spectroscopy, providing a higher identification rate and incidentally a higher discriminating power. The fact that it prevents fluorescence and gives an extra kick in sensitivity

(a)   (b)   (c)

Figure 10.8 Small pieces of Agarose gel can be put in contact with ink entries and left a couple of minutes to extract the dyes (a). The procedure is almost non-destructive, leaving only a wavy surface where the wet gel was in contact with the paper (b). The amount of dye collected is largely sufficient for SERS analysis and appears as faint strokes on the collection gel (c).

gives more detailed spectra and many additional peaks corresponding to other minor colorants components.

Printers and toner inks are a bit more challenging than pen inks for obtaining good SERS spectra. It is mostly due to the prevalence of black inks whose main component the carbon black is essentially uninformative in Raman and SERS with only two large peaks [32]. Printer inks also contain a significant amount of pigments which are insoluble and mostly nonreactive with the metal substrates. Bringing them in contact with the silver colloids can be challenging without wet-chemistry principles and relies then mostly on a physical collection and then bringing them at a reasonable distance from the colloids in order to be able to benefit from the increase in electromagnetic field. The enhancement factors for pigments are not always impressive (10^3 at most) but remain possible in certain conditions. SERS was nonetheless proven successful for sensitively identifying the inkjet dyes. SERS being a *surface* technique it provides a localized excitation of the dyes in contact with the colloids and greatly limits the interaction from the support (paper and fillers). Standard Raman techniques, and even FT-Raman at 1064 nm, cannot avoid obtaining a signal from the paper due to the larger interaction volume [36].

SERS has shown a great deal of interest for security inks development. The capacity to encapsulate a silver core and a SERS taggant shell into silica nanoparticles turns them into nanometer-sized "signature" taggants that can only be detected with appropriate equipment [37]. These nanotags can be added (hidden) in different proportions and combinations to any kind of ink (government sticker, stamp, barcode, and QR code) and provide replacement alternatives to common lanthanides and other rare earth taggants.

10.4.2 Explosives

The detection of explosives is a major concern of homeland security agencies, and a domain in which many different analytical techniques are currently available (e.g. colorimetric tests, dogs sniffing, ion mobility spectrometry, or X-ray scanning). Compared to these techniques, SERS presents the advantage of versatility for a very sensitive early detection of the explosives in various environments with portable Raman spectrometers. The analysis of explosives by SERS has recently been covered by numerous papers and is one of the most prolific fields of SERS analyses in forensic science [38, 39]. The trace detection of explosives, either in the form of vapor, liquid, or solid has been very successful.

For most of the explosives, the use of gold represents the best SERS substrate for providing a strong and specific binding of the molecule to the surface. Nitroaromatic explosives such as TNT or dinitrotoluene (DNT) have a great affinity to gold since the nitro groups form bidentate chelate with the surface through their oxygen atoms. This specific binding with gold has been at the forefront of the development of numerous types of sensors [15, 16, 40–52]. All these substrates were proven successful due to their rough surface capable of sustaining greater SPR. Various strategies for collecting explosive traces have been proposed. Collection swabs are the most convenient for surfaces supposed to have been in contact with explosives. Gold nanoparticles are either deposited on standard filter papers by inkjet printing, or prepared by coating the extremity of cotton swabs [44, 45, 53]. Both solutions are convenient, providing simple and effective means for trace collection and subsequent analysis. Vapors of explosives can be detected by SERS using adapted captation chambers or microfluidic chips down to concentrations in the range of sub-ppb [50, 51, 54]. The probes are designed around a roughened metal surface and a fan to carry the vapors. A few drawbacks currently limit the vapor detection, mostly the low vapor pressure of common explosives at room temperature and the difficulty of obtaining a stable bond to metals other than gold. Gold thus remains the best alternative for detecting explosives, even though a few examples of application were proposed using either silver [53, 55, 56], nickel [57], or copper substrates [58]. Overall, the various types of explosives (TNT, PETN, RDX, and TNB) do not have a sufficient affinity with silver substrates, which results in a low adsorption and low enhancement factor (if any).

10.4.3 Fibers

The use of SERS for fiber analysis has a long history but most of the literature comes from art and cultural heritage objects field rather than forensic science. Microscopic examination is the most powerful technique for identifying the class and subclass of fibers, while infrared and Raman based techniques are very valuable next in the sequence of examination for respectively confirming the composition or identifying dyes and pigments [59]. The most convenient way of analyzing fibers by SERS is the in situ deposition of silver colloids, by adding a few drops directly on the samples [60–63]. The main difficulty using this procedure is the colloid coating of the fibers that is often insufficient to obtain good spectra or to quench completely the fluorescence of the substrate. One way to get around this problem is to concentrate the colloids with successive centrifugation steps until getting a colloidal paste [62], one that is much denser and provides a better coverage of the fiber.

As opposed to in situ analyses of dyes, it is sometimes preferable to treat the fiber in order to facilitate the migration of the dyes into the drop. Two different kinds of treatments should be distinguished depending on the nature of the dyes, mordant, or non-mordant. Non-mordant dyes can rapidly be extracted by common solvents. In a systematic review of extraction solvent mixtures, pyridine: water (4: 3) was the most commonly cited across a range of fiber and dye chemistries

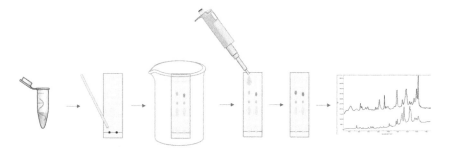

Figure 10.9 Schematic procedure of the extraction of non-mordant dyes from fibers, thin-layer chromatography (TLC) separation, followed by the addition of silver colloids on the revealed spots, and their identification by surface-enhanced Raman spectroscopy (SERS).

[64]. The extraction aliquot can be followed by a separation if multiple dyes are expected to be present. The most convenient separation technique is the TLC [60]. It is particularly interesting for SERS analyses because the various colored spots corresponding to the dyes can be identified directly on the plate under a Raman microscope (Figure 10.9). The TLC plates can be preloaded with silver for large volume analyses, or AgNPs can be added on a case-by-case basis on the migrated spots according to the needs of the analysis. One important feature to consider is that the chromatographic plates sometimes are loaded with fluorescent indicators such as GFP (green fluorescence protein). These can overpower the SERS signal and mask all relevant bands in the spectrum. Nonfluorescent plates are preferred.

Mordant dyes are a particular category of dyes which are covalently bond to the fibers and need to be hydrolyzed for better aggregation with the colloids. The most common procedure requires a pretreatment which consists of a hydrolysis by hydrofluoric acid (HF) or hydrochloric acid (HCl) [65]. Single fibers are placed for two minutes in a sealed chamber with a single drop of HF. The standard sequence of extraction/separation can then be applied in its entirety.

10.4.4 Paint

Similar to the fibers, the use of Raman and SERS in paint analysis mostly aims at identifying the dyes and pigments used to give them their color and appearance. Raman and SERS have a unique position among the sequence of examination and are most likely the best choice for a precise identification of the colorants [32], as opposed to techniques aiming at the other paint constituents (FTIR, Py-GC/MS, and elemental analysis). Pigments are usually providing very detailed and intense spectra with standard Raman spectroscopy. Employing a multilasers Raman instrument allows to move away from fluorescence, and obtain complementary information about pigment mixtures under different excitation sources. In most cases, the further use of SERS is not necessarily required. SERS for paint analysis is also rendered difficult by the nature of pigments that do not naturally bind with the metals due to their low solubility. Although some dyes might sometimes be used in the composition of paints, their color is mostly given by the pigment components. A particular type of dyes, the Lake dyes, is complexed with metal salts and represents the equivalent of fibers mordant dyes. Lake dyes are sometimes considered under the pigments category (and even called lake pigments) and require an extraction to be able to analyze them without contribution from the binder or other pigments/dyes. While the extraction of dyes is possible by various techniques presented above for the fibers (gels, films, and hydrolysis), these will extract the lake dyes almost exclusively, and would not permit a complete physicochemical characterization of the remaining components. Additional techniques would still be required in sequence.

One area which attracted a lot of attention is the development of laser ablation coupled to SERS (LA-SERS). It makes use of a pulsed laser that can precisely select which areas to collect from [66, 67]. The ablated particles are directly trapped onto a silver surface on top of them. Although destructive at a microscopic level, LA-SERS has the enormous advantage of being able to target individual layers within a paint cross-section and does not require additional sample handling.

10.4.5 Fingermarks

The analysis of fingermarks by Raman spectroscopy has two main fields of application: (i) the punctual analysis at a given spot on the fingermark to determine the secretions composition or detect a contamination residue (i.e. information other than the identification of fingermark source) or (ii) the chemical imaging of whole fingermark ridges in order to try to improve the contrast of standard techniques used previously.

Detection of contaminants in fingermarks using Raman spectroscopy has been documented for explosives [45], and drugs of abuse [68, 69]. SERS has also been reported using Raman probes that bind to specific proteins within latent fingermark using antibody–silver nanoparticles [70]. In a proof-of-concept laboratory experiment, the explosive DNT was still detected after more than 20 consecutive depletions [45]. Unfortunately, the proposed methodology requires a complete swabbing of the fingermark surface, making it incompatible with a dactyloscopic examination for identification. The detection of different drugs of abuse and common adulterants was made possible through contactless measurements on latent fingermark [68]. While the spectral quality was satisfactory, the most complicated step was to efficiently locate the contaminants over the surface. Chemical imaging might provide an interesting solution to this problem, with multiple replicated measurements over the whole surface. Once long and laborious, the acquisition of thousands of spectra over a grid of a few centimeters can now be realized in a couple of hours. This opened up great opportunities for introducing SERS in the possibilities of visualization for latent fingermarks [71–73]. Although it will probably never be considered as the first-choice technique in a sequence of fingermark revelation, the fact that numerous revelation techniques already employ nanoparticles – that are efficient and proven SERS substrates – makes it a solid complementary technique to improve contrast on difficult surfaces. Multimetal deposition (MMD), single-metal deposition (SMD), and small particle reagent (SPR) are all based on silver or gold nanoparticles. The particle size is typically 1–100 nm, which is very well adapted for targeting the SPR or resonance SERS effects. Examples of application are not abundant in the literature at the time of this review but show promising new applications [73]. One difficulty has been to image full-sized fingermarks, limited in part by the motorized stage amplitude, and by long acquisition time. Examples in the literature often demonstrate fingermark images over a few ridges at most.

Other semiconductors nanoparticles or quantum dots used for fingermark revelation can offer Raman resonance conditions and provide alternatives to standard metal substrates. Among the advantages it allows for a chemical imaging of the fingermark, which can be of interest for traces on difficult substrates with a background giving poor contrast. Most importantly, it does not necessitate to further reveal the trace using chemicals, as it simply adds a visualization alternative to already existing and accepted techniques. As it has already been done with standard Raman spectroscopy, SERS might also be able to detect contaminants in the fingermarks, either with functionalized nanoparticles or by taking advantage of the SERS enhancement on these molecules as well.

10.4.6 Fire Accelerants

The detection of accelerants from fire debris is extremely challenging on a scene due to the extreme conditions sustained (e.g. temperature, structural destruction, and fire extinguishing). While vapors might still be present on burnt floors, carpets, or furniture, the most common situation encountered is a complete consumption of the fire accelerants. The detection of vapors from fire debris through passive absorption on charcoal followed by analysis by gas chromatography is the current technique of choice. Standard Raman spectroscopy has been reported for the characterization of gasoline or other ignitable liquids used in arsons [74–76]. Since it does not benefit from the increased sensitivity allowed by SERS, the analyses cannot be performed on degraded traces or vapors. Examples of SERS detection might eventually prove useful using the same strategies as explosives vapor detection, but currently lack the possibility to detect all types of accelerant liquids with a single substrate, due to the specific resonance conditions required (i.e. white spirit, gasoline, and diesel might independently present better resonances for silver or gold substrates at different excitation wavelengths). The most promising avenue of research is toward developing specific sensors with substrates that are adapted to a particular type of fire accelerant through functionalization. Previous examples of SERS use was for identifying the fraudulent use of adulterated gasoline [77], or identifying polycyclic aromatic hydrocarbons (PAHs) [78–80].

10.4.7 Gunshot Residues

The detection of GSRs on a shooter's hands is the primary evidence of firearm use. The identification of GSR is usually demonstrated by looking at particles of Pb, Ba, and Sn, produced from the combustion of the bullet's primer. The presence of these three elements in a single solidified particle is characteristic of firearm usage. This analysis is almost exclusively done by electron microscopy (SEM/EDX) which permits both counting and characterizing the particles. While being capable of demonstrating firearm usage, SEM/EDX; however, cannot analyze the organic fraction of GSR, and eventually infer ammunition or firearm type, or associate the residues with a recovered cartridge. Raman has been successfully used for analyzing both organic and inorganic components of GSR [81–83]. It was proven useful for determining firearm type and caliber [82], comparison of unfired and GSR [84], or collection and chemical mapping of the residues over a surface [83]. The inorganic faction of GSR and its characteristic metal ions were successfully attributed in a Raman spectra [85]. No references report the use of SERS for characterizing GSRs yet. The increase in sensitivity permitted by SERS will probably further open up the possibilities to detect minor constituents of the residues. Combining Raman spectroscopy with swabbing procedures using a SERS active substrate has not been tested yet for GSRs but might provide very valuable results for direct on-field characterization of GSRs.

10.4.8 Cosmetic Products

Cosmetic products are common types of physical evidence encountered in forensic laboratories, ranging from lipsticks to glitters and shimmers from body creams, shampoos, or hair products. Cosmetic products can be found on pieces of cloth, bed sheets and pillows, vehicle airbags, or on any touched surface through contamination. Raman spectroscopy is particularly useful for contactless in situ measurement of smears on various textiles. Similar to other type of traces, the pigments composition will be the main contributor to the Raman signal. Raman spectroscopy can provide information about the polymer composition of glitters and eventually carry out a depth profiling through layers, but overall a microscopic examination will be much more effective at discriminating glitter particles.

Raman spectroscopy has been a proven technique for analyzing lipstick smears on various fabrics [86–88]. The lipstick matrix (oil and wax) is often responsible for strong fluorescence. One study in particular demonstrated the usefulness of using several lasers for moving further away from the fluorescence and taking advantage of their complementarity [87]. Traces of lipsticks can be analyzed by SERS in situ with the simple addition of silver colloids over the smear [89]. This strategy proved useful and effectively decreased the fluorescence background while providing spectra with well resolved peaks. While a few samples still had some residual fluorescence, the strategy used for fibers by using a colloidal paste can be a solution of choice.

Raman spectroscopy has been employed for identifying tattoo pigments [90]. Of particular interest for forensic scientists is the ability to gain information on the pigment once injected under the skin, for the purpose of improving visualization of the tattoo on deceased persons whose bodies have begun to degrade.

10.4.9 Other Types of Physical Evidence

Various other fields of forensic science have found niche examples of applications of SERS, sometimes restricted to a specific type of samples or techniques. Most of these applications concern the analysis of dyes in various matrices.

A good example of the potential of SERS analysis is the analysis of vehicle headlights. Multilayered original equipment manufacturer (OEM) paint systems represent the best alternative in hit-and-run accidents when the suspected vehicle has fled the scene. Access to database often permits to find a list of potential brands, models, and years of production with great discrimination. Headlights were never exploited that way because of its very low discriminating power. About 90% of headlights are currently composed of polymethylmethacrylate (PMMA) and no other components permit to further discriminate the population at the exception of the dyes for orange and red headlights. It was shown that SERS was a technique of choice, capable of identifying both the polymer and the dye composition, either in situ or through extraction [91]. The successful identification of solvent dye mixtures improved the discriminating power from around 30–50% to over 90%, opening up new opportunities to exploit vehicle headlights in comparing unknown and reference samples or creating databases of profiles.

SERS has also been used to analyze shoe polish smears in cases of aggression and kicking. Even black or brownish, the most common shoe polishes may contain a mixture of various colorants, usually solvent dyes [92]. It was first tested on cotton fabrics and glass slide [92] or as part of mixtures [93].

References

1 Fleischmann, M., Hendra, P.J., and McQuillan, A.J. (1974). Raman spectra of pyridine adsorbed at a silver electrode. *Chem. Phys. Lett.* 26 (2): 163–166.

2 Jeanmaire, D.L. and Van Duyne, R.P. (1977). Surface Raman electrochemistry part I. Heterocyclic, aromatic, and aliphatic amines adsorbed on the anodized silver electrode. *J. Electroanal. Chem.* 84: 1–20.

3 Albrecht, M.G. and Creighton, J.A. (1977). Anomalously intense Raman spectra of pyridine at a silver electrode. *J. Am. Chem. Soc.* 99 (15): 5215–5217.

4 Lombardi, J.R. and Birke, R.L. (2008). A unified approach to surface-enhanced Raman spectroscopy. *J. Phys. Chem.* 112: 5605–5617.

5 Lombardi, J.R. and Birke, R.L. (2014). Theory of surface-enhanced Raman scattering in semiconductors. *J. Phys. Chem.* 118: 11120–11130.

6 Kneipp, K., Wang, Y., Kneipp, H. et al. (1997). Single molecule detection using surface enhanced Raman scattering (SERS). *Phys. Rev. Lett.* 78: 1667–1670.

7 Lombardi, J.R., Birke, R.L., and Haran, G. (2011). Single molecule SERS spectral blinking and vibronic coupling. *J. Phys. Chem. C* 115 (11): 4540–4545.

8 Canamares, M.V., Chenal, C., Birke, R.L. et al. (2008). DFT, SERS, and single-molecule SERS of crystal violet. *J. Phys. Chem. C* 112: 20295–20300.

9 Wang, Z. and Rothberg, L.J. (2005). Origins of blinking in single-molecule Raman spectroscopy. *J. Phys. Chem. B* 109: 3387–3391.

10 Wang, Z., Pan, S., Krauss, T.D. et al. (2003). The structural basis for giant enhancement enabling single-molecule Raman scattering. *Proc. Natl. Acad. Sci.* 100 (15): 8638–8643.

11 Lux, C., Lubio, A., Ruediger, A. et al. (2019). Optimizing the analysis of dyes by surface-enhanced Raman spectroscopy (SERS) using a conventional-microwave silver nanoparticles synthesis. *Forensic Chem.* 16: 100186.

12 Li, W., Zhao, X., Yi, Z. et al. (2017). Plasmonic substrates for surface enhanced Raman scattering. *Anal. Chim. Acta* 984: 19–41.

13 Tao, W., Zhao, A., Sun, H. et al. (2014). Periodic silver nanodishes as sensitive and reproducible surface-enhanced Raman scattering substrates. *RSC Adv.* 4: 3487–3493.

14 Wang, Y., Lu, N., Wang, W. et al. (2013). Highly effective and reproducible surface-enhanced Raman scattering substrates based on Ag pyramidal arrays. *Nano Res.* 6 (3): 159–166.

15 Demeritte, T., Kanchanapally, R., Fan, Z. et al. (2012). Highly efficient SERS substrate for direct detection of explosive TNT using popcorn-shaped gold nanoparticle-functionalized SWCNT hybrid. *Analyst* 137: 5041–5045.

16 Xu, J.Y., Wang, J., Kong, L.T. et al. (2011). SERS detection of explosive agent by macrocyclic compound functionalized triangular gold nanoprisms. *J. Raman Spectrosc.* 42: 1728–1735.

17 Yang, Q., Deng, M., Li, H. et al. (2015). Highly reproducible SERS arrays directly written by inkjet printing. *Nanoscale* 7: 421–425.

18 Yu, W.W. and White, I.M. (2013). Inkjet-printed paper-based SERS dipsticks and swabs for trace chemical detection. *Analyst* 138: 1020–1025.

19 Andreou, C., Hoonejani, M.R., Barmi, M.R. et al. (2013). Rapid detection of drugs of abuse in saliva using surface enhanced Raman spectroscopy and microfluidics. *ACS Nano* 7 (8): 7157–7164.

20 Vo-Dinh, T. and Stokes, D.L. (1999). Surface-enhanced Raman detection of chemical vapors with the use of personal dosimeters. *Field Anal. Chem. Technol.* 3 (6): 346–356.

21 Huang, Y., Fang, Y., Zhang, Z. et al. (2014). Nanowire-supported plasmonic waveguide for remote excitation of surface-enhanced Raman scattering. *Light: Sci. Appl.* 3: 199.

22 Muehlethaler, C., Considine, C.R., Menon, V. et al. (2016). Ultrahigh Raman enhancement on monolayer MoS_2. *ACS Photon.* 3 (7): 1164–1169.

23 Ji, W., Zhao, B., and Ozaki, Y. (2016). Semiconductor materials in analytical applications of surface-enhanced Raman scattering. *J. Raman Spectrosc.* 47 (1): 51–58.

24 Ma, S., Livingstone, R., Zhao, B. et al. (2011). Enhanced Raman spectroscopy of nanostructured semiconductor phonon modes. *J. Phys. Chem. Lett.* 2: 671–674.

25 McNay, G., Eustace, D., Smith, W.E. et al. (2011). Surface-enhanced Raman scattering (SERS) and surface-enhanced resonance Raman scattering (SERRS): a review of applications. *Appl. Spectrosc.* 65 (8): 825–837.

26 Muehlethaler, C., Leona, M., and Lombardi, J.R. (2016). Review of surface enhanced Raman scattering applications in forensic science. *Anal. Chem.* 88 (1): 152–169.

27 Suzuki, E.M. and Buzzini, P. (2018). Applications of Raman spectroscopy in forensic science. II: Analysis considerations, spectral interpretation, and examination of evidence. *Forensic Sci. Rev.* 30: 137–169.

28 Muro, C.K., Doty, K.C., Bueno, J. et al. (2015). Vibrational spectroscopy: recent developments to revolutionize forensic science. *Anal. Chem.* 87: 306–327.

29 Muehlethaler, C., Leona, M., and Lombardi, J.R. (2016). Towards a validation of surface-enhanced Raman scattering (SERS) for use in forensic science: repeatability and reproducibility experiments. *Forensic Sci. Int.* 268: 1–13.

30 Braz, A., Lopez-Lopez, M., and Garcia-Ruiz, C. (2013). Raman spectroscopy for forensic analysis of inks in questioned documents. *Forensic Sci. Int.* 232: 206–212.

31 Carcerrada, M. and Garcia-Ruiz, C. (2015). Analysis of questioned documents: a review. *Anal. Chim. Acta* 853: 143–166.

32 Buzzini, P. and Suzuki, E. (2015). Forensic applications of Raman spectroscopy for the in situ analyses of pigments and dyes in ink and paint evidence. *J. Raman Spectrosc.* 47: 16–27.

33 Geiman, I., Leona, M., and Lombardi, J.R. (2009). Application of Raman spectroscopy and surface-enhanced Raman scattering to the analysis of synthetic dyes found in ballpoint pen inks. *J. Forensic Sci.* 54 (4): 947–952.

34 Seifar, R.M., Verheul, J.M., Ariese, F. et al. (2001). Applicability of surface-enhanced resonance Raman scattering for the direct discrimination of ballpoint pen inks. *Analyst* 126 (8): 1418–1422.

35 Pozzi, F., Shibayama, N., Leona, M. et al. (2013). TLC-SERS study of Syrian rue (Peganum harmala) and its main alkaloid constituents. *J. Raman Spectrosc.* 44: 102–107.

36 Rodger, C., Dent, G., Watkinson, J. et al. (2000). Surface-enhanced resonance Raman scattering and near-infrared fourier transform Raman scattering as in situ probes of ink jet dyes printed on paper. *Appl. Spectrosc.* 54 (11): 1567–1576.

37 Golightly, R.S., DOering, W.E., and Natan, M.J. (2009). Surface-enhanced Raman spectroscopy and homeland security: a perfect match? *ACS Nano* 3 (10): 2859–2869.

38 Hakonen, A., Wu, K., Schmidt, M.S. et al. (2018). Detecting forensic substances using commercially available SERS substrates and handheld Raman spectrometers. *Talanta* 189: 649–652.

39 Liszewska, M.B., Bartosewicz, B., Budner, B. et al. (2019). Evaluation of selected SERS substrates for trace detection of explosive materials using portable Raman systems. *Vib. Spectrosc.* 100: 79–85.

40 Botti, S., Cantarini, L., Almaviva, S. et al. (2014). Assessment of SERS activity and enhancement factors for highly sensitive gold coated substrates probed with explosive molecules. *Chem. Phys. Lett.* 592: 277–281.

41 Chen, T.F., Lu, S.H., Wang, A.J. et al. (2014). Detection of explosives by surface enhanced Raman scattering using substrate with a monolayer of ordered Au nanoparticles. *Appl. Surf. Sci.* 317: 940–945.

42 Chou, A., Jaatinen, E., Buividas, R. et al. (2012). SERS substrate for detection of explosives. *Nanoscale* 4: 7419–7424.

43 Fang, X. and Ahmad, S.R. (2009). Detection of explosive vapour using surface-enhanced Raman spectroscopy. *Appl. Phys. B* 97: 723–726.

44 Fierro-Mercado, P.M. and Hernandez-Rivera, S.P. (2012). Highly sensitive filter paper substrate for SERS trace explosives detection. *Int. J. Spect.* 2012: 1–7.

45 Gong, Z., Du, H., Cheng, F. et al. (2014). Fabrication of SERS swab for direct detection of trace explosives in fingerprints. *ACS Appl. Mater. Interfaces* 6: 21931–21937.

46 Liu, X.C., Zhao, L., Shen, H. et al. (2011). Ordered gold nanoparticle arrays as surface-enhanced Raman spectroscopy substrates for label-free detection of nitroexplosives. *Talanta* 83: 1023–1029.

47 Ma, R.-M., Ota, S., Li, Y. et al. (2014). Explosives detection in a lasing plasmon nanocavity. *Nat. Nanotechnol.* 9: 600–604.

48 Moore, D.S. (2007). Recent advances in trace explosive detection instrumentation. *Sense Image* 8: 9–38.

49 Nuntawong, N., Eiamchai, P., Limwichean, S. et al. (2013). Trace detection of perchlorate in industrial-grade emulsion explosive with portable surface-enhanced Raman spectroscopy. *Forensic Sci. Int.* 233: 174–178.

50 Piorek, B.D., Lee, S.J., Moskovits, M. et al. (2012). Free-surface microfluidics/surface-enhanced Raman spectroscopy for real-time trace vapor detection of explosives. *Anal. Chem.* 84: 9700–9705.

51 Tamane, S., Topal, C.O., and Kalkan, A.K. (2011). Vapor phase SERS sensor for explosives detection. *IEEE International Conference on Nanotechnology* 301–306.

52 Wackerbarth, H., Salb, C., Gundrum, L. et al. (2010). Detection of explosives based on surface-enhanced Raman spectroscopy. *Appl. Opt.* 49 (23): 4362–4366.

53 Liu, J., Si, T., and Zhang, Z. (2019). Mussel-inspired immobilization of silver nanoparticles toward sponge for rapid swabbing extraction and SERS detection of trace inorganic explosives. *Talanta* 204: 189–197.

54 Khaing Oo, M.K., Chang, C.-F., Sun, Y. et al. (2011). Rapid, sensitive DNT vapor detection with UV-assisted photo-chemically synthesized gold nanoparticle SERS substrates. *Analyst* 136 (13): 2811–2817.

55 Xu, Z., Hao, J., Braida, W. et al. (2011). Surface-enhanced Raman scattering spectroscopy of explosive 2,4-dinitroanisole using modified silver nanoparticles. *Langmuir* 27: 13773–13779.

56 Zhou, H., Zhang, Z., Jiang, C. et al. (2011). Trinitrotoluene explosive lights up ultrahigh Raman scattering of nonresonant molecule on a top-closer silver nanotube array. *Anal. Chem.* 83: 6913–6917.

57 Sajanlal, P.R. and Pradeep, T. (2012). Functional hybrid nickel nanostructures as recyclable SERS substrates: detection of explosives and biowarfare agents. *Nanoscale* 4: 3427–3437.

58 Hamad, S., Podagatlapalli, G.K., Mohiddon, M.A. et al. (2014). Cost effective nanostructured copper substrates prepared with ultrafast laser pulses for explosives detection using surface enhanced Raman scattering. *Appl. Phys. Lett.* 104 (263104).

59 Buzzini, P. and Massonnet, G. (2013). The discrimination of colored acrylic, cotton, and wool textile fibers using micro-Raman spectroscopy. Part 1: in situ detection and characterization of dyes. *J. Forensic Sci.* 58 (6): 1593–1600.

60 Brosseau, C.L., Gambardella, A., Casadio, F. et al. (2009). Ad-hoc surface-enhanced Raman spectroscopy methodologies for the detection of artist dyestuffs: thin layer chromatography-surface enhanced Raman spectroscopy and in situ on the fiber analysis. *Anal. Chem.* 81: 3056–3062.

61 Casadio, F., Leona, M., Lombardi, J.R. et al. (2010). Identification of organic colorants in fibers, paints, and glazes by surface enhanced Raman spectroscopy. *Acc. Chem. Res.* 43 (6): 782–791.

62 Idone, A., Gulmini, M., Henry, A.-H. et al. (2013). Silver colloidal pastes for dye analysis of references and historical textile fibers using direct, extractionless, non-hydrolysis surface-enhanced Raman spectroscopy. *Analyst* 138: 5895.

63 Jurasekova, Z., Del Puerto, E., Bruno, G. et al. (2010). Extractionless non-hydrolysis surface-enhanced Raman spectroscopic detection of historical mordant dyes on textile fibers. *J. Raman Spectrosc.* 41 (11): 1455–1461.

64 Groves, E., Palenik, C.S., and Palenik, S. (2016). A survey of extraction solvents in the forensic analysis of textile dyes. *Forensic Sci. Int.* 268: 139–144.

65 Pozzi, F., Lombardi, J.R., Bruni, S. et al. (2012). Sample treatment considerations in the analysis of organic colorants by surface-enhanced Raman scattering. *Anal. Chem.* 84: 3751–3757.

66 Londero, P.S., Lombardi, J.R., and Leona, M. (2013). Laser ablation surface-enhanced Raman microspectroscopy. *Anal. Chem.* 85 (11): 5463–5467.

67 Cesaratto, A., Leona, M., Lombardi, J.R. et al. (2014). Detection of organic colorants in historical painting layers using UV laser ablation surface-enhanced Raman microspectroscopy. *Angew. Chem. Int. Ed.* 53: 14373–14377.

68 Day, J.S., Edwards, H.G.M., Dobrowski, S.A. et al. (2004). The detection of drugs of abuse in fingerprints using Raman spectroscopy I: latent fingerprints. *Spectrochim. Acta. Part A* 60: 563–568.

69 Day, J.S., Edwards, H.G.M., Dobrowski, S.A. et al. (2004). The detection of drugs of abuse in fingerprints using Raman spectroscopy II: cyanoacrylate-fumed fingerprints. *Spectrochim. Acta. Part A* 60: 1725–1730.

70 Song, W.H., Mao, Z., Liu, X. et al. (2012). Detection of protein deposition within latent fingerprints by surface-enhanced Raman spectroscopy imaging. *Nanoscale* 4 (7): 2333–2338.

71 Connatser, M., Prokes, S.M., Glembocki, O.J. et al. (2010). Toward surface-enhanced Raman imaging of latent fingerprints. *J. Forensic Sci.* 55 (6): 1462–1470.

72 Ferguson, L., S. Francese, R. Wolstenholme, et al. (2010). *Gold nanoparticles for SERS in fingermark identification.* Application Note Perkin Elmer.

73 Kolhatkar, G., Parisien, C., Ruediger, A. et al. (2019). Latent fingermark imaging by single-metal deposition of gold nanoparticles and surface enhanced Raman spectroscopy. *Front. Chem.* 7: 1–8.

74 Gonzalez-Rodriguez, J., Sissons, N., and Robinson, S. (2011). Fire debris analysis by Raman spectroscopy and chemometrics. *J. Anal. Appl. Pyrolysis* 91: 210–218.

75 Li, S. and Dai, L.-K. (2012). Classification of gasoline brand and origin by Raman spectroscopy and a novel R-weighted LSSVM algorithm. *Fuel* 96: 146–152.

76 Zhang, X., Qi, X., Zou, M. et al. (2012). Rapid detection of gasoline by a portable Raman spectrometer and chemometrics. *J. Raman Spectrosc.* 43: 1487–1491.

77 White, P. and Wilkinson, T. (2013). Surface-enhanced Raman scattering (SERS) spectroscopy identifies fraudulent uses of fuels. *Spectrosc. Eur.* 25 (2): 18–22.

78 Costa, J.C.S., Sant'Ana, A.C., Corio, P. et al. (2006). Chemical analysis of polycyclic aromatic hydrocarbons by surface-enhanced Raman spectroscopy. *Talanta* 70: 1011–1016.

79 Gu, H., Zhang, Y., and Cao, L. (2014). Fabrication of silver nanoparticle-coated silica substrate for SERS detection of polycyclic aromatic hydrocarbons in fire scene. In: *Proceedings of 11th International GeoRaman Conference*. St.Louis, USA.

80 Xu, J., Du, J., Jing, C. et al. (2014). Facile detection of polycyclic aromatic hydrocarbons by a surface-enhanced Raman scattering sensor based on the Au coffee ring effect. *ACS Appl. Mater. Interfaces* 6: 6891–6897.

81 Bueno, J. and Lednev, I.K. (2013). Advanced statistical analysis and discrimination of gunshot residue implementing combined Raman and FT-IR data. *Anal. Methods* 5: 6292–6296.

82 Bueno, J., Sikirzhytski, V., and Lednev, I.K. (2012). Raman spectroscopic analysis of gunshot residue offering great potential for caliber differentiation. *Anal. Chem.* 84: 4334–4339.

83 Bueno, J. and Lednev, I.K. (2014). Raman microspectroscopic chemical mapping and chemometric classification for the identification of gunshot residue on adhesive tape. *Anal. Bioanal.Chem.* 406: 4595–4599.

84 Lopez-Lopez, M., Delgado, J.J., and Garcia-Ruiz, C. (2012). Ammunition identification by means of the organic analysis of gunshot residues using Raman spectroscopy. *Anal. Chem.* 84: 3581–3585.

85 Stich, S., Bard, D., Gros, L. et al. (1998). Raman microscopic identification of gunshot residues. *J. Raman Spectrosc.* 29: 787–790.

86 López-López, M., Özbek, N., and García-Ruiz, C. (2014). Confocal Raman spectroscopy to trace lipstick with their smudges on different surfaces. *Talanta* 123: 135–139.

87 Gardner, P., Bertino, M.F., Weimer, R. et al. (2013). Analysis of lipsticks using Raman spectroscopy. *Forensic Sci. Int.* 232: 67–72.

88 Salahioglu, F. and Went, M.J. (2012). Differentiation of lipsticks by Raman spectroscopy. *Forensic Sci. Int.* 223: 148–152.

89 Rodger, C., Rutherford, V., Broughton, D. et al. (1998). The in-situ analysis of lipsticks by surface enhanced resonance Raman scattering. *Analyst* 123: 1823–1826.

90 Yakes, B.J., Michael, T.J., Perez-Gonzalez, M. et al. (2017). Investigation of tattoo pigments by Raman spectroscopy. *J. Raman Spectrosc.* 48 (5): 736–743.

91 Muehlethaler, C., Muehlethaler, C., Lombardi, J.R. et al. (2018). Contribution of Raman and surface enhanced Raman spectroscopy (SERS) to the analysis of vehicle headlights: dye(s) characterization. *Forensic Sci. Int.* 287: 98–107.

92 White, P.C. (2000). SERRS spectroscopy – a new technique for forensic science? *Sci. Justice* 40 (2): 113–119.

93 Muehlethaler, C. et al. (2017). Raman and SERS characterization of solvent dyes: an example of shoe polish analysis. *Dyes Pigm.* 137: 539–552.

Index

Leading Edge Techniques in Forensic Trace Evidence Analysis: More New Trace Analysis Methods,
First Edition. Edited by Robert D. Blackledge.
© 2023 John Wiley & Sons, Inc. Published 2023 by John Wiley & Sons, Inc.

Printed and bound by CPI Group (UK) Ltd, Croydon, CR0 4YY

16/04/2025

14658348-0001